Plate Bending Analysis with Boundary Elements

Advances in Boundary Elements Series

Objectives

The continuing interest in the application of the Boundary Element Method has generated a series of books and numerous scientific papers. In spite of all their advantages, the need exists for a serial publication in which the most recent advances in the method are demonstrated in a more complete form.

Each volume in the Series comprises authored or edited books written by leading researchers in the field. The volumes are all self contained and cover a particular topic in sufficient detail for the analyst to understand the subject. Some books report on practical applications of the technique.

The Series covers topics such as:

Fluid Mechanics
Heat Transfer
Fracture Mechanics
Stress Analysis
Contact Mechanics
Structural Dynamics
Inelastic Problems
Optimization and Sensitivity
Plate Bending
Mathematical and Computational Aspects
Acoustics
Cathodic Protection
Electrical and Electromagnetic Problems
High Performance Computing
Sparse Methods
Numerical Integration
Industrial Applications
Basic Principles

Series Editors

C A Brebbia
Wessex Institute of Technology
Ashurst Lodge, Ashurst, Southampton, SO40 7AA, UK

M H Aliabadi
Queen Mary College, University of London,
Mile End, London, E1 4NS, UK

Associate Editors

C Alessandri
Universita di Ferrara
Facolta di Architettura
via Quartieri 8
44100 Ferrara
Italy

D E Beskos
Department of Civil Engineering
University of Patras
GR 261 10 Patras
Greece

M Bonnet
Ecole Polytechnique
Lab de Mecanique des Solides
91128 Palaiseau Cedex
France

M Bush
The University of Western Australia
Department of Mechanical Engineering
Nedland
W Australia 6009
Australia

A D Cheng
University of Delaware
Department of Civil Engineering
355 Dupont Hall
Newark, DE 19716
USA

D E Cormack
University of Toronto
Department of Chemical Engineering
200 College Street
Toronto, Ontario, M5S 1A4
Canada

T A Cruse
Department of Mechanical Engineering
Vanderbilt University
Nashville
TN 37235
USA

G De Mey
Ghent State University
Lab of Electronics
Sint Pietersnieuwstraat 41
9000 Ghent
Belgium

J Dominguez
Escuela Sup. de Ing. Indust.
University of Seville
41012 Seville
Spain

Q H Du
Tsinghua University
Mechanical Engineering
Beijing 100084
China

P Fedelinski
Department of Engineering Mechanics
Silesian Technical University
44-100 Gliwice
Poland

J I Frankel
University of Tennessee
Mech. & Aerospace Engineering Dept.
Knoxville, TN 37996-2210
USA

L Gaul
Universitat Stuttgart
Institut A fur Mechanik
Pfaffenwalring 9
D 70550 Stuttgart
Germany

G S Gipson
Oklahoma State University
Department of Civil Engineering
Engineering South 314
Stillwater, OK 74078-0327
USA

M Golberg
2025 University Circle
Las Vegas
NV 89119
USA

S Grilli
University of Rhode Island
Department of Ocean Engineering
Kingston, RI 02881
USA

D B Ingham
Department of Applied Mathematics
The University of Leeds
Leeds, LS2 9JT
United Kingdom

N Kamiya
Nagoya University
Department of Informatics & Sciences
School of Inform. & Sciences
Nagoya 464-01
Japan

D L Karabalis
University of Patras
Department of Civil Engineering
26110 Patras
Greece

A J Kassab
University of Central Florida
Department of Mechanical & Aero Engg.
Orlando
FL 32816-2993
USA

J T Katsikadelis
Dept of Civil Engineering
National Technical University of Athens
Institute of Structural Analysis
GR-157 73 Athens
Greece

V Leitao
Department de Engg.Civil
Inst. Superior Tecnico
Av Rovisco Pais
1096 Lisboa Codex
Portugal

W J Mansur
COPPE/UFRJ
Prog. Engineering Civil
CXP 68506
Rio de Janeiro 21945
Brazil

R A Meric
Istanbul Technical University
Faculty of A & A
Maslak
Istanbul 80626
Turkey

K Onishi
Science University of Tokyo
Department of Mathematics II
26 Wakamiya-cho, Shinjuku-ku
Tokyo 162
Japan

D Ouazar
EMI
BP 765
Agdal
Rabat
Morocco

F Paris
Escuela Sup. de Ing. Indust.
University of Seville
E S I I
41012 Sevilla
Spain

M Predeleanu
Univ. Paris VI
ENS Cachan
41 ave de President Wilson
94235 Cachan Cedex
France

J J Rencis
Worcester Polytechnic Institute
100 Institute Road
Worcester
MA 01609-2280
USA

T J Rudolphi
Iowa State University of Science & Technology
Department of Engg. Science & Mech.
Ames, Iowa 50011
USA

A P Selvadurai
McGill University
Department of Civil Engineering & Applied Mathematics
817 Sherbrooke St West
Montreal QC H3A 2K6
Canada

R P Shaw
Department of Civil Engineering
University of Buffalo, SUNY
212 Ketter Hall
Buffalo, NY 14260,
USA

P Skerget
Department of Mechanical Engineering
University of Maribor
POB 224
Smetanova 17
62000 Maribor,
Slovenia

V Sladek
Slovak Academy of Sciences
Inst. of Construction & Architecture
842 20 Bratislava
Slovakia

R N Smith
RMCS
Shrivenham
Wiltshire
SN6 8LA
United Kingdom

M Tanaka
Shinshu University
Department of Mechanical Engineering
500 Wakasato
Nagano 380
Japan

J C F Telles
COPPE/UFRJ
Prog. Engineering Civil
CP 68506
21945 Rio de Janeiro
Brazil

N Tosaka
Nihon University
Dept. of Mathematical Engineering
1-2-1 Narashino
Chiba 275
Japan

T Trans-Cong
Faculty of Engineering and Surveying
The University of Southern Queensland
New South Wales 2006
Australia

W S Venturini
University of Sao Paulo
13560 Campus de Sao Carlos
Sao Carlos, SP
Brazil

J L Wearing
Department of Mechanical Engineering
University of Sheffield
Mappin Street
Sheffield, S1 3JD
United Kingdom

J C Wu
Georgia Institute of Technology
48967 Ventura Drive
Fremont, CA 94539
USA

Plate Bending Analysis with Boundary Elements

EDITOR

M. H. Aliabadi
Queen Mary College, University of London, UK

Computational Mechanics Publications
Southampton, UK and Boston, USA

M.H. Aliabadi
Queen Mary College
University of London
Mile End
London E1 4NS
UK

Published by

Computational Mechanics Publications
Ashurst Lodge, Ashurst, Southampton, SO40 7AA, UK
Tel: 44 (0)1703 293223; Fax: 44 (0)1703 292853
Email: cmp@cmp.co.uk
http://www.cmp.co.uk

For USA, Canada and Mexico

Computational Mechanics Inc
25 Bridge Street, Billerica, MA 01821, USA
Tel: 978 667 5841; Fax: 978 667 7582
Email: cmina@ix.netcom.com

British Library Cataloguing-in-Publication Data

A Catalogue record for this book is available
from the British Library

ISBN 1 85312 531 8 Computational Mechanics Publications, Southampton
ISSN 1460-1419

Library of Congress Catalog Card Number 97-067408

Printed and bound in Great Britain by Bookcraft Ltd., Bath.

CONTENTS

Preface

Thin walled plate structures are widely used in engineering practice for the design of aircraft, spacecraft and ground structures. Accordingly the study of their behaviour when subjected to different loadings is essential.

This book presents a boundary element formulation for linear and non-linear problems in plate bending, providing readers with a detailed formulation and implementation of the method for plate bending analysis. The book contains ten chapters, covering the application of the boundary element method to foundation plates, stability analysis, and fracture mechanics. Also presented are hypersingular formulations and adaptive techniques.

The editor is grateful to the authors for their excellent contributions.

M H Aliabadi
University of London
1998

Chapter 1

The boundary element method for Reissner plates resting on elastic foundations

Youssef F. Rashed[a], M.H. Aliabadi[b], C.A. Brebbia[c]
[a]*Department of Structural Engineering, Cairo University, Giza, Egypt*
[b]*Department of Engineering, Queen Mary College, University of London, Mile End, London, E1 4NS, UK*
[c]*Wessex Institute of Technology, Ashurst, Southampton, SO40 7AA, UK*

Abstract

In this chapter, the boundary element theory of plates resting on elastic foundation is presented. The Reissner plate theory is used to model the plate bending behavior. The foundation is modeled using either the Winkler or the Pasternak models. The fundamental solutions are derived and the related kernels are given. Domain integrals due to inform loads are performed using equivalent boundary integrals. The necessary particular solutions required for the equivalent boundary integrals are derived. The internal points kernels are derived and listed. Several numerical examples are presented including all types of boundary conditions. The results are compared to analytical and previously published fininte element and boundary element results for both thin and thick plate theories. The results obtained demonstrate the efficacy and accuracy of the proposed formulations.

1 Introduction

Plates resting on elastic foundations have many applications in structural engineering such as building foundations and road pavements. A precise mathematical modeling of such structures would require a three-dimensional analysis. However, the three-dimensional analysis usually requires a considerable computational effort and data preparation. An alternative to three-dimensional analysis is the two-dimensional theories of plates. There are two main categories in the modeling of plates, namely: the classical thin plate theory [1] and the thick plate theories such as Reissner plate theory [2]. From the computational point of view, the classical plate theory presents the problem in terms of two independent degrees of freedom. Additional corner unknowns have to be added due to the Kirchhoff shear jump. This theory ignores the transverse shear deformations which, sometimes, makes modeling real structures difficult, particularly for problems of stress concentrations and composites. The Reissner plate theory on the other hand takes into account the effect of the shear deformation. Rashed *et al.* [3] have presented wide range of applications of Reissner's theory to real structural members.

Modeling of the underneath foundation as a half space would requires dealing with complicated mathematical functions. An alternative to this model is the simplified foundation models (see for example, Kerr [4]) that widely used in the literature. The simplest foundation model was proposed by Winkler [4]. In the Winkler model (refereed as the one-parameter model), the soil is represented by infinite number of springs. However these springs are not connected to each other, which results in a discontinuous settlement between the plate and the foundation along any free boundary. A more realistic example to the Winkler model is a liquid base. The Winkler foundation model is sufficient in analyzing building foundations [5]. A more refined model than the Winkler model, is the two-parameter Pasternak [4] model. In the Pasternak model, the Winkler springs are connected to each other via a shear layer. This leads to a more accurate model and overcomes the discontinuity of the settlement problem that appears in the Winkler model. The Pasternak model is usually used in modeling road pavements.

The boundary element method (BEM) has become a powerful numerical technique in the last decade. Many applications are considered and the BEM shows its superiority to domain type methods. One of the early applications of the boundary element method to thin plates on a Pasternak foundation is by Balaš *et al.* [6], who derived the fundamental solution based on the Yu [7] formulation. They also considered three different cases of the fundamental solution. The coupling integral equation for the free edge boundary condition was also derived, however, no example was presented for a free edge plate.

Wang *et al.* [8] extended the work of Balaš *et al.* [6] to deal with a thick plate resting on a Pasternak foundation. Similarly, they showed that the fundamental solution has three different cases depending on the parameters of the plate and the foundation. However, their work is incomplete as only one of the three possible cases for the fundamental solutions was considered. Wang *et al.* [8] used a cell discretization technique to compute the domain integral due to the body force and no examples were solved using a free edge boundary condition. Fadhil and El-Zafrany [9] derived an alternative BEM formulation for thick plate resting on a Pasternak foundation. The formulation presented in [9] contains a higher order of singularity ($O(1/r^2)$) and does not account for one of the three cases of the fundamental solutions. A regular BEM analysis was used in Ref. [9] to avoid the treatment of such a singularity. However, the regular BEM formulation is not a reliable method as its accuracy is highly dependent on the location of the collocation points and the type of the problem. Fadhil and El-Zafrany [9] solved problems with different boundary conditions using constant elements.

In this chapter, the boundary element formulation for Reissner plates resting on elastic foundation is presented. Both Winkler and Pasternak models are considered. The fundamental solutions are derived and the necessary kernels and their relevant derivatives are given. As it will be seen the present formulation has the same order of singularity as in two-dimensional elasticity problems, that is $O(\ln r)$ and $O(1/r)$. Domain integrals due to uniform domain loading are transformed to equivalent boundary integrals. The necessary particular solutions are derived and the explicit forms of the internal point kernels are presented. Quadratic isoparametric elements are used to approximate the distribution of the boundary unknowns. A discussion on the arrangement of the BEM system matrices is provided, particularly for the case of the free edge boundary condition. Several examples, including all possible types of boundary conditions, are presented to demonstrate the accuracy and the validity of the present formulation.

The source of information for this chapter is the work by Rashed *et al.* [10-14].

2 Governing equations

In this section, the basic equations for a Reissner plate resting on elastic foundations are reviewed. Throughout this chapter, the indicial notation is used. Comma denotes differentiation ($(\cdots)_{,\theta} = \frac{\partial(\cdots)}{\partial x_\theta}$) and $(\cdots)_{,n}$ denotes the derivative with respect to the normal n. Greek indices will vary from 1 to 2, whereas Roman indices vary from 1 to 3.

Consider an arbitrary plate domain Ω with boundary Γ. The plate has a thickness h and the $x_1 - x_2$ plane is assumed to be the middle surface $x_3 = 0$. The plate domain Ω is surrounded by an infinite foundation domain Ω_f (exterior problem) with boundary Γ_f, where $\Gamma = -\Gamma_f$.

2.1 The plate

The generalized displacements are denoted as u_i, where u_α denotes rotations (ϕ_{x_1} and ϕ_{x_2}) and u_3 denotes the transverse deflection w in x_3 direction. According to Reissner [2], the stress resultants-displacement relationships are given by:

$$\begin{aligned} M_{\alpha\beta} &= D\frac{1-\nu}{2}\left(u_{\alpha,\beta} + u_{\beta,\alpha} + \frac{2\nu}{1-\nu}u_{\gamma,\gamma}\delta_{\alpha\beta}\right) \\ Q_\alpha &= D\frac{1-\nu}{2}\lambda^2(u_\alpha + u_{3,\alpha}) \qquad \alpha,\beta = 1,2 \end{aligned} \tag{1}$$

where $M_{\alpha\beta}$ and Q_α are the bending and shear stress resultants respectively, and $D = Eh^3/12(1-\nu^2)$ is the plate flexural rigidity, E is Young's modulus, ν is Poisson's ratio and $\lambda = \sqrt{10}/h$ is the shear factor. It has to be noted that the effect of the normal stresses and the foundation reaction on the bending moment is ignored from eqn (1). The equilibrium equations can be written as follows:

$$\begin{aligned} & M_{\alpha\beta,\beta} - Q_\alpha = 0 \\ & Q_{\alpha,\alpha} + q - p = 0 \end{aligned} \tag{2}$$

where the term q is the distributed load per unit area, and p is an interface pressure between the plate and the foundation which is given by [4]:

$$p = k_f u_3 - G_f \nabla^2 u_3 \tag{3}$$

where ∇^2 is the two-dimensional Laplace operator and k_f, G_f denote the modulus of sub grade reaction and the shear modulus for the foundation respectively. For the Winkler model $G_f = 0$. Eqn (2) can be rewritten in terms of the generalized displacements as follows:

$$L_{ij}u_j + b_i = 0 \tag{4}$$

where

$$\begin{aligned} L_{\alpha\beta} &= \left(D\frac{(1-\nu)}{2}\nabla^2 - C \right) \delta_{\alpha\beta} + D\frac{(1+\nu)}{2}\partial_\alpha \partial_\beta \\ L_{\alpha 3} &= -C\partial_\alpha \\ L_{3\beta} &= C\partial_\beta \\ L_{33} &= (C+G_f)\nabla^2 - k_f \end{aligned} \tag{5}$$

and

$$C = D\frac{1-\nu}{2}\lambda^2 \tag{6}$$

where L_{ij} is the generalized Navier differential operator and b_i is the generalized body force components. The generalized tractions at a boundary point can be defined as:

$$\begin{aligned} p_\alpha &= M_{\alpha\beta} n_\beta \\ p_3 &= Q_\alpha n_\alpha \end{aligned} \tag{7}$$

where n_β are the components of the outward normal vector to the plate boundary Γ.

2.2 The foundation

The governing equation for the foundation settlement in the case of the Pasternak model is given by Kerr [4] as:

$$\left(\nabla^2 - \frac{k_f}{G_f} \right) u_f(\mathbf{X}'') = 0 \tag{8}$$

where u_f denotes the foundation settlement and $\mathbf{X}'' \in \Omega_f$.

3 Boundary conditions

The possible types of boundary conditions can be summarized as follows:

Clamped boundary:

$$u_t = 0, \qquad u_n = 0, \qquad u_3 = 0 \tag{9}$$

The subscripts n, t denotes the normal and the tangential directions, in which

$$u_t = u_\alpha t_\alpha, \qquad u_n = u_\alpha n_\alpha \tag{10}$$

where t_α are the tangential direction cosines.

Simply-supported boundary:

$$u_t = 0, \qquad u_3 = 0, \qquad M_n = 0 \tag{11}$$

where

$$M_n = p_\beta n_\beta \tag{12}$$

Guided boundary:

$$u_n = 0, \qquad u_t = 0, \qquad p_3 = 0 \tag{13}$$

Free boundary:

$$M_n = 0, \qquad M_t = 0, \qquad p_3 = 0 \tag{14}$$

where

$$M_t = p_\alpha t_\alpha \tag{15}$$

In the case of the free boundary an additional unknown shear traction p_f appears along the free edge due to the discontinuity between the plate and the foundation normal slopes. The unknown shear p_f has the same direction as p_3 and it can be defined as follows [15]:

$$p_f = G_f(u_{f,n} - u_{3,n}) \tag{16}$$

where $u_{3,n}$ can be defined as follows (recall eqn (1)):

$$u_{3,n} = \frac{p_f}{C} - u_n \tag{17}$$

4 Integral representation

In order to establish the integral representation of eqn (2), the following integral identity is considered (see Rashed *et al.* [12]):

$$\int_{\Omega} \left[(M_{\alpha\beta,\beta} - Q_{\alpha})\ U_{\alpha}^{*} + (Q_{\alpha,\alpha} + q - k_f u_3 + G_f \nabla^2 u_3)\ U_3^{*} \right] \mathrm{d}\Omega = 0 \quad (18)$$

where U_i^* $(i = \alpha, 3)$ are weighting functions. Integrating by parts (i.e. applying Green's second identity) and making use of eqn (7) the final integral representation is obtained as:

$$\begin{aligned} \int_{\Gamma} P_j^* u_j \mathrm{d}\Gamma + G_f \int_{\Gamma} U_{3,n}^* u_3 \mathrm{d}\Gamma &= \int_{\Gamma} U_j^* p_j \mathrm{d}\Gamma + \int_{\Gamma} U_3^* G_f u_{3,n} \mathrm{d}\Gamma + \int_{\Omega} U_3^* q \mathrm{d}\Omega \\ &+ \int_{\Omega} \left[(M_{\alpha\beta,\beta}^* - Q_{\alpha}^*)\ u_{\alpha} + (Q_{\alpha,\alpha}^* - k_f U_3^* + G_f \nabla^2 U_3^*)\ u_3 \right] \mathrm{d}\Omega \end{aligned} \quad (19)$$

where $(i = \alpha, 3)$. Eqn (19) represents the generalized Betti reciprocal theory for Reissner plates on elastic foundations between the first state (u, p) and the second state (U^*, P^*). Introducing arbitrary directions i of generalized loads applied at the point $\mathbf{X}' \in \Omega$, then eqn (19) can be re-written as:

$$\begin{aligned} &\int_{\Gamma} P_{ij}^*(\mathbf{X}', \mathbf{x}) u_j(\mathbf{x}) \mathrm{d}\Gamma(\mathbf{x}) + G_f \int_{\Gamma} U_{i3,n}^*(\mathbf{X}', \mathbf{x}) u_3(\mathbf{x}) \mathrm{d}\Gamma(\mathbf{x}) \\ = &\int_{\Gamma} U_{ij}^*(\mathbf{X}', \mathbf{x}) p_j(\mathbf{x}) \mathrm{d}\Gamma(\mathbf{x}) + \int_{\Gamma} U_{i3}^*(\mathbf{X}', \mathbf{x}) G_f u_{3,n}(\mathbf{x}) \mathrm{d}\Gamma(\mathbf{x}) \\ + &\int_{\Omega} U_{i3}^*(\mathbf{X}', \mathbf{X}) q(\mathbf{X}) \mathrm{d}\Omega(\mathbf{X}) + \int_{\Omega} \left[L_{ij}^{adj} U_{kj}^*(\mathbf{X}', \mathbf{X}) u_k(\mathbf{X}) \right] \mathrm{d}\Omega(\mathbf{X}) \end{aligned} \quad (20)$$

where $\mathbf{x} \in \Gamma$, $\mathbf{X} \in \Omega$ are boundary and internal field points and L_{ij}^{adj} is the adjoint operator of the original operator L_{ij} and given by:

$$\begin{aligned} L_{\alpha\beta}^{adj} &= \left(D\frac{(1-\nu)}{2} \nabla^2 - C \right) \delta_{\alpha\beta} + D\frac{(1+\nu)}{2} \partial_{\alpha}\partial_{\beta} \\ L_{\alpha 3}^{adj} &= C\partial_{\alpha} \\ L_{3\beta}^{adj} &= -C\partial_{\beta} \\ L_{33}^{adj} &= (C + G_f)\nabla^2 - k_f \end{aligned} \quad (21)$$

If the $(\cdot)^*$ state represents the fundamental state (two concentrated bending moments and concentrated shear force applied at the point $\mathbf{X}'$ in an infinite plate) i.e.

$$L_{ij}^{adj} U_{kj}^*(\mathbf{X}', \mathbf{X}) + \delta(\mathbf{X}', \mathbf{X})\delta_{ki} = 0 \tag{22}$$

where $\delta(\mathbf{X}', \mathbf{x})$ is the Dirac delta, then eqn (20) can be written as:

$$\begin{aligned} u_i(\mathbf{X}') &+ \int_\Gamma P_{ij}^*(\mathbf{X}', \mathbf{x})u_j(\mathbf{x})d\Gamma(\mathbf{x}) + G_f \int_\Gamma U_{i3,n}^*(\mathbf{X}', \mathbf{x})u_3(\mathbf{x})d\Gamma(\mathbf{x}) \\ &= \int_\Gamma U_{ij}^*(\mathbf{X}', \mathbf{x})p_j(\mathbf{x})d\Gamma(\mathbf{x}) + \int_\Gamma U_{i3}^*(\mathbf{X}', \mathbf{x})G_f u_{3,n}(\mathbf{x})d\Gamma(\mathbf{x}) \\ &+ \int_\Omega U_{i3}^*(\mathbf{X}', \mathbf{X})q(\mathbf{X})d\Omega(\mathbf{X}) \end{aligned} \tag{23}$$

where U_{ij}^* and P_{ij}^* are the two-point fundamental solution kernels and it represents the displacement and traction at the point $\mathbf{x}$ in the direction j due to generalized impulse load at the point $\mathbf{x}'$ in the direction i. Eqn (23) represents the generalized Somigliana's identity for Reissner plates.

5 Fundamental solutions

An operator decoupling process via Hörmander method [16] can be used to obtain the relevant fundamental solutions (see Rashed *et al.* [10]). Eqn (22) can be rewritten as:

$$U_{ij}^*(\mathbf{X}', \mathbf{X}) = {}^{co}L_{ji}^{adj} \phi^*(\mathbf{X}', \mathbf{X}) \tag{24}$$

where $\phi^*(\mathbf{X}', \mathbf{X})$ is an unknown scalar potential and ${}^{co}L_{ij}^{adj}$ is the cofactor matrix of the operator L_{ij}^{adj} and given by:

$$\begin{aligned} {}^{co}L_{\alpha\beta}^{adj} &= D(C+G_f)[\nabla^4 - \alpha\nabla^2 + \beta]\delta_{\alpha\beta} \\ &- \partial_\alpha\partial_\beta\left[D(C+G_f)\frac{1+\nu}{2}\nabla^2 + C^2 - Dk_f\frac{1+\nu}{2}\right] \\ {}^{co}L_{\alpha 3}^{adj} &= C\partial_\alpha\left(D\frac{1-\nu}{2}\nabla^2 - C\right) \\ {}^{co}L_{3\beta}^{adj} &= -C\partial_\beta\left(D\frac{1-\nu}{2}\nabla^2 - C\right) \\ {}^{co}L_{33}^{adj} &= (D\nabla^2 - C)\left(D\frac{1-\nu}{2}\nabla^2 - C\right) \end{aligned} \tag{25}$$

where

$$\alpha = \frac{Dk_f + CG_f}{D(C+G_f)} \quad \text{and} \quad \beta = \frac{Ck_f}{D(C+G_f)} \tag{26}$$

Now the potential $\phi^*(\mathbf{x}', \mathbf{x})$ can be evaluated as follows:

$$det[{}^{co}L_{ij}^{adj}]\phi^*(\mathbf{X}', \mathbf{X}) = -\delta(\mathbf{X}', \mathbf{X}) \tag{27}$$

To evaluate the determinant of ${}^{co}L_{ij}^{adj}$, it is sufficient to deduce the sum of the multiplication of the elements of the two corresponding columns or rows in L_{ij}^{adj} and ${}^{co}L_{ij}^{adj}$. In doing such an operation the following relationship can be discovered:

$$\left[\frac{1+\nu}{2}\lambda^2 + \frac{C^2}{D(C+G_f)} - \frac{k_f\frac{1+\nu}{2}}{C+G_f}\right] = \frac{\lambda^4 - \alpha\lambda^2 + \beta}{\lambda^2} \tag{28}$$

Eqn (28) can be used to obtain the final form of the fundamental solution in a more compact form. Now eqn (27) can be rewritten as follows:

$$\frac{\mathcal{F}}{4}(\nabla^4 - \alpha\nabla^2 + \beta)(\nabla^2 - \lambda^2)\phi^*(\mathbf{X}', \mathbf{X}) = -\delta(\mathbf{X}', \mathbf{X}) \tag{29}$$

where

$$\mathcal{F} = D^3(1-\nu)^2\lambda^2\alpha_3 \quad \text{and} \quad \alpha_3 = \left(1 + \frac{G_f}{C}\right) \tag{30}$$

In eqn (29), the first operator $(\nabla^4 - \alpha\nabla^2 + \beta)$ is the same as the operator for thin plates on elastic foundation [6] and it can be written as:

$$\frac{\mathcal{F}}{4}(\nabla^2 - d^2)(\nabla^2 - e^2)\mathcal{A}^*(\mathbf{X}', \mathbf{X}) = -\delta(\mathbf{X}', \mathbf{X}) \tag{31}$$

and

$$d^2 = \frac{\kappa + \sqrt{\kappa^2 - 1}}{\ell^2} \quad \text{and} \quad e^2 = \frac{\kappa - \sqrt{\kappa^2 - 1}}{\ell^2} \tag{32}$$

where $\kappa = \alpha\ell^2/2$ and $\ell^4 = 1/\beta$.

The fundamental solution $\mathcal{A}^*(\mathbf{X}', \mathbf{X})$ is the same as that of thin plates on elastic foundation, and it can be evaluated by either using Fourier integral transform [6] or using the so-called difference trick [17] as follows:

$$\mathcal{A}^*(\mathbf{X}', \mathbf{X}) = \frac{-\ell^2}{\pi \mathcal{F} S}[K_0(er) - K_0(dr)] \quad \text{and} \quad S = \sqrt{\kappa^2 - 1} \tag{33}$$

where $K_0(\cdot)$ is modified Bessel function [18]. Eqn (33) represent the first case for the fundamental solution and it has to be noted that this solution is valid only for $\kappa > 1$. When $\kappa < 1$, the parameter d and e are complex conjugate [6] (i.e. $e = \overline{d}$). Defining [6]:

$$e = \overline{d} = \zeta = \frac{\exp(i(\frac{\pi}{2} + \psi))}{\ell} \quad \text{and} \quad i = \sqrt{-1} \tag{34}$$

where $\kappa = \cos 2\psi$, $\sqrt{1 - \kappa^2} = \sin 2\psi$ and $\psi \in [0, \pi/4]$. Using the relationship $\overline{K_i(z)} = K_{\overline{i}}(\overline{z})$ [18] and substituting in eqn (33) yields:

$$\mathcal{A}^*(\mathbf{X}', \mathbf{X}) = \frac{\ell^2}{\mathcal{F} \sin 2\psi} \mathbf{Re}[H_0^{(1)}(\zeta r)] \tag{35}$$

where $H_0^{(1)}(\cdot)$ is Hankel function [18]. Eqn (35) represent another case for the fundamental solution. In this chapter, this case is referred to as case 3.

Case 2 of the fundamental solution occurred when $e = d$ then $\kappa = 1$. This case should be analyzed separately. By using Fourier integral transform yield [6]:

$$\mathcal{A}^*(\mathbf{X}', \mathbf{X}) = \frac{-\ell}{\pi \mathcal{F}} r K_1(\frac{r}{\ell}) \tag{36}$$

where $K_1(\cdot)$ is a modified Bessel function [18].

Returning to equation(29), the second operator $(\nabla^2 - \lambda^2)$ is a modified Helmholtz operator, and its fundamental solution can be written as follows [8]:

$$\mathcal{B}^*(\mathbf{X}', \mathbf{X}) = \frac{2}{\pi \mathcal{F}} K_0(z) \tag{37}$$

where $z = \lambda r$.

Now the scalar potential $\phi^*(\mathbf{x}', \mathbf{x})$ can be expressed as a linear combination of the two fundamental solutions $\mathcal{A}^*(\mathbf{x}', \mathbf{x})$ and $\mathcal{B}^*(\mathbf{x}', \mathbf{x})$ as follows [8]:

$$\phi^*(\mathbf{X}', \mathbf{X}) = \frac{(\alpha - \lambda^2 - \nabla^2)\mathcal{A}^*(\mathbf{x}', \mathbf{x}) + \mathcal{B}^*(\mathbf{x}', \mathbf{x})}{\lambda^4 - \alpha\lambda^2 + \beta} \tag{38}$$

The explicit forms for this potential are given as follows:

Case 1 $(\kappa > 1)$

$$\phi^*(\mathbf{X}', \mathbf{X}) = \frac{\ell^2(\lambda^2 - \alpha)\Upsilon_0 + \ell^2\Upsilon_2 + 2K_0(z)S}{\pi\mathcal{F}(\lambda^4 - \alpha\lambda^2 + \beta)S} \tag{39}$$

where

$$\Upsilon_j = \begin{cases} e^j K_0(er) - d^j K_0(dr) & j\text{: even number} \\ e^j K_1(er) - d^j K_1(dr) & j\text{: odd number} \end{cases} \tag{40}$$

Case 2 $(\kappa = 1)$

$$\phi^*(\mathbf{X}', \mathbf{X}) = \frac{\ell^4}{\pi\mathcal{F}(\ell^2\lambda^2 - 1)^2}[2K_0(\frac{r}{\ell}) + (\frac{r}{\ell})K_1(\frac{r}{\ell})(1 - \alpha\ell^2 + \lambda^2\ell^2) + 2K_0(z)] \tag{41}$$

Case 3 $(\kappa < 1)$

$$\phi^*(\mathbf{X}', \mathbf{X}) = \frac{\ell^2\pi(\alpha - \lambda^2)\Psi_0 + \ell^2\pi\Psi_2 + 2K_0(z)\sin 2\psi}{\pi\mathcal{F}(\lambda^4 - \alpha\lambda^2 + \beta)\sin 2\psi} \tag{42}$$

$$\Psi_j = \begin{cases} 2\mathbf{Re}[\zeta^j H_0^{(1)}(\zeta r)] & j\text{: even number} \\ 2\mathbf{Re}[\zeta^j H_1^{(1)}(\zeta r)] & j\text{: odd number} \end{cases} \tag{43}$$

5.1 Displacement kernels

The fundamental solution kernels can be computed using eqn (24). It has to be noted that Wang *et al.* [8] derived the fundamental solution for the case 3. Rashed *et al.* [10] derived the other two cases. Substituting for ϕ^* into eqn (24) and using the relationship in eqn (28), the fundamental solution can be obtained as follows:

Case 1 ($\kappa > 1$)

$$
\begin{aligned}
U^*_{\alpha\beta} &= \frac{1}{\pi D(1-\nu)}[B_1(z)\delta_{\alpha\beta} - A_1(z)r_{,\alpha}r_{,\beta}] \\
&+ \frac{\ell^2}{2\pi D(1-\nu)\lambda^2 S}\left\{\frac{1}{r}[\Upsilon_3 - \alpha_2\Upsilon_1]\delta_{\alpha\beta}\right. \\
&- \left.\left[(\Upsilon_4 - \alpha_2\Upsilon_2) + \frac{2}{r}(\Upsilon_3 - \alpha_2\Upsilon_1)\right] r_{,\alpha}r_{,\beta}\right\} \\
U^*_{\alpha 3} &= \frac{-\ell^2}{4\pi D\alpha_3 S} r_{,\alpha}\Upsilon_1 \\
U^*_{3\alpha} &= -U^*_{\alpha 3} \\
U^*_{33} &= \frac{\ell^2}{4\pi D\alpha_3 S}\left[\frac{-2}{(1-\nu)\lambda^2}\Upsilon_2 + \Upsilon_0\right]
\end{aligned}
\tag{44}
$$

where

$$
A_1(\cdot) = K_0(\cdot) + \frac{2}{(\cdot)}K_1(\cdot) \quad \text{and} \quad B_1(\cdot) = K_0(\cdot) + \frac{1}{(\cdot)}K_1(\cdot) \tag{45}
$$

and $\alpha_2 = \alpha - \frac{1-\nu}{2}\lambda^2$.

Case 2 ($\kappa = 1$)

$$
\begin{aligned}
U^*_{\alpha\beta} &= \frac{1}{\pi D(1-\nu)}\left\{[B_1(z)\delta_{\alpha\beta} - A_1(z)r_{,\alpha}r_{,\beta}]\right. \\
&+ \frac{1}{\lambda^2\ell^2}\left[A_1(\frac{r}{\ell})r_{,\alpha}r_{,\beta} - (\frac{\ell}{r})K_1(\frac{r}{\ell})\delta_{\alpha\beta}\right] \\
&+ \left.\frac{\alpha_4}{4}\left[K_0(\frac{r}{\ell})\delta_{\alpha\beta} - (\frac{r}{\ell})K_1(\frac{r}{\ell})r_{,\alpha}r_{,\beta}\right]\right\} \\
U^*_{\alpha 3} &= \frac{-1}{4\pi D\alpha_3} rK_0(\frac{r}{\ell})r_{,\alpha} \\
U^*_{3\alpha} &= -U^*_{\alpha 3} \\
U^*_{33} &= \frac{1}{4\pi D(1-\nu)\lambda^2\alpha_3}\left[4K_0(\frac{r}{\ell}) + \ell^2\lambda^2\alpha_4(\frac{r}{\ell})K_1(\frac{r}{\ell})\right]
\end{aligned}
\tag{46}
$$

where $\alpha_4 = (1-\nu) - \frac{2}{\ell^2\lambda^2}$

Case 3 ($\kappa < 1$)

$$
\begin{aligned}
U^*_{\alpha\beta} &= \frac{1}{\pi D(1-\nu)}[B_1(z)\delta_{\alpha\beta} - A_1(z)r_{,\alpha}r_{,\beta}] \\
&+ \frac{\ell^2}{4D(1-\nu)\lambda^2 \sin 2\psi}\left\{\frac{1}{r}[\Psi_3 + \alpha_2\Psi_1]\delta_{\alpha\beta}\right. \\
&+ \left.\left[(\Psi_4 + \alpha_2\Psi_2) - \frac{2}{r}(\Psi_3 + \alpha_2\Psi_1)\right] r_{,\alpha}r_{,\beta}\right\} \\
U^*_{\alpha 3} &= \frac{\ell^2}{8D\alpha_3 \sin 2\psi} r_{,\alpha}\Psi_1 \\
U^*_{3\alpha} &= -U^*_{\alpha 3} \\
U^*_{33} &= \frac{-\ell^2}{8D\alpha_3 \sin 2\psi}\left[\frac{2}{(1-\nu)\lambda^2}\Psi_2 + \Psi_0\right]
\end{aligned}
\tag{47}
$$

The necessary normal derivatives of the displacement fundamental solutions are given in Appendix A.

5.2 Traction kernels

The traction fundamental solution can be computed by substituting the displacement fundamental solution into equations (1) and (7), to give:

$$
\begin{aligned}
P^*_{i\alpha} &= \frac{D(1-\nu)}{2}\left(U^*_{i\alpha,\beta} + U^*_{i\beta,\alpha} + \frac{2\nu}{1-\nu}U^*_{i\theta,\theta}\delta_{\alpha\beta}\right) n_\beta \\
P^*_{i3} &= \frac{D(1-\nu)}{2}\lambda^2\left(U^*_{i\beta} + U^*_{i3,\beta}\right) n_\beta
\end{aligned}
\tag{48}
$$

It has to be noted that the differentiation should be carried out with respect to the coordinate of the field point **X**. The explicit forms the final kernels are given as follows (Rashed *et al.* [10]):

Case 1 ($\kappa > 1$)

$$
\begin{aligned}
P^*_{\gamma\alpha} &= \frac{1}{2\pi r}[zK_1(z)(2r_{,\gamma}r_{,\alpha}r_{,n} - \delta_{\gamma\alpha}r_{,n} - r_{,\alpha}n_\gamma) \\
&+ 2A_1(z)(4r_{,\gamma}r_{,\alpha}r_{,n} - \delta_{\gamma\alpha}r_{,n} - r_{,\alpha}n_\gamma - r_{,\gamma}n_\alpha)] \\
&+ \frac{\ell^2}{4\pi\lambda^2 S}\left\{(\Upsilon_5 - \alpha_2\Upsilon_3)\left[\frac{2\nu}{1-\nu}n_\alpha r_{,\gamma} + 2r_{,\alpha}r_{,\gamma}r_{,n}\right]\right.
\end{aligned}
$$

$$
\begin{aligned}
&+ \left[\frac{4}{r^2}(\Upsilon_3 - \alpha_2\Upsilon_1) + \frac{2}{r}(\Upsilon_4 - \alpha_2\Upsilon_2)\right] \\
&\times \left.(4r_{,\gamma}r_{,\alpha}r_{,n} - \delta_{\gamma\alpha}r_{,n} - r_{,\alpha}n_\gamma - r_{,\gamma}n_\alpha)\right\} \\
P^*_{\gamma 3} &= \frac{\lambda^2}{2\pi}[B_1(z)n_\gamma - A_1(z)r_{,\gamma}r_{,n}] \\
&+ \frac{\ell^2}{4\pi S}\left\{\frac{1}{r}(\Upsilon_3 - \alpha_1\Upsilon_1)(n_\gamma - 2r_{,\gamma}r_{,n}) - (\Upsilon_4 - \alpha_1\Upsilon_2)r_{,\gamma}r_{,n}\right\} \\
P^*_{3\alpha} &= \frac{-\ell^2}{4\pi\alpha_3 S}\left\{\Upsilon_2[(1-\nu)r_{,\alpha}r_{,n} + \nu n_\alpha] - \frac{1-\nu}{r}\Upsilon_1(n_\alpha - 2r_{,\alpha}r_{,n})\right\} \\
P^*_{33} &= \frac{\ell^2}{4\pi\alpha_3 S}\Upsilon_3 r_{,n}
\end{aligned}
\tag{49}
$$

where $r_{,n} = r_{,\alpha}n_\alpha$ and $\alpha_1 = \frac{k_f}{C+G_f}$.

Case 2 ($\kappa = 1$)

$$
\begin{aligned}
P^*_{\gamma\alpha} &= \frac{1}{2\pi r}[zK_1(z)(2r_{,\gamma}r_{,\alpha}r_{,n} - \delta_{\gamma\alpha}r_{,n} - r_{,\alpha}n_\gamma) \\
&+ 2A_1(z)(4r_{,\gamma}r_{,\alpha}r_{,n} - \delta_{\gamma\alpha}r_{,n} - r_{,\alpha}n_\gamma - r_{,\gamma}n_\alpha)] \\
&+ \left[\frac{\alpha_4}{4\pi\ell}K_1(\frac{r}{\ell}) - \frac{1}{\pi\lambda^2\ell^2 r}A_1(\frac{r}{\ell})\right](4r_{,\gamma}r_{,\alpha}r_{,n} - \delta_{\gamma\alpha}r_{,n} - r_{,\alpha}n_\gamma - r_{,\gamma}n_\alpha) \\
&+ \left[\frac{\alpha_4}{4\pi\ell}(\frac{r}{\ell})K_0(\frac{r}{\ell}) - \frac{1-\nu}{2\pi\ell}K_1(\frac{r}{\ell})\right]\left(\frac{\nu}{1-\nu}n_\alpha r_{,\gamma} + r_{,\alpha}r_{,\gamma}r_{,n}\right) \\
P^*_{\gamma 3} &= \frac{\lambda^2}{2\pi}[B_1(z)n_\gamma - A_1(z)r_{,\gamma}r_{,n}] - \frac{1}{4\pi\ell^2}\left\{2(\frac{\ell}{r})K_1(\frac{r}{\ell})(n_\gamma - 2r_{,\gamma}r_{,n})\right. \\
&+ \left.\left[(\frac{r}{\ell})K_1(\frac{r}{\ell})r_{,\gamma}r_{,n} - K_0(\frac{r}{\ell})n_\gamma\right](1 - \ell^2\alpha_1) - 2K_0(\frac{r}{\ell})r_{,\gamma}r_{,n}\right\} \\
P^*_{3\alpha} &= \frac{1}{4\pi\alpha_3}\left\{(1+\nu)K_0(\frac{r}{\ell})n_\alpha - (\frac{r}{\ell})K_1(\frac{r}{\ell})(\nu n_\alpha + (1-\nu)r_{,\alpha}r_{,n})\right\} \\
P^*_{33} &= \frac{r_{,n}}{4\pi\ell\alpha_3}\left[(\frac{r}{\ell})K_0(\frac{r}{\ell}) - 2K_1(\frac{r}{\ell})\right]
\end{aligned}
\tag{50}
$$

Case 3 ($\kappa < 1$)

$$
\begin{aligned}
P^*_{\gamma\alpha} &= \frac{1}{2\pi r}[zK_1(z)(2r_{,\gamma}r_{,\alpha}r_{,n} - \delta_{\gamma\alpha}r_{,n} - r_{,\alpha}n_\gamma) \\
&+ 2A_1(z)(4r_{,\gamma}r_{,\alpha}r_{,n} - \delta_{\gamma\alpha}r_{,n} - r_{,\alpha}n_\gamma - r_{,\gamma}n_\alpha)] \\
&+ \frac{\ell^2}{8\lambda^2\sin 2\psi}\left\{ - (\Psi_5 + \alpha_2\Psi_3)\left[\frac{2\nu}{1-\nu}n_\alpha r_{,\gamma} + 2r_{,\alpha}r_{,\gamma}r_{,n}\right]\right.
\end{aligned}
$$

$$
\begin{aligned}
&+ \left[\frac{4}{r^2}(\Psi_3 + \alpha_2\Psi_1) - \frac{2}{r}(\Psi_4 + \alpha_2\Psi_2)\right] \\
&\times \left.(4r_{,\gamma}r_{,\alpha}r_{,n} - \delta_{\gamma\alpha}r_{,n} - r_{,\alpha}n_\gamma - r_{,\gamma}n_\alpha)\right\} \\
P^*_{\gamma 3} &= \frac{\lambda^2}{2\pi}[B_1(z)n_\gamma - A_1(z)r_{,\gamma}r_{,n}] \\
&+ \frac{\ell^2}{8\sin 2\psi}\left\{\frac{1}{r}(\Psi_3 + \alpha_1\Psi_1)(n_\gamma - 2r_{,\gamma}r_{,n}) + (\Psi_4 + \alpha_1\Psi_2)r_{,\gamma}r_{,n}\right\} \\
P^*_{3\alpha} &= \frac{-\ell^2}{8\alpha_3\sin 2\psi}\left\{\Psi_2[(1-\nu)r_{,\alpha}r_{,n} + \nu n_\alpha] + \frac{1-\nu}{r}\Psi_1(n_\alpha - 2r_{,\alpha}r_{,n})\right\} \\
P^*_{33} &= \frac{\ell^2}{8\alpha_3\sin 2\psi}\Psi_3 r_{,n} \qquad (51)
\end{aligned}
$$

A discussion on the singularity of the fundamental solution kernels are given in Appendix B.

6 Boundary integral equations

Taking the limiting form of eqn (23) as $\mathbf{X}'$ approaches the boundary at the position $\mathbf{x}'$, gives (see Rashed *et al.* [12]):

$$
\begin{aligned}
c_{ij}(\mathbf{x}')u_j(\mathbf{x}') &+ \fint_\Gamma P^*_{ij}(\mathbf{x}',\mathbf{x})u_j(\mathbf{x})\mathrm{d}\Gamma(\mathbf{x}) + G_f\int_\Gamma U^*_{i3,n}(\mathbf{x}',\mathbf{x})u_3(\mathbf{x})\mathrm{d}\Gamma(\mathbf{x}) \\
&= \int_\Gamma U^*_{ij}(\mathbf{x}',\mathbf{x})p_j(\mathbf{x})\mathrm{d}\Gamma(\mathbf{x}) \\
&+ \int_\Gamma U^*_{i3}(\mathbf{x}',\mathbf{x})[p_f(\mathbf{x}) + G_f u_{3,n}(\mathbf{x})]\mathrm{d}\Gamma(\mathbf{x}) \\
&+ \int_\Omega U^*_{i3}(\mathbf{x}',\mathbf{X})q(\mathbf{X})\mathrm{d}\Omega(\mathbf{X}) \qquad (52)
\end{aligned}
$$

where $\fint$ denotes a Cauchy principal value integral and c_{ij} is the jump term matrix. It has to be noted that, in eqn (52), the unknown p_f (in case of free edge plate on Pasternak foundation) is separated from p_3.

In order to evaluate the jump term c_{ij}, the terms of $O(1/r)$ and $O(r_{,n}/r)$ are separated from the relevant kernels, as follows:

$$
P^*_{\gamma\alpha} \rightarrow \frac{1}{2\pi r}\left\{ -\delta_{\gamma\alpha}r_{,n} + \frac{1-\nu}{2}(r_{,\gamma}n_\alpha - r_{,\alpha}n_\gamma)\right.
$$

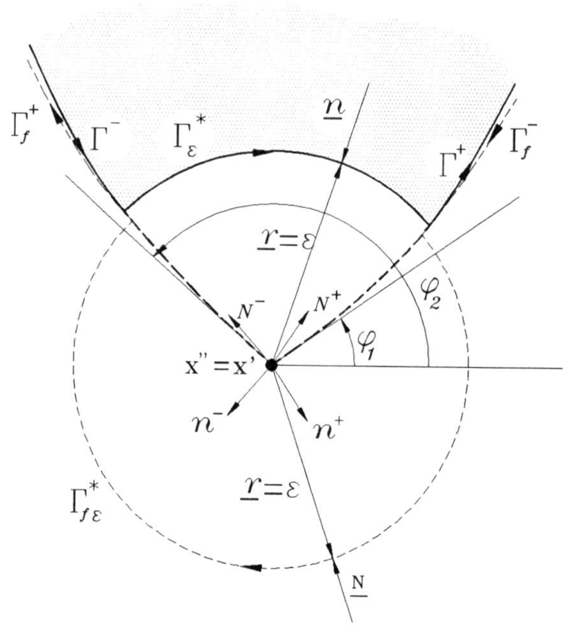

Figure 1. Geometrical representation of a corner point.

$$
\begin{aligned}
& + && \frac{1+\nu}{2}(\delta_{\gamma\alpha}r_{,n} - 2r_{,\alpha}r_{,\gamma}r_{,n})\Big\} \\
P_{33}^* & \rightarrow && \frac{-1}{2\pi\alpha_3 r}r_{,n} \\
U_{33,n}^* & \rightarrow && \frac{-1}{2\pi\alpha_3 C r}r_{,n}
\end{aligned}
\tag{53}
$$

The former terms of $O(1/r)$ and $O(r_{,n}/r)$ appear in each case of the fundamental solution (Independent of κ). From Figure 1, the term c_{ij} can be defined as follows:

$$
c_{ij} = \lim_{\varepsilon \to 0} \int_{\Gamma_\varepsilon^*} [P_{ij}^* + G_f U_{i3,n}^* \delta_{j3}] \mathrm{d}\Gamma \tag{54}
$$

and the following relationships can be defined:

$$
r = \varepsilon, \qquad r_{,n} = -1, \qquad \mathrm{d}\Gamma = \varepsilon \mathrm{d}\varphi,
$$

$$
r_{,1} = -n_1 = \cos\varphi, \qquad r_{,2} = -n_2 = \sin\varphi,
$$

and

$$\int_{\Gamma_\varepsilon^*} \cdots \mathrm{d}\Gamma = \int_{\varphi_1}^{\varphi_2} \cdots \varepsilon \mathrm{d}\varphi \tag{55}$$

Using the above relationships and taking a limiting form of the integral in eqn (54) as $\varepsilon \to 0$, the following jump term can be obtained:

$$c_{ij} = \frac{1}{2\pi} \begin{bmatrix} \Delta\varphi + ss & -cc & 0 \\ -cc & \Delta\varphi - ss & 0 \\ 0 & 0 & \Delta\varphi \end{bmatrix} \tag{56}$$

where

$$\begin{aligned} ss &= \left(\frac{1+\nu}{4}\right)(\sin 2\varphi_2 - \sin 2\varphi_1) \\ cc &= \left(\frac{1+\nu}{4}\right)(\cos 2\varphi_2 - \cos 2\varphi_1) \end{aligned} \tag{57}$$

in which $\Delta\varphi = \varphi_2 - \varphi_1$. For a smooth boundary, $\varphi_1 = 0$ and $\varphi_2 = \pi$, therefore, $c_{ij} = \frac{1}{2}\delta_{ij}$.

For Winkler model ($p_f = 0$ and $G_f = 0$), so that the previous integral eqn can be simplified as follows:

$$\begin{aligned} c_{ij}(\mathbf{x}')u_j(\mathbf{x}') \; + \; & ⨍_\Gamma P_{ij}^w(\mathbf{x}', \mathbf{x})u_j(\mathbf{x})\mathrm{d}\Gamma(\mathbf{x}) = \\ & \int_\Gamma U_{ij}^w(\mathbf{x}', \mathbf{x})p_j(\mathbf{x})\mathrm{d}\Gamma(\mathbf{x}) + \int_\Omega U_{ij}^w(\mathbf{x}', \mathbf{X})b_j(\mathbf{X})\mathrm{d}\Omega(\mathbf{X}) \end{aligned} \tag{58}$$

where $(\cdot)^w = (\cdot)^*\|_{G_f=0}$ and the term c_{ij} in eqn (58) is the same as in eqn (52) as it is independent of G_f.

It has to be noted that eqn (52) contains a set of three integral equations (two for rotation and one for deflection). However, it has four unknowns (u_i and p_f). This equation is sufficient for solving problems with a clamped, simply supported, or guided boundary as ($p_f = 0$). For the case of free boundary, however, an additional integral equation for Pasternak foundation is required. The additional integral equation can be generated from eqn (8) as given in [6] as:

$$\begin{aligned} c_f(\mathbf{x}'')u_f(\mathbf{x}'') &= \int_{\Gamma_f} U_f^*(\mathbf{x}'', \mathbf{x})u_{f,N}(\mathbf{x})\mathrm{d}\Gamma_f(\mathbf{x}) \\ &- \int_{\Gamma_f} U_{f,N}^*(\mathbf{x}'', \mathbf{x})u_f(\mathbf{x})\mathrm{d}\Gamma_f(\mathbf{x}) \end{aligned} \tag{59}$$

where $\mathbf{x}'' \in \Gamma_f$, N is the normal to Γ_f and U_f^* is the two-point fundamental solution of the modified Helmholtz equation in eqn (8) and is given by:

$$U_f^* = \frac{1}{2\pi}K_0(\lambda_1 r) \tag{60}$$

in which $\lambda_1 = \sqrt{\frac{k_f}{G_f}}$.

The expression of the free term c_f can be obtained in similar way as the free term c_{ij}. From Figure 1:

$$c_f = \lim_{\varepsilon \to 0} \int_{\Gamma_{f\varepsilon}^*} U_{f,N}^* \mathrm{d}\Gamma = 1 - \frac{\Delta\varphi}{2\pi} \tag{61}$$

For a smooth boundary $c_f = 1/2$.

Along the boundary, $u_f = u_3$, and from equations (16) and (17), the following relationship can be obtained:

$$u_{f,n} = \frac{p_f}{G_f}\alpha_3 - u_\alpha n_\alpha \tag{62}$$

Substituting eqn (62) into eqn (59) and moving from Γ_f to Γ, gives:

$$\begin{aligned} c_f(\mathbf{x}')u_3(\mathbf{x}') = {} & \int_\Gamma U_{f,n}^*(\mathbf{x}',\mathbf{x})u_3(\mathbf{x})\mathrm{d}\Gamma(\mathbf{x}) + \int_\Gamma U_f^*(\mathbf{x}',\mathbf{x})u_\alpha(\mathbf{x})n_\alpha(\mathbf{x})\mathrm{d}\Gamma(\mathbf{x}) \\ & - \frac{\alpha_3}{G_f}\int_\Gamma U_f^*(\mathbf{x}',\mathbf{x})p_f(\mathbf{x})\mathrm{d}\Gamma(\mathbf{x}) \end{aligned} \tag{63}$$

with

$$U_{f,n}^* = \frac{-\lambda_1}{2\pi}K_1(\lambda_1 r)r_{,n} \tag{64}$$

Using eqn (63) together with eqn (52), free boundary condition can now be solved.

7 Treatment of the body force domain integrals

The last integral in eqn (52) is a domain integral and can be transformed to the boundary for the case of uniform domain load q as follows (Rashed *et al.* [11]):

$$\int_\Omega U^*_{i3}(\mathbf{X}'',\mathbf{X})q\mathrm{d}\Omega(\mathbf{X}) = q\int_\Gamma V^*_{i,n}(\mathbf{X}'',\mathbf{x})\mathrm{d}\Gamma(\mathbf{x}) \tag{65}$$

and

$$\nabla^2 V^*_i = U^*_{i3} \tag{66}$$

where V^*_i are unknown particular solutions.

7.1 Evaluation of the particular solutions

The three different fundamental solutions corresponding to the three cases will be considered individually to compute the relevant particular solutions V^*_i:

Case 1 ($\kappa > 1$)

Eqn (66) can be expanded as follows:

$$\begin{aligned}\nabla^2 V^*_\alpha &= U^*_{\alpha 3} = \frac{-\ell^2}{4\pi D\alpha_3 S} r_{,\alpha}\Upsilon_1\\ \nabla^2 V^*_3 &= U^*_{33} = \frac{\ell^2}{4\pi D\alpha_3 S}\left[\frac{-2}{(1-\nu)\lambda^2}\Upsilon_2 + \Upsilon_0\right]\end{aligned} \tag{67}$$

The expressions for the particular solutions V^*_α and V^*_3 can be obtained as follows:

$$\begin{aligned}V^*_\alpha &= \frac{-\ell^2}{4\pi D\alpha_3 S} r_{,\alpha}\Upsilon_{-1}\\ V^*_3 &= \frac{\ell^2}{4\pi D\alpha_3 S}\left[\frac{-2}{(1-\nu)\lambda^2}\Upsilon_0 + \Upsilon_{-2}\right]\end{aligned} \tag{68}$$

and their derivatives with respect to the normal are:

$$\begin{aligned}V^*_{\alpha,n} &= \frac{-\ell^2}{4\pi D\alpha_3 S}\left[\frac{\Upsilon_{-1}}{r}(n_\alpha - 2r_{,\alpha}r_{,n}) - \Upsilon_0 r_{,\alpha}r_{,n}\right]\\ V^*_{3,n} &= \frac{-\ell^2 r_{,n}}{4\pi D\alpha_3 S}\left[\frac{-2}{(1-\nu)\lambda^2}\Upsilon_1 + \Upsilon_{-1}\right]\end{aligned} \tag{69}$$

Case 2 ($\kappa = 1$)

Similar to case 1, the expressions for the particular solutions V_α^* and V_3^* can be obtained as follows:

$$
\begin{aligned}
V_\alpha^* &= \frac{-\ell^3 r_{,\alpha}}{4\pi D\alpha_3}\left[(\frac{r}{\ell})K_0(\frac{r}{\ell}) + 2K_1(\frac{r}{\ell})\right] \\
V_3^* &= \frac{\ell^2}{4\pi D\lambda^2(1-\nu)\alpha_3}\left[2(2+\ell^2\lambda^2\alpha_4)K_0(\frac{r}{\ell}) + \ell^2\lambda^2\alpha_4(\frac{r}{\ell})K_1(\frac{r}{\ell})\right]
\end{aligned} \tag{70}
$$

and their derivatives with respect to the normal are:

$$
\begin{aligned}
V_{\alpha,n}^* &= \frac{-\ell^2}{4\pi D\alpha_3}\left[A_1(\frac{r}{\ell})(n_\alpha - 2r_{,\alpha}r_{,n}) - (\frac{r}{\ell})K_1(\frac{r}{\ell})r_{,\alpha}r_{,n}\right] \\
V_{3,n}^* &= \frac{-r_{,n}\ell}{4\pi D(1-\nu)\lambda^2\alpha_3}\left[4K_1(\frac{r}{\ell}) + \ell^2\lambda^2\alpha_4\left(2K_1(\frac{r}{\ell}) + (\frac{r}{\ell})K_0(\frac{r}{\ell})\right)\right]
\end{aligned} \tag{71}
$$

Case 3 ($\kappa < 1$)

Similar to case 1 and 2, the expressions for the particular solutions V_α^* and V_3^* can be obtained as follows:

$$
\begin{aligned}
V_\alpha^* &= \frac{-\ell^2}{8D\alpha_3\sin 2\psi}\Psi_{-1}r_{,\alpha} \\
V_3^* &= \frac{\ell^2}{8D\alpha_3\sin 2\psi}\left[\frac{2}{(1-\nu)\lambda^2}\Psi_0 + \Psi_{-2}\right]
\end{aligned} \tag{72}
$$

and their derivatives with respect to the normal are:

$$
\begin{aligned}
V_{\alpha,n}^* &= \frac{-\ell^2}{8D\alpha_3\sin 2\psi}\left[\frac{\Psi_{-1}}{r}(n_\alpha - 2r_{,\alpha}r_{,n}) + \Psi_0 r_{,\alpha}r_{,n}\right] \\
V_{3,n}^* &= \frac{-\ell^2 r_{,n}}{8D\alpha_3\sin 2\psi}\left[\frac{2}{(1-\nu)\lambda^2}\Psi_1 + \Psi_{-1}\right]
\end{aligned} \tag{73}
$$

7.2 Singular terms

Expanding Bessel functions and Hankel functions for small argument [18], it can be seen that, the same singular terms are found in the kernel $V_{i,n}^*$ for the three cases of the fundamental solution. The singular terms can be written as:

$$
\begin{aligned}
V^{s*}_{\alpha,n} &= \frac{-\ell^4}{2\pi D\alpha_3 r^2}(n_\alpha - 2r_{,\alpha}r_{,n}) \\
V^{s*}_{3,n} &= \frac{-\ell^4 r_{,n}}{2\pi D\alpha_3 r}
\end{aligned}
\tag{74}
$$

where the superscript $(\cdot)^s$ denotes the singular term of $(\cdot)$.

As can be seen that $V^*_{\alpha,n}$ contains a hypersingular term $(O(1/r^2))$ and a strong singular term $(O(r_{,n}/r^2))$; whereas the kernel $V^*_{3,n}$ contains weakly singular terms of order $(O(r_{,n}/r))$.

7.3 Boundary and internal collocation

As mentioned in the previous section, the high order of singularity in the kernel $V^*_{i,n}$ results in a discontinuity in the equivalent boundary integrals. These boundary integrals (as discussed earlier) can be evaluated using Gauss-Legendre formulae when the collocation point is placed outside the boundary. In the next sections, the evaluation of the integrals when the collocation point is placed on the boundary or at an internal point is discussed.

7.3.1 Boundary collocation

In the following sections the two relevant integrals ($i = \alpha$ and $i = 3$) will be considered individually as the collocation point $\mathbf{X}''$ approaches the boundary at $\mathbf{x}'$. Define I_i as:

$$
\begin{aligned}
I_\alpha &= q \, ⨎_\Gamma V^*_{\alpha,n}(\mathbf{x}', \mathbf{x}) d\Gamma(\mathbf{x}) \\
I_3 &= q \int_\Gamma V^*_{3,n}(\mathbf{x}', \mathbf{x}) d\Gamma(\mathbf{x})
\end{aligned}
\tag{75}
$$

Noting that the integral I_α are expressed in Hadamard finite part (⨎) [19] as the kernel $V^*_{\alpha,n}$ contains hypersingularity.

The integral I_α

The integral expression for I_α can be rewritten as:

$$I_\alpha = q\left[\int_\Gamma V^{r*}_{\alpha,n}(\mathbf{x}',\mathbf{x})\mathrm{d}\Gamma(\mathbf{x}) + \oint_\Gamma V^{s*}_{\alpha,n}(\mathbf{x}',\mathbf{x})\mathrm{d}\Gamma(\mathbf{x})\right] \tag{76}$$

where the superscripts $(\cdot)^r$ and $(\cdot)^s$ represent the regular part and the singular part of $(\cdot)$ respectively.

In eqn (76), the first integral is regular and can be evaluated using the Gauss-Legendre scheme. The second integral is hypersingular. However, this second integral can be shown to vanish. If the singular part of the kernel V^*_α is considered, as follows:

$$V^{s*}_\alpha = -\frac{\ell^4}{2\pi D\alpha_3}\frac{r_{,\alpha}}{r} \tag{77}$$

It can be shown that:

$$\nabla^2 V^{s*}_\alpha = 0 \tag{78}$$

If this function is integrated over the a closed domain, it gives:

$$\int_\Omega \nabla^2 V^{s*}_\alpha \mathrm{d}\Omega = 0 \tag{79}$$

Using eqn (65) with $q = 1$, one can write:

$$\oint_\Gamma V^{s*}_{\alpha,n}\mathrm{d}\Gamma = 0 \tag{80}$$

Therefore eqn (76) can be written as:

$$I_\alpha = q\int_\Gamma V^{r*}_{\alpha,n}(\mathbf{x}',\mathbf{x})\mathrm{d}\Gamma(\mathbf{x}) \tag{81}$$

In order to carry out the numerical evaluation of I_α using the Gauss-Legendre scheme, the kernel $V^{s*}_{\alpha,n}$ must be introduced to the final integral to ensure the cancellation of the hypersingular terms. So that:

$$I_\alpha = q\int_\Gamma [V^{*}_{\alpha,n}(\mathbf{x}',\mathbf{x}) - V^{s*}_{\alpha,n}(\mathbf{x}',\mathbf{x})]\mathrm{d}\Gamma(\mathbf{x}) \tag{82}$$

The integral I_3

In this section the second integral I_3 will be considered. If a half circular region Γ_ε^* of radius ε is constructed around collocation point $\mathbf{x}'$ at a corner as shown in Figure 1, I_3 can be written as follows:

$$I_3 = \lim_{\varepsilon \to 0} q \left[\int_{\Gamma-\Gamma_\varepsilon} V_{3,n}^*(\mathbf{x}', \mathbf{x}) \mathrm{d}\Gamma(\mathbf{x}) + \int_{\Gamma_\varepsilon^*} V_{3,n}^*(\mathbf{x}', \mathbf{x}) \mathrm{d}\Gamma(\mathbf{x}) \right] \tag{83}$$

Using eqn (74), the following expression can be written:

$$\lim_{\varepsilon \to 0} q \int_{\Gamma_\varepsilon^*} V_{3,n}^*(\mathbf{x}', \mathbf{x}) \mathrm{d}\Gamma(\mathbf{x}) = \frac{-\ell^4 q}{2\pi D \alpha_3} \lim_{\varepsilon \to 0} \int_{\Gamma_\varepsilon^*} \frac{r_{,n}}{r} \mathrm{d}\Gamma \tag{84}$$

Using the relationships in eqn (55) and taking the limiting form of the integral in eqn (84) as $\varepsilon \to 0$, the following jump term can be obtained:

$$\frac{-\ell^4 q}{2\pi D \alpha_3} \lim_{\varepsilon \to 0} \int_{\Gamma_\varepsilon^*} \frac{r_{,n}}{r} \mathrm{d}\Gamma = \frac{\ell^4 q}{\alpha_3 D} \frac{\Delta\varphi}{2\pi} \tag{85}$$

The integral I_3 can be written as follows:

$$I_3 = q \int_\Gamma V_{3,n}^*(\mathbf{x}', \mathbf{x}) \mathrm{d}\Gamma(\mathbf{x}) + \frac{\ell^4 q}{\alpha_3 D} \frac{\Delta\varphi}{2\pi} \tag{86}$$

and for smooth boundary $\varphi_2 - \varphi_1 = \pi$, it simplifies to:

$$I_3 = q \int_\Gamma V_{3,n}^*(\mathbf{x}', \mathbf{x}) \mathrm{d}\Gamma(\mathbf{x}) + \frac{\ell^4 q}{2\alpha_3 D} \tag{87}$$

Now the equivalent transformed kernels for a boundary collocation point $\mathbf{x}'$ can be written as follows:

$$\begin{aligned} \int_\Omega U_{i3}^*(\mathbf{x}', \mathbf{X}) q \mathrm{d}\Omega(X) &= q \int_\Gamma [V_{i,n}^*(\mathbf{x}', \mathbf{x}) - V_{i,n}^{s*}(\mathbf{x}', \mathbf{x})(1 - \delta_{i3})] \mathrm{d}\Gamma(\mathbf{x}) \\ &+ \frac{\ell^4 q}{\alpha_3 D} \frac{\Delta\varphi}{2\pi} \delta_{i3} \qquad \text{(no summation on } i\text{)} \end{aligned} \tag{88}$$

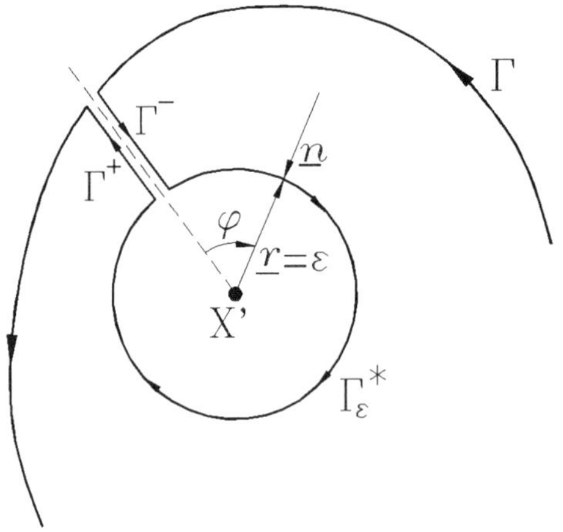

Figure 2. Internal collocation.

7.3.2 Internal collocation

Eqn (65) will be considered as the collocation point $\mathbf{X}''$ moves inside the domain to $\mathbf{X}'$ position. A circular region Γ_ε^* of radius ε is constructed around $\mathbf{X}'$ as shown in Figure 2. Now eqn (65) can be rewritten as follows:

$$\begin{aligned}\int_\Omega U_{i3}^*(\mathbf{X}',\mathbf{X})q\mathrm{d}\Omega(\mathbf{X}) &= \lim_{\varepsilon\to 0} q\Big[\int_\Gamma V_{i,n}^*(\mathbf{X}',\mathbf{x})\mathrm{d}\Gamma(\mathbf{x}) + \int_{\Gamma-} V_{i,n}^*(\mathbf{X}',\mathbf{x})\mathrm{d}\Gamma(\mathbf{x}) \\ &+ \int_{\Gamma+} V_{i,n}^*(\mathbf{X}',\mathbf{x})\mathrm{d}\Gamma(\mathbf{x}) + \int_{\Gamma_\varepsilon^*} V_{i,n}^*(\mathbf{x}',\mathbf{x})\mathrm{d}\Gamma(\mathbf{x})\Big] \end{aligned} \quad (89)$$

Similar to the boundary collocation, by considering the relationships in eqn (55) and noting that:

$$\int_{\Gamma+} \cdots \mathrm{d}\Gamma = -\int_{\Gamma-} \cdots \mathrm{d}\Gamma$$

and

$$\int_{\Gamma_\varepsilon^*} \cdots \mathrm{d}\Gamma_\varepsilon^* = \int_0^{2\pi} \cdots \varepsilon \mathrm{d}\varphi \quad (90)$$

eqn (89) can be written as follows:

$$\int_\Omega U_{i3}^*(\mathbf{X}',\mathbf{X})q\mathrm{d}\Omega(\mathbf{X}) = q\int_\Gamma V_{i,n}^*(\mathbf{X}',\mathbf{x})\mathrm{d}\Gamma(\mathbf{x}) + \frac{\ell^4 q}{\alpha_3 D}\delta_{i3} \quad (91)$$

8 Numerical implementation

In this chapter, quadratic isoparametric elements are used to discretize the boundary of the plate. The local positions of the element nodes are left arbitrary, to allow for the use of continuous or discontinuous elements (see Rashed *et al.* [12]).

After the discretization of the boundary into N_n nodes (N_f of which have free boundary), eqn (52) and eqn (63) can be written in a matrix form as:

$$[H]\{u\} = [G]\{p\} + \{Q\} \tag{92}$$

where $[H]$ and $[G]$ are the boundary element influence matrices, $\{u\}$ and $\{p\}$ are the vectors of the boundary displacements and tractions respectively, and $\{Q\}$ is the domain load vector. After applying the boundary conditions, eqn (92) can be written in the following form:

$$[A]_{3N_n+N_f \times 3N_n+N_f} \{x\}_{3N_n+N_f \times 1} = \{b\}_{3N_n+N_f \times 1} \tag{93}$$

where $[A]$ is the coefficient matrix, $\{x\}$ is the vector of the unknown boundary values, and $\{b\}$ is the vector of the prescribed values. In case of using the Winkler foundation, $N_f = 0$. For simplicity in programing, eqn (92) can be assembled assuming that $p_f = 0$ then applying the boundary conditions to give:

$$[A_0]_{3N_n \times 3N_n} \{x_0\}_{3N_n \times 1} = \{b_0\}_{3N_n \times 1} \tag{94}$$

The new unknowns due to the free boundary conditions can be included in the final system of equations, to give:

$$\begin{bmatrix} [A_0] & [H_f] \\ & \\ [U_f] & [P_f] \end{bmatrix} \begin{Bmatrix} \{x_0\} \\ \\ \{p_f\} \end{Bmatrix} = \begin{Bmatrix} \{b_0\} \\ \\ \{0\} \end{Bmatrix} \tag{95}$$

which is an expanded form of eqn (93). The sub-matrices in eqn (95) are the coefficients of the following integrals (recall equations (52) and (63)):

$$\begin{aligned}
[H_f]_{3N_n \times N_f} &\rightarrow \int_\Gamma U^*_{i3} p_f(\mathbf{x}) \mathrm{d}\Gamma \\
[U_f]_{N_f \times 3N_n} &\rightarrow c_f u_3 - \int_\Gamma U^*_{f,n} u_3 \mathrm{d}\Gamma - \int_\Gamma U^*_f u_\alpha n_\alpha \mathrm{d}\Gamma \\
[P_f]_{N_f \times N_f} &\rightarrow \frac{\alpha_3}{G_f} \int_\Gamma U^*_f p_f \mathrm{d}\Gamma
\end{aligned} \tag{96}$$

The weak singularity is evaluated using non-linear coordinate transformation proposed by Telles [20]. Strong singularity, on the other hand, is evaluated based on semi-analytical process using the Taylor series expansion (see Aliabadi *et al.* [21]).

9 Internal point kernels

After the solution for the boundary values, the displacement at any internal point $\mathbf{X}'$ can be evaluated using eqn (52) with $c_{ij} = \delta_{ij}$ and $\Delta\varphi = 2\pi$. The settlement at an exterior point $\mathbf{X}''$ can be also evaluated using eqn (63) with $c_f = 1$.

The stress resultants at an internal point $\mathbf{X}'$ can be evaluated by differentiating eqn (52) with the respect to the coordinate of the source point and substituting in eqn (1) to give:

$$\begin{aligned} M_{\alpha\beta}(\mathbf{X}') &= \int_\Gamma U^*_{\alpha\beta j}(\mathbf{X}',\mathbf{x})p_j(\mathbf{x})\mathrm{d}\Gamma(\mathbf{x}) \\ &+ \int_\Gamma U^*_{\alpha\beta 3}(\mathbf{X}',\mathbf{x})[p_f(\mathbf{x}) + G_f u_{3,n}(\mathbf{x})]\mathrm{d}\Gamma(\mathbf{x}) \\ &- \int_\Gamma P^*_{\alpha\beta j}(\mathbf{X}',\mathbf{x})u_j(\mathbf{x})\mathrm{d}\Gamma(\mathbf{x}) - G_f\int_\Gamma U^*_{\alpha\beta 3,n}(\mathbf{X}',\mathbf{x})u_3(\mathbf{x})\mathrm{d}\Gamma(\mathbf{x}) \\ &+ q\int_\Gamma W^*_{\alpha\beta}(\mathbf{X}',\mathbf{x})\mathrm{d}\Gamma(\mathbf{x}) \qquad (97) \\ Q_\beta(\mathbf{X}') &= \int_\Gamma U^*_{3\beta j}(\mathbf{X}',\mathbf{x})p_j(\mathbf{x})\mathrm{d}\Gamma(\mathbf{x}) \\ &+ \int_\Gamma U^*_{3\beta 3}(\mathbf{X}',\mathbf{x})[p_f(\mathbf{x}) + G_f u_{3,n}(\mathbf{x})]\mathrm{d}\Gamma(\mathbf{x}) \\ &- \int_\Gamma P^*_{3\beta j}(\mathbf{X}',\mathbf{x})u_j(\mathbf{x})\mathrm{d}\Gamma(\mathbf{x}) - G_f\int_\Gamma U^*_{3\beta 3,n}(\mathbf{X}',\mathbf{x})u_3(\mathbf{x})\mathrm{d}\Gamma(\mathbf{x}) \\ &+ q\int_\Gamma W^*_{3\beta}(\mathbf{X}',\mathbf{x})\mathrm{d}\Gamma(\mathbf{x}) \qquad (98) \end{aligned}$$

where

$$\begin{aligned} U^*_{\alpha\beta j} &= \frac{D(1-\nu)}{2}\left[U^*_{\alpha j,\beta} + U^*_{\beta j,\alpha} + \frac{2\nu}{1-\nu}U^*_{\gamma j,\gamma}\delta_{\alpha\beta}\right] \\ U^*_{3\beta j} &= \frac{D(1-\nu)\lambda^2}{2}[U^*_{\beta j} + U^*_{3j,\beta}] \qquad (99) \end{aligned}$$

$$
\begin{aligned}
P^*_{\alpha\beta j} &= \frac{D(1-\nu)}{2}\left[P^*_{\alpha j,\beta} + P^*_{\beta j,\alpha} + \frac{2\nu}{1-\nu}P^*_{\gamma j,\gamma}\delta_{\alpha\beta}\right] \\
P^*_{3\beta j} &= \frac{D(1-\nu)\lambda^2}{2}[P^*_{\beta j} + P^*_{3j,\beta}]
\end{aligned} \tag{100}
$$

$$
\begin{aligned}
W^*_{\alpha\beta} &= \frac{D(1-\nu)}{2}\left[V^*_{\alpha,n\beta} + V^*_{\beta,n\alpha} + \frac{2\nu}{1-\nu}V^*_{\gamma,n\gamma}\delta_{\alpha\beta}\right] \\
W^*_{3\beta} &= \frac{D(1-\nu)\lambda^2}{2}[V^*_{\beta,n} + V^*_{3,n\beta}]
\end{aligned} \tag{101}
$$

The expressions for the kernels U^*_{ijk}, $U^*_{i\beta 3,n}$,P^*_{ijk} and $W^*_{i\beta}$ are given by Rashed *et al.* [12] and listed in Appendix C.

10 Examples

In this section, several numerical examples will be presented to demonstrate the accuracy and efficiency of the present formulation. The examples will cover all possible types of boundary conditions. The Gauss-Legendre integration scheme is used to evaluate the integrals. Defining:

$$
\overline{k_f} = \frac{k_f a^4}{D}, \qquad \overline{G_f} = \frac{G_f a^2}{D}, \qquad \overline{w} = \frac{wD}{qa^4} \quad \text{and} \quad \overline{M} = \frac{M}{qa^2} \tag{102}
$$

to represent the normalized values of the modulus of sub grade reaction, the shear modulus, the deflection and the moment.

10.1 Comparison to the thin plate theory

In this example, a circular plate of 10 in radius is considered. The plate has a modulus of rigidity $D = 10^6$ *in.lb* and under uniform pressure $q = 1000$ psi. The modulus of sub-grade reaction k_f is taken to be 2×10^4 lb/in^3. Yu [7] presented the analytical solution of that problem based on the thin plate theory. In [7] three cases were considered for the Pasternak model and one case for the Winkler model. The corresponding dimensionless parameter that distinguishes between the cases for thin plate theory is κ_t and defined [6] as follows:

$$
\kappa_t = \frac{G_f}{2\sqrt{k_f D}} \tag{103}
$$

Unlike thick plate theory, thin plate theory has only one case of the fundamental solution for the Winkler model (as $\kappa_t = 0$). Table 1 lists the parametric studies considered for this example.

Table 1. Parametric studies for the circular plate

G_f	κ_t	Thin plate case	κ	Thick plate case
3×10^6	10.607	1*	7.798	1
2.82843×10^5	1	2*	0.981	3
2.88557×10^5	1.041	1	1	2
3×10^4	0.106	3*	0.126	3
0	0	Winkler*	0.020	3

* Studies considered by Yu [7].

Figures 3 and 4 show the deflection and the radial moment along the central line for the simply-supported and the clamped boundary conditions. As can be seen the BEM results are in good agreement with the analytical results given by Yu [7].

10.2 Comparison to Case 3

A simply supported square plate of side length a is considered in this example. A uniform load q is applied over the whole of the plate domain. The plate is analyzed using 16 boundary elements. The following parametric studies are carried out:

1. A comparison between the results obtained by Wang *et al.* [8] using 32 linear elements and the present solution for different plate thickness is shown in Table 2. Poisson's ratio is chosen to be 0.25 and $\overline{k_f} = 200$ as in Ref. [8]. The results obtained are in good agreement with those reported in Ref. [8] with a smaller number of elements, as the quadratic element is more suitable for this type of problem.

2. Here, the above problem is considered, with k_f selected to give $0.75 \leq \kappa \leq 1.25$. The plate thickness is fixed as $0.3a$ and Poisson's ratio $= 0.3$. Figures 5 and 6 represent the variation of $\overline{w_c} \times 10^4$ and $\overline{M_c} \times 100$ against

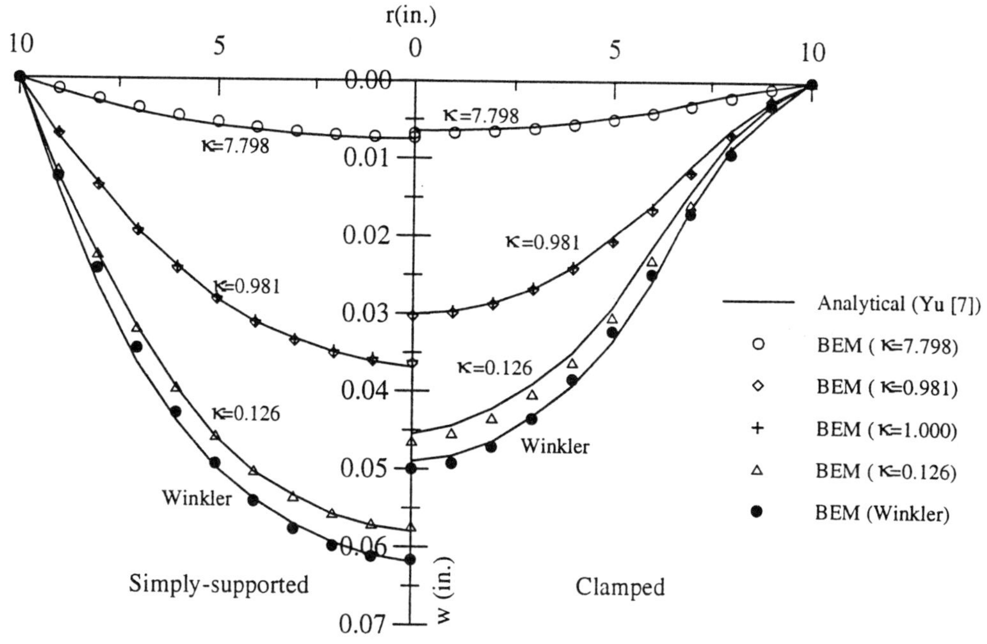

Figure 3. Deflections for the circular plate.

κ for different values of $\overline{G_f}$. From these figures, it can be seen that the agreement between the three cases is excellent, particularly in the vicinity of $\kappa = 0.99, 1.00, 1.01$.

10.3 Free edge circular plate

A circular plate of 1 m diameter (d) is resting on a Pasternak foundation ($k_f = 6.48 \times 10^7$ N/m^3 and $G_f = 2.25 \times 10^6$ N/m). The plate material properties are: $\nu = 0.3$ and $E = 2.1 \times 10^{11}$ N/m^2. Table 3 lists different cases considered for the plate thickness. Figures 7 and 8 show the results of the present model for the radial deflection, the foundation settlement and the radial bending moment together with results given in Ref. [9] for $h/d = 0.01, 0.1$. It has to be noted that the present results for $h/d = 0.01$ are close to the results of $h/d = 0.010046$, as κ is nearly the same; however they are computed based on two different fundamental solutions. That confirms the agreement between the fundamental solution cases

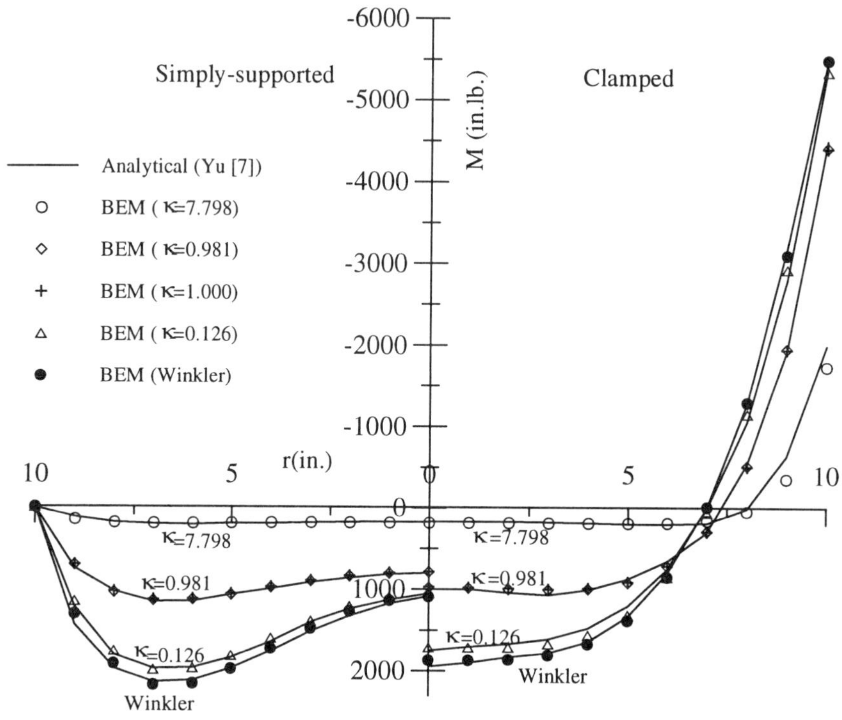

Figure 4. Radial bending moments for the circular plate.

in the transition zone between them.

Figure 9 shows the variation of boundary shear with different values of h/d together with the results reported in Ref. [9]. It can be seen that h/d increase as the plate rigidity increases which results in high discontinuity between the plate and the foundation and higher boundary shear.

Figure 10 shows the variation of the edge deflection for different values of h/d. Results are presented for the Pasternak and the Winkler model. For Pasternak model, results are compared against results obtained from Ref. [9], whereas for the Winkler model ($G_f = 0$) results are compared against the analytical solution which is equal to the elongation of a one-dimensional bar under axial load q $(= q/k_f)$.

10.4 Comparison to finite element analysis

In this example, a simple reinforced concrete combined footing given in Ref. [5] and shown in Figure 11 is considered. The combined footing dimensions are

Table 2. Results for the simply supported plate

$\overline{G_f}$	h/a	$\overline{w_c}[8] \times 100$	$\overline{w_c}$ Present $\times 100$	$\overline{M_c}[8] \times 100$	$\overline{M_c}$ Present $\times 100$
5	0.005	0.22496	0.22414	2.40655	2.40590
	0.100	0.23031	0.23034	2.35212	2.35519
	0.200	0.24409	0.24455	2.19879	2.20423
20	0.005	0.15598	0.15565	1.60640	1.60870
	0.100	0.15816	0.15827	1.56331	1.56675
	0.200	0.16373	0.16403	1.44464	1.44873

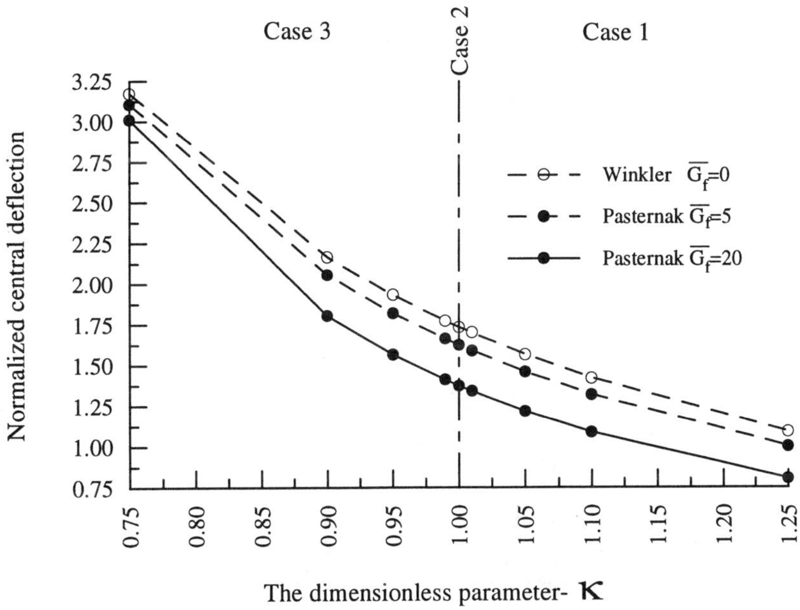

Figure 5: Normalized central defection $\overline{w_c} \times 10^4$ against κ for the simply supported square plate.

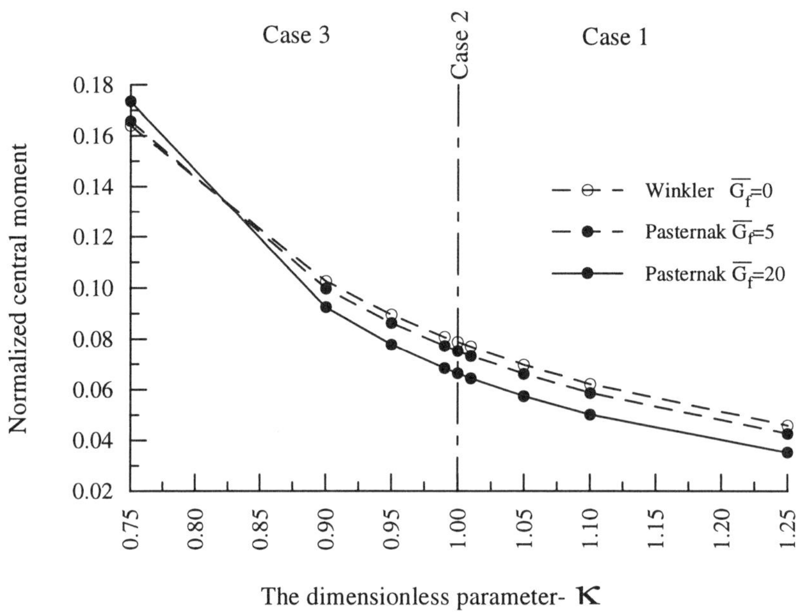

Figure 6. Normalized central moment $\overline{M_c}\times 100$ against κ for the simply supported square plate.

Table 3. Considered cases for the free edge plate

h/d	κ	Case
0.01	1.007	1
0.010046	1.000	2
0.1	0.034	3

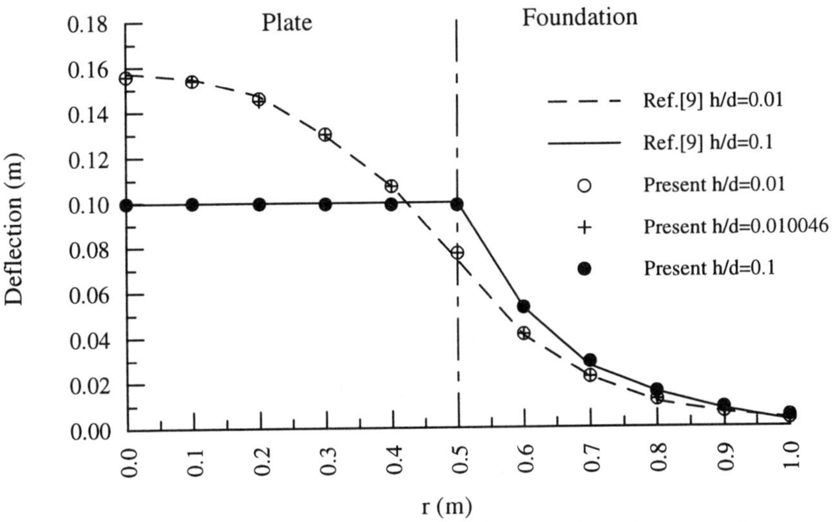

Figure 7. Radial distribution of the deflection for the free edge plate.

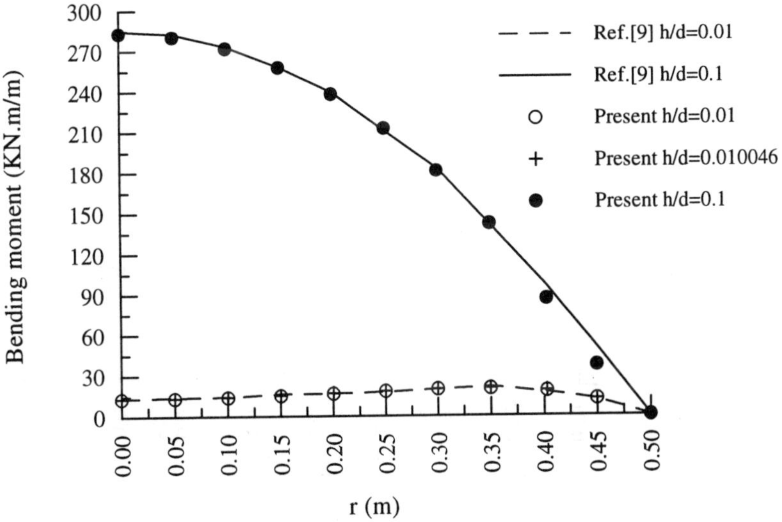

Figure 8: Radial distribution of the bending moment for the free edge plate.

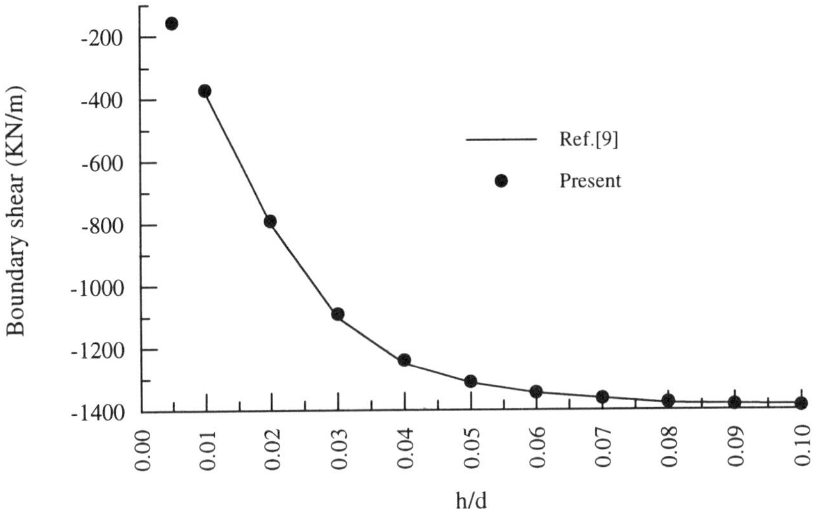

Figure 9: Boundary shear for different plate thicknesses for the free edge plate.

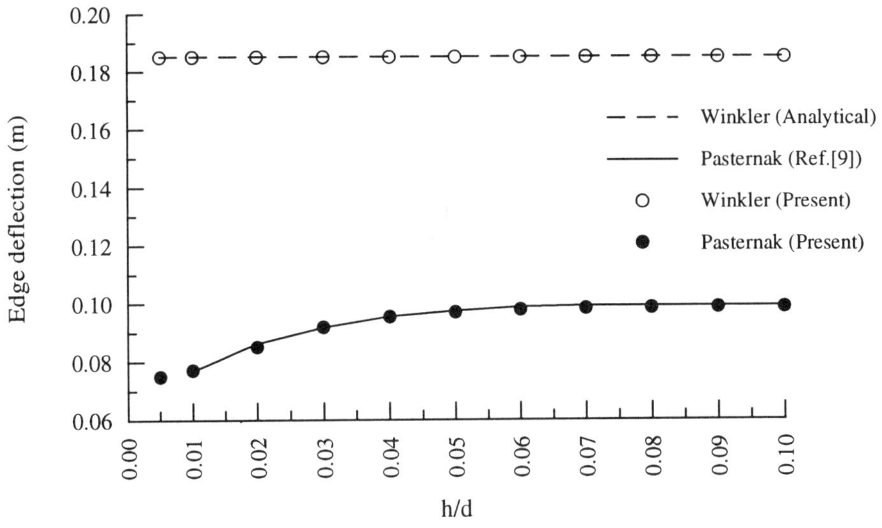

Figure 10. Edge deflection for different plate thicknesses for the free edge plate.

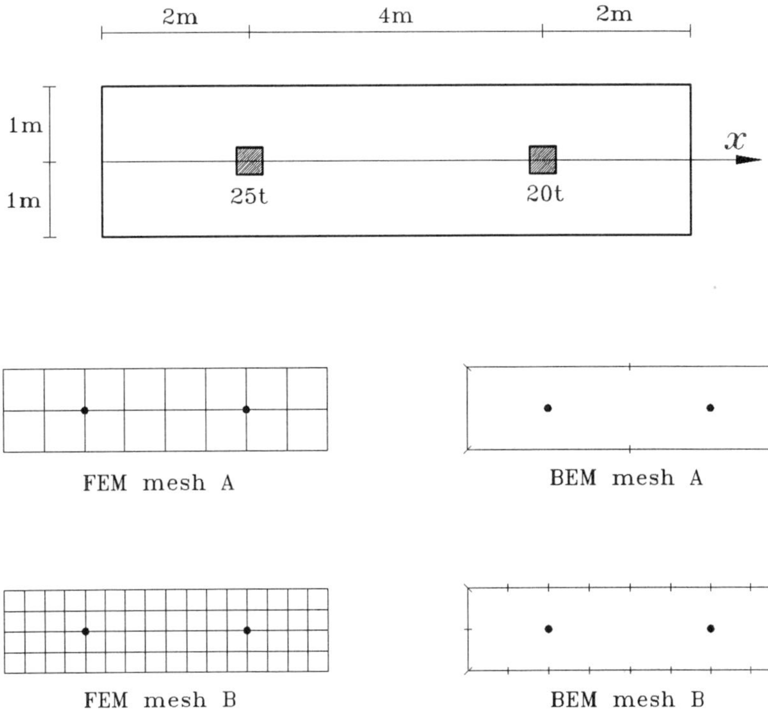

Figure 11. The combined footing under consideration.

$8m \times 2$ m and has thickness of 0.4 m. Two column loads are applied: 25 t and 20 t. The material properties are: $E = 2.1 \times 10^6$ t/m^2 and $\nu = 1/6$. The modulus of sub grade reaction is chosen for soft clay soil and equal to 1000 t/m^3. The problem is analyzed using the present formulation via the BEM and using the FEM (The FEM analysis was performed using a commercial package). Two meshes are considered in the BEM analysis (see Figure 11). The applied load is represented by two square cells 0.1 m $\times$ 0.1 m in each of the meshes. In the FEM, shell elements (thin plate elements) are used. The soil is represented by spring elements acting at the element nodes. The applied load is represented by concentrated loads applied at the element nodes. Two meshes were considered in the FEM analysis (see Figure 11).

Figures 12-15 show the deflection, rotation, bending moment and the shearing force distribution along the x axis. It can be seen that the results of the BEM and the FEM for the deflection and the rotation are close and are independent of the mesh discretization. The results of the bending moment are accurate in the

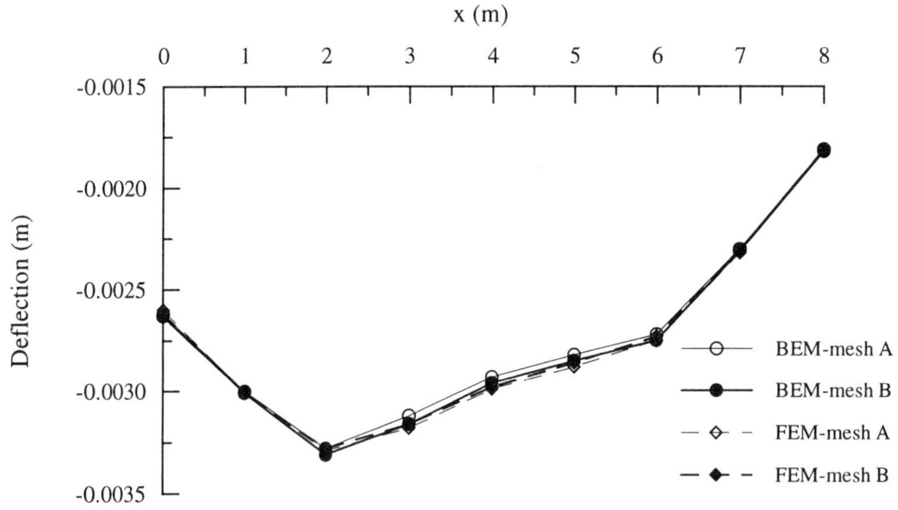

Figure 12. Deflection along the x axis for the combined footing.

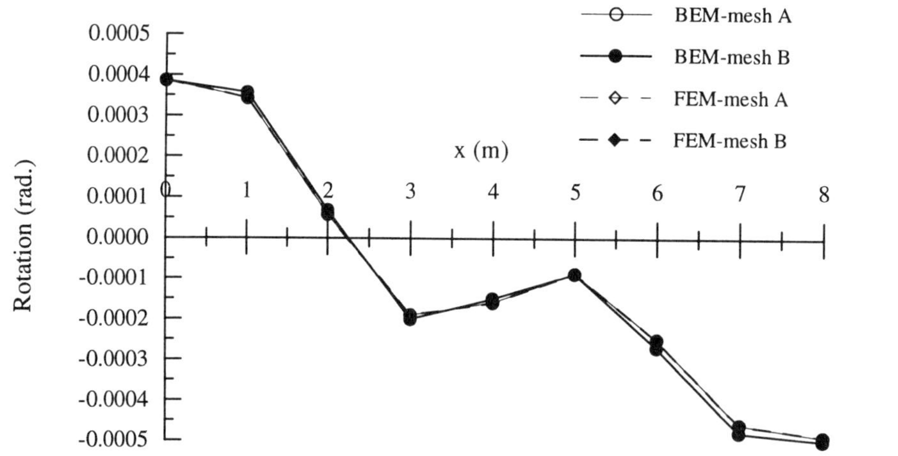

Figure 13. Rotation along the x axis for the combined footing.

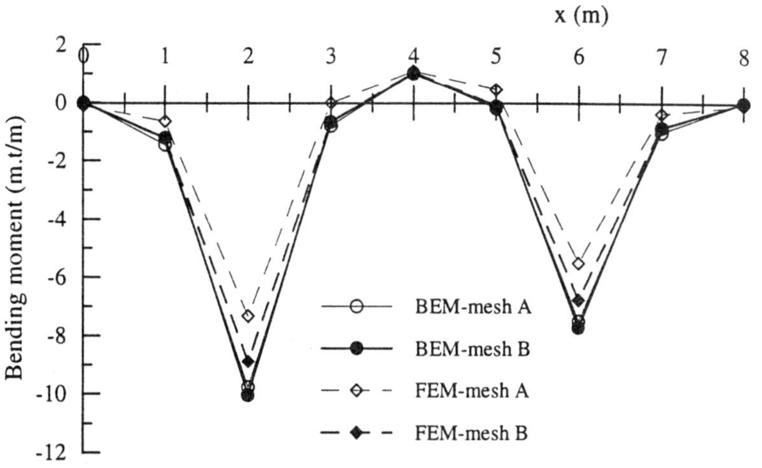

Figure 14. Bending moment along the x axis for the combined footing.

BEM, with only few elements (mesh A). It has to be noted that only the BEM results are represented for the shearing force in Figure 15, as the FEM package used does not give the shear force as output. Also, it was found that the BEM formulation gives zero stresses on the free boundary. For example in the FEM, the bending moment at $x = 0$ is -0.110865 m.t/m for mesh A and -0.0017469 m.t/m for mesh B, whereas in the BEM this value is zero.

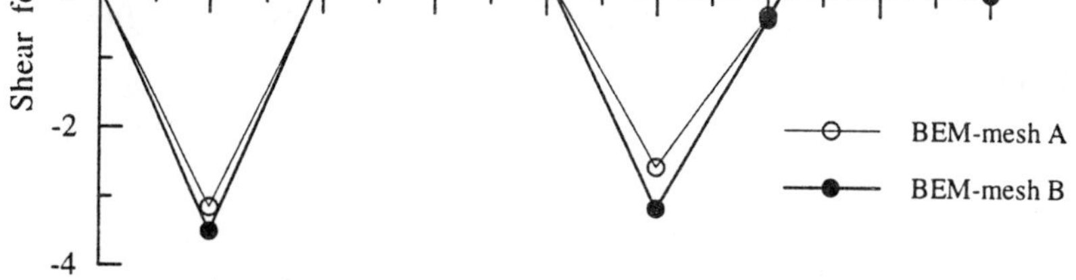

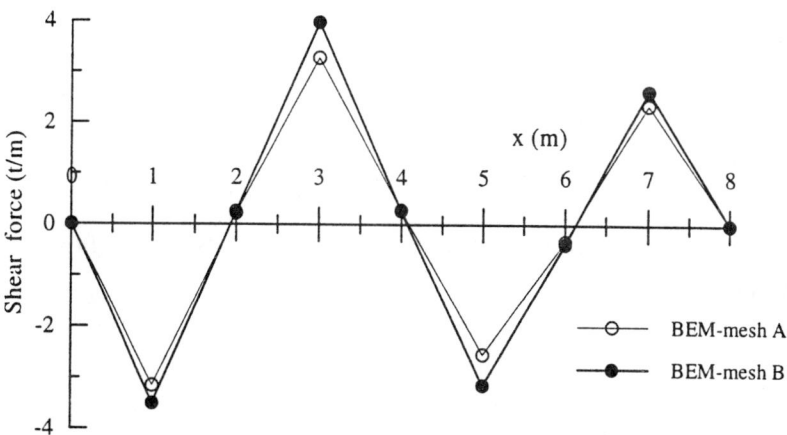

Figure 15. Shear force along the x axis for the combined footing.

References

[1] Timoshenko, S. & Woinowsky-Krieger, S. *Theory of Plates and Shells*, McGraw-Hill, New York, 1959.

[2] Reissner, E. On bending of elastic plates, *Quart. Applied Mathematics*, **5**, 55-68, 1947.

[3] Rashed, Y.F., Aliabadi, M.H. & Brebbia, C.A. Stress and displacement boundary integral formulations for shear-deformable plate bending problems, *BEM 18*, Portugal, 1996.

[4] Kerr, A.D. Elastic and viscoelastic foundation models, *J. Applied Mechanics*, **31**, 491-498, 1964.

[5] Zaher, M.N. *Notes on FEM Analysis-Part 2: Plates and Shells*, Moharram-Bakhoum Ltd., Cairo, Egypt, 1990.

[6] Balaš, J., Sládek, V. & Sládek, J. The boundary integral equation method for plates resting on a two-parameter foundation, *ZAMM*, **64**, 137-146, 1984.

[7] Yu, Y.Y. Axisymmetrical bending of circular plates under simultaneous action of lateral load, force in the middle plane, and elastic foundation, *J. Applied Mechanics*, **24**, 141-143, 1957.

[8] Wang, J., Wang, X. & Huang, M. A boundary integral equation formulation for the Reissner's plates resting on two-parameter foundation, *Acta Mechanica Sinica*, **5**(1), 85-98, 1992.

[9] Fadhil, S. & El-Zafrany, A. Boundary element analysis of thick Reissner plates on two-parameter foundation, *Int. J. Solids Structures*, **31**(21), 2901-2917, 1994.

[10] Rashed, Y.F., Aliabadi, M.H. & Brebbia, C.A. Fundamental solutions for thick foundation plates, *Mech. Res. Comm.*, **24**(3), 331-340, 1997.

[11] Rashed, Y.F., Aliabadi, M.H. & Brebbia, C.A. Transformation of the domain integrals in BEM for thick foundation plates, *J. Applied Mechanics*, (accepted).

[12] Rashed, Y.F., Aliabadi, M.H. & Brebbia, C.A. A boundary element formulation for Reissner plates on a Pasternak foundation, (submitted).

[13] Rashed, Y.F., Aliabadi, M.H. & Brebbia, C.A. The BEM for thick plates on an elastic foundation, *BETECH 97*, Knoxville, USA, 1997.

[14] Rashed, Y.F., Aliabadi, M.H. & Brebbia, C.A. Boundary element analysis of building foundation plates, (submitted).

[15] Selvadurai, A.P.S. *Elastic Analysis of Soil-foundation Interaction*, Elsevier, New York, 1979.

[16] Hörmander, L. *Linear Partial Differential Operators*, Springer Verlag, Berlin, 1963.

[17] Ortner, N. Methods of construction of fundamental solutions of decomposable linear differential operators, *BEM XI*, Vol. 1, Ed. C.A. Brebbia, W.L. Wendland and G. Kuhn, CMP, 1987.

[18] Abramowitz, M. & Stegun, I.A., ed. *Handbook of Mathematical Functions*, Dover, New York, 1965.

[19] Hadamard, J. *Lectures on Cauchy's Problem in Linear Partial Differential Equations*, New Haven, Yale University press, 1923.

[20] Telles, J.C.F. A self-adaptive coordinate transformation for efficient numerical evaluation of general boundary element integrals, *Int. J. Num. Methods Engineering*, **24**, 959-973, 1987.

[21] Aliabadi, M.H., Hall, W.S., & Phemister, T.G. Taylor expansion for singular kernels in the boundary element method. *Int. J. Num. Methods Engineering*, **21**, 2221-2236, 1985.

Appendix A: The kernel $U^*_{i3,n}$

The expressions for $U_{i3,n}$ are as follows (Rashed *et al.* [12]):

Case 1 ($\kappa > 1$)

$$
\begin{aligned}
U^*_{\alpha 3,n} &= \frac{-\ell^2}{4\pi D\alpha_3 S}\left\{\frac{\Upsilon_1}{r}(n_\alpha - 2r_{,\alpha}r_{,n}) - \Upsilon_2 r_{,\alpha}r_{,n}\right\} \\
U^*_{33,n} &= \frac{-\ell^2 r_{,n}}{4\pi D\alpha_3 S}\left\{\Upsilon_1 - \frac{2}{(1-\nu)\lambda^2}\Upsilon_3\right\}
\end{aligned}
\tag{104}
$$

Case 2 ($\kappa = 1$)

$$
\begin{aligned}
U^*_{\alpha 3,n} &= \frac{-1}{4\pi D\alpha_3}\left\{K_0(\frac{r}{\ell})n_\alpha - (\frac{r}{\ell})K_1(\frac{r}{\ell})r_{,\alpha}r_{,n}\right\} \\
U^*_{33,n} &= \frac{-r_{,n}}{4\pi D\alpha_3(1-\nu)\lambda^2}\left\{\frac{4}{\ell}K_1(\frac{r}{\ell}) + \alpha_4 r\lambda^2 K_0(\frac{r}{\ell})\right\}
\end{aligned}
\tag{105}
$$

Case 3 ($\kappa < 1$)

$$
\begin{aligned}
U^*_{\alpha 3,n} &= \frac{\ell^2}{8D\alpha_3 \sin 2\psi}\left\{\frac{\Psi_1}{r}(n_\alpha - 2r_{,\alpha}r_{,n}) + \Psi_2 r_{,\alpha}r_{,n}\right\} \\
U^*_{33,n} &= \frac{\ell^2 r_{,n}}{8D\alpha_3 \sin 2\psi}\left\{\Psi_1 + \frac{2}{(1-\nu)\lambda^2}\Psi_3\right\}
\end{aligned}
\tag{106}
$$

Appendix B: Singular terms

It can be seen (after the expansion of the modified Bessel functions and Hankel functions for small argument [18]) that the kernel U^*_{ij} is weakly singular ($O(\ln r)$) and the kernel P^*_{ij} contains weakly singular terms and strongly singular ($O(1/r)$).

The strongly singular terms do not depend on κ and are given by:

$$
P^*_{\gamma\alpha} \rightarrow \frac{1-\nu}{4\pi r}(r_{,\gamma}n_\alpha - r_{,\alpha}n_\gamma)
\tag{107}
$$

The expressions of the weak singular terms are given as follows (Rashed *et al.* [13]):

Case 1 ($\kappa > 1$)

$$
\begin{aligned}
U^*_{\alpha\beta} &\rightarrow \frac{1}{\pi D(1-\nu)}\left[-\frac{1}{2}\ln(\frac{z}{2}) + \frac{\ell^2}{8S}(1-\nu)\Lambda_1 - \frac{1}{4\lambda^2\ell^2 S}\Lambda_{-1}\right]\delta_{\alpha\beta} \\
&+ \frac{\ell^2}{4\pi D(1-\nu)S}\left[(1-\nu)[\Lambda_2 - \Lambda_1] - \frac{2}{\lambda^2\ell^4}[\Lambda_1 - \Lambda_{-1}]\right] r_{,\alpha} r_{,\beta} \\
U^*_{33} &\rightarrow \frac{\ell^2}{4\pi D\alpha_3 S}\left[\frac{2}{\lambda^2(1-\nu)}\Lambda_2 - \Lambda_0\right]
\end{aligned}
\tag{108}
$$

$$
U^*_{\alpha 3,n} \rightarrow -\frac{\ell^2}{8\pi D S\alpha_3} n_\alpha \Lambda_1 \tag{109}
$$

$$
\begin{aligned}
P^*_{\gamma 3} &\rightarrow -\frac{\lambda^2}{4\pi}\ln(\frac{z}{2})n_\gamma + \frac{\ell^2}{4\pi S}\left[\frac{1}{2}(\Lambda_4 - \Lambda_1\Lambda_1)(n_\gamma - 2r_{,\gamma}r_{,n})\right. \\
&+ \left.(\Lambda_4 - \alpha_1\Lambda_2)r_{,\gamma}r_{,n}\right] \\
P^*_{3\alpha} &\rightarrow \frac{\ell^2}{4\pi S\alpha_3}\Lambda_2[(1-\nu)r_{,\alpha}r_{,n} + \nu n_\alpha]
\end{aligned}
\tag{110}
$$

where

$$
\Lambda_i = e^i \ln(\frac{er}{2}) - d^i \ln(\frac{dr}{2})
$$

Case 2 ($\kappa = 1$)

$$
\begin{aligned}
U^*_{\alpha\beta} &\rightarrow -\frac{1}{2\pi D(1-\nu)}\left[\ln(\frac{z}{2}) + \frac{1-\nu}{2}\ln(\frac{r}{2\ell})\right]\delta_{\alpha\beta} \\
U^*_{33} &\rightarrow -\frac{1}{\pi D(1-\nu)\lambda^2\alpha_3}\ln(\frac{r}{2\ell})
\end{aligned}
\tag{111}
$$

$$
U^*_{\alpha 3,n} \rightarrow \frac{n_\alpha}{4\pi D\alpha_3}\ln(\frac{r}{2\ell}) \tag{112}
$$

$$
\begin{aligned}
P^*_{\gamma 3} &\rightarrow -\frac{\lambda^2}{4\pi}\left[\ln(\frac{z}{2}) + \frac{1}{\lambda^2}(2-\alpha_1)\ln(\frac{r}{2\ell})\right] n_\gamma \\
P^*_{3\alpha} &\rightarrow -\frac{1}{4\pi\alpha_3}(1+\nu)\ln(\frac{r}{2\ell})n_\alpha
\end{aligned}
\tag{113}
$$

Case 3 ($\kappa < 1$)

$$
\begin{aligned}
U^*_{\alpha\beta} &\rightarrow -\frac{1}{2\pi D(1-\nu)}\left[\ln(\frac{z}{2}) + (1-\nu)\ln(\frac{r}{\ell})\right]\delta_{\alpha\beta} \\
U^*_{33} &\rightarrow -\frac{1}{2\pi D\alpha_3}\ln(\frac{r}{\ell})
\end{aligned}
\tag{114}
$$

$$
U^*_{\alpha 3,n} \rightarrow \frac{n_\alpha}{4\pi D\alpha_3}\ln(\frac{r}{\ell})
\tag{115}
$$

$$
\begin{aligned}
P^*_{\gamma 3} &\rightarrow -\frac{\lambda^2}{4\pi}\left[\frac{1}{2}\ln(\frac{z}{2}) + \frac{1}{2\lambda^2}(1-\alpha_1)\ln(\frac{r}{\ell})\right]n_\gamma \\
P^*_{3\alpha} &\rightarrow -\frac{1}{4\pi\alpha_3}(1+\nu)\ln(\frac{r}{\ell})n_\alpha
\end{aligned}
\tag{116}
$$

Appendix C: Internal points kernels

Case 1 ($\kappa > 1$)

$$
\begin{aligned}
U^*_{\alpha\beta\gamma} &= \frac{-1}{2\pi r}\{zK_1(z)[2r_{,\alpha}r_{,\beta}r_{,\gamma} - \delta_{\alpha\gamma}r_{,\beta} - \delta_{\gamma\beta}r_{,\alpha}] \\
&+ 2A_1(z)[4r_{,\alpha}r_{,\beta}r_{,\gamma} - \delta_{\alpha\gamma}r_{,\beta} - \delta_{\gamma\beta}r_{,\alpha} - \delta_{\alpha\beta}r_{,\gamma}]\} \\
&- \frac{\ell^2}{4\pi\lambda^2 S}\left\{[\Upsilon_5 - \alpha_2\Upsilon_3]\left[2r_{,\alpha}r_{,\beta}r_{,\gamma} + \frac{2\nu}{1-\nu}\delta_{\alpha\beta}r_{,\gamma}\right]\right. \\
&+ \left[\frac{4}{r^2}(\Upsilon_3 - \alpha_2\Upsilon_1) + \frac{2}{r}(\Upsilon_4 - \alpha_2\Upsilon_2)\right] \\
&\times \left.[4r_{,\alpha}r_{,\beta}r_{,\gamma} - \delta_{\alpha\gamma}r_{,\beta} - \delta_{\gamma\beta}r_{,\alpha} - \delta_{\alpha\beta}r_{,\gamma}]\right\} \\
U^*_{3\beta\alpha} &= \frac{\lambda^2}{2\pi}[B_1(z)\delta_{\alpha\beta} - A_1(z)r_{,\alpha}r_{,\beta}] \\
&+ \frac{\ell^2}{4\pi S}\left\{\frac{1}{r}[\Upsilon_3 - \alpha_1\Upsilon_1](\delta_{\alpha\beta} - 2r_{,\alpha}r_{,\beta}) - [\Upsilon_4 - \alpha_1\Upsilon_2]r_{,\alpha}r_{,\beta}\right\} \\
U^*_{\alpha\beta 3} &= \frac{-\ell^2}{4\pi\alpha_3 S}\left\{\Upsilon_2\left[(1-\nu)r_{,\alpha}r_{,\beta} + \nu\delta_{\alpha\beta}\right] - \frac{1-\nu}{r}\Upsilon_1[\delta_{\alpha\beta} - 2r_{,\alpha}r_{,\beta}]\right\} \\
U^*_{3\beta 3} &= \frac{-\ell^2}{4\pi\alpha_3 S}\Upsilon_3 r_{,\beta}
\end{aligned}
\tag{117}
$$

$$
\begin{aligned}
U^*_{\alpha\beta 3,n} &= \frac{\ell^2}{4\pi\alpha_3 S}\left\{\Upsilon_3[\nu\delta_{\alpha\beta}r_{,n}+(1-\nu)r_{,\alpha}r_{,\beta}r_{,n}]+\frac{1-\nu}{r}\left[\Upsilon_2+\frac{2}{r}\Upsilon_1\right]\right.\\
&\times \left.(4r_{,\alpha}r_{,\beta}r_{,n}-\delta_{\alpha\beta}r_{,n}-n_\beta r_{,\alpha}-n_\alpha r_{,\beta})\right\}\\
U^*_{3\beta 3,n} &= \frac{-\ell^2}{4\pi\alpha_3 S}\left\{\frac{\Upsilon_3}{r}[n_\beta-2r_{,\beta}r_{,n}]-\Upsilon_4 r_{,\beta}r_{,n}\right\} \qquad (118)
\end{aligned}
$$

$$
\begin{aligned}
P^*_{\alpha\beta\gamma} &= \frac{D(1-\nu)}{4\pi r^2}\{z^2K_0(z)(4r_{,\alpha}r_{,\beta}r_{,\gamma}r_{,n}-\delta_{\alpha\gamma}r_{,\beta}r_{,n}\\
&- r_{,\gamma}r_{,\beta}n_\alpha-\delta_{\beta\gamma}r_{,\alpha}r_{,n}-n_\beta r_{,\alpha}r_{,\gamma})\\
&+ zK_1(z)(32r_{,\alpha}r_{,\beta}r_{,\gamma}r_{,n}+2\delta_{\gamma\beta}n_\alpha+2\delta_{\gamma\alpha}n_\beta-4n_\gamma r_{,\alpha}r_{,\beta}-6n_\beta r_{,\alpha}r_{,\gamma}\\
&- 6n_\alpha r_{,\beta}r_{,\gamma}-6\delta_{\beta\gamma}r_{,\alpha}r_{,n}-6\delta_{\alpha\gamma}r_{,\beta}r_{,n}-4\delta_{\alpha\beta}r_{,\gamma}r_{,n})\\
&+ 2A_1(z)(48r_{,\alpha}r_{,\beta}r_{,\gamma}r_{,n}-8\delta_{\alpha\gamma}r_{,\beta}r_{,n}-8r_{,\gamma}r_{,\beta}n_\alpha-8r_{,\alpha}r_{,\beta}n_\gamma\\
&- 8\delta_{\alpha\beta}r_{,\gamma}r_{,n}-8\delta_{\beta\gamma}r_{,\alpha}r_{,n}-8r_{,\alpha}r_{,\gamma}n_\beta+2\delta_{\alpha\gamma}n_\beta+2\delta_{\beta\gamma}n_\alpha+2\delta_{\alpha\beta}n_\gamma)\}\\
&+ \frac{D(1-\nu)\ell^2}{8\pi\lambda^2 S}\left\{[\Upsilon_6-\alpha_2\Upsilon_4]\left[4r_{,\alpha}r_{,\beta}r_{,\gamma}r_{,n}\right.\right.\\
&+ \left.\frac{4\nu}{1-\nu}\left(\delta_{\alpha\beta}r_{,\gamma}r_{,n}+n_\gamma r_{,\alpha}r_{,\beta}+\frac{\nu}{1-\nu}\delta_{\alpha\beta}n_\gamma\right)\right]\\
&- \frac{1}{r}[\Upsilon_5-\alpha_2\Upsilon_3]\left[4n_\gamma r_{,\alpha}r_{,\beta}+4n_\beta r_{,\alpha}r_{,\gamma}+4n_\alpha r_{,\beta}r_{,\gamma}-32r_{,\alpha}r_{,\beta}r_{,\gamma}r_{,n}\right.\\
&+ 4\delta_{\beta\gamma}r_{,\alpha}r_{,n}+4\delta_{\alpha\gamma}r_{,\beta}r_{,n}+4\delta_{\alpha\beta}r_{,\gamma}r_{,n}\\
&+ \left.\frac{4\nu}{1-\nu}(2\delta_{\alpha\beta}n_\gamma-2n_\gamma r_{,\alpha}r_{,\beta}-2\delta_{\alpha\beta}r_{,\gamma}r_{,n})\right]\\
&+ \frac{4}{r^2}\left[\frac{2}{r}(\Upsilon_3-\alpha_2\Upsilon_1)+(\Upsilon_4-\alpha_2\Upsilon_2)\right]\\
&\times (24r_{,\alpha}r_{,\beta}r_{,\gamma}r_{,n}-4\delta_{\alpha\gamma}r_{,\beta}r_{,n}-4r_{,\gamma}r_{,\beta}n_\alpha-4r_{,\alpha}r_{,\beta}n_\gamma-4\delta_{\alpha\beta}r_{,\gamma}r_{,n}\\
&- \left.4\delta_{\beta\gamma}r_{,\alpha}r_{,n}-4r_{,\alpha}r_{,\gamma}n_\beta+\delta_{\alpha\gamma}n_\beta+\delta_{\beta\gamma}n_\alpha+\delta_{\alpha\beta}n_\gamma)\right\}\\
P^*_{3\beta\alpha} &= \frac{D(1-\nu)\lambda^2}{4\pi r}\{zK_1(z)[2r_{,\alpha}r_{,\beta}r_{,n}-n_\beta r_{,\alpha}-\delta_{\alpha\beta}r_{,n}]\\
&+ 2A_1(z)[4r_{,\alpha}r_{,\beta}r_{,n}-\delta_{\alpha\beta}r_{,n}-n_\beta r_{,\alpha}-r_{,\beta}n_\alpha]\}\\
&+ \frac{D(1-\nu)\ell^2}{8\pi S}\left\{\left(\frac{4}{r^2}[\Upsilon_3-\alpha_1\Upsilon_1]+\frac{2}{r}[\Upsilon_4-\alpha_1\Upsilon_2]\right)\right.\\
&\times [4r_{,\alpha}r_{,\beta}r_{,n}-\delta_{\alpha\beta}r_{,n}-n_\beta r_{,\alpha}-r_{,\beta}n_\alpha]\\
&+ \left.[\Upsilon_5-\alpha_1\Upsilon_3]\left(2r_{,\alpha}r_{,\beta}r_{,n}+\frac{2\nu}{1-\nu}r_{,\beta}n_\alpha\right)\right\}
\end{aligned}
$$

$$\begin{aligned}
P^*_{\alpha\beta 3} &= \frac{D(1-\nu)\lambda^2}{4\pi r}\{zK_1(z)[n_\beta r_{,\alpha} + n_{,\alpha} r_{,\beta} - 2r_{,\alpha} r_{,\beta} r_{,n}] \\
&+ 2A_1(z)[n_\beta r_{,\alpha} + n_\alpha r_{,\beta} + \delta_{\alpha\beta} r_{,n} - 4r_{,\alpha} r_{,\beta} r_{,n}]\} \\
&+ \frac{D(1-\nu)\ell^2}{4\pi S}\Big\{ - [\Upsilon_5 - \alpha_1 \Upsilon_3]\left[r_{,\alpha} r_{,\beta} r_{,n} + \frac{\nu}{1-\nu}\delta_{\alpha\beta} r_{,n}\right] \\
&+ \left[\frac{2}{r^2}(\Upsilon_3 - \alpha_1 \Upsilon_1) + \frac{1}{r}(\Upsilon_4 - \alpha_1 \Upsilon_2)\right] \\
&\times \ [n_\beta r_{,\alpha} + n_\alpha r_{,\beta} + \delta_{\alpha\beta} r_{,n} - 4r_{,\alpha} r_{,\beta} r_{,n}]\Big\} \\
P^*_{3\beta 3} &= \frac{D(1-\nu)\lambda^4}{4\pi}[B_1(z) n_\beta - A_1(z) r_{,\beta} r_{,n}] \\
&+ \frac{D(1-\nu)\lambda^2\ell^2}{8\pi S}\Big\{\frac{1}{r}[\alpha_5 \Upsilon_3 - \alpha_1 \Upsilon_1](n_\beta - 2r_{,\beta} r_{,n}) \\
&- \ [\alpha_5 \Upsilon_4 - \alpha_1 \Upsilon_2] r_{,\beta} r_{,n}\Big\}
\end{aligned} \tag{119}$$

where $\alpha_5 = 1 - \frac{1}{\alpha_3}$.

$$\begin{aligned}
W^*_{\alpha\beta} &= \frac{\ell^2}{4\pi\alpha_3 S}\Big\{r_{,n}\Upsilon_1[\nu\delta_{\alpha\beta} + (1-\nu) r_{,\alpha} r_{,\beta}] \\
&+ \frac{1-\nu}{r}\left[\frac{2\Upsilon_{-1}}{r} + \Upsilon_0\right](4r_{,\alpha} r_{,\beta} r_{,n} - \delta_{\alpha\beta} r_{,n} - n_\beta r_{,\alpha} - n_\alpha r_{,\beta})\Big\} \\
W^*_{3\beta} &= \frac{\ell^2}{4\pi\alpha_3 S}\Big\{\Upsilon_2 r_{,\beta} r_{,n} - \frac{\Upsilon_1}{r}[n_\beta - 2r_{,\beta} r_{,n}]\Big\}
\end{aligned} \tag{120}$$

Case 2 ($\kappa = 1$)

$$\begin{aligned}
U^*_{\alpha\beta\gamma} &= \frac{-1}{2\pi r}\{zK_1(z)[2r_{,\alpha} r_{,\beta} r_{,\gamma} - \delta_{\alpha\gamma} r_{,\beta} - \delta_{\gamma\beta} r_{,\alpha}] \\
&+ 2A_1(z)[4r_{,\alpha} r_{,\beta} r_{,\gamma} - \delta_{\alpha\gamma} r_{,\beta} - \delta_{\gamma\beta} r_{,\alpha} - \delta_{\alpha\beta} r_{,\gamma}]\} \\
&- \left[\frac{\alpha_4}{4\pi\ell}K_1(\frac{r}{\ell}) - \frac{1}{\pi\lambda^2\ell^2 r}A_1(\frac{r}{\ell})\right](4r_{,\alpha} r_{,\beta} r_{,\gamma} - \delta_{\alpha\gamma} r_{,\beta} - \delta_{\gamma\beta} r_{,\alpha} - \delta_{\alpha\beta} r_{,\gamma}) \\
&- \left[\frac{\alpha_4}{4\pi\ell}(\frac{r}{\ell})K_0(\frac{r}{\ell}) - \frac{(1-\nu)}{2\pi\ell}K_1(\frac{r}{\ell})\right]\left(\frac{\nu}{1-\nu}\delta_{\alpha\beta} r_{,\gamma} + r_{,\alpha} r_{,\beta} r_{,\gamma}\right) \\
U^*_{3\beta\alpha} &= \frac{\lambda^2}{2\pi}[B_1(z)\delta_{\alpha\beta} - A_1(z) r_{,\alpha} r_{,\beta}] - \frac{1}{4\pi\ell^2}\Big\{2(\frac{\ell}{r})K_1(\frac{r}{\ell})(\delta_{\alpha\beta} - 2r_{,\alpha} r_{,\beta}) \\
&+ (1-\alpha_1\ell^2)\left[(\frac{r}{\ell})K_1(\frac{r}{\ell}) r_{,\alpha} r_{,\beta} - K_0(\frac{r}{\ell})\delta_{\alpha\beta}\right] - 2K_0(\frac{r}{\ell}) r_{,\alpha} r_{,\beta}\Big\} \\
U^*_{\alpha\beta 3} &= \frac{1}{4\pi\alpha_3}\Big\{K_0(\frac{r}{\ell})(1+\nu)\delta_{\alpha\beta} - (\frac{r}{\ell})K_1(\frac{r}{\ell})[\nu\delta_{\alpha\beta} + (1-\nu) r_{,\alpha} r_{,\beta}]\Big\}
\end{aligned}$$

$$U^*_{3\beta 3} = \frac{r_{,\beta}}{4\pi\ell\alpha_3}\left[2K_1(\frac{r}{\ell}) - (\frac{r}{\ell})K_0(\frac{r}{\ell})\right] \tag{121}$$

$$\begin{aligned}
U^*_{\alpha\beta 3,n} &= \frac{1-\nu}{4\pi\ell\alpha_3}\left\{(\frac{r}{\ell})K_0(\frac{r}{\ell})\left[\frac{\nu}{1-\nu}\delta_{\alpha\beta}r_{,n} + r_{,\alpha}r_{,\beta}r_{,n}\right]\right.\\
&+ \left. K_1(\frac{r}{\ell})\left[2r_{,\alpha}r_{,\beta}r_{,n} - \left(\frac{1+\nu}{1-\nu}\right)\delta_{\alpha\beta}r_{,n} - n_\beta r_{,\alpha} - n_\alpha r_{,\beta}\right]\right\}\\
U^*_{3\beta 3,n} &= \frac{1}{4\pi\ell^2\alpha_3}\left\{2(\frac{\ell}{r})K_1(\frac{r}{\ell})[n_\beta - 2r_{,\beta}r_{,n}] - K_0(\frac{r}{\ell})[n_\beta + 2r_{,\beta}r_{,n}]\right.\\
&+ \left.(\frac{r}{\ell})K_1(\frac{r}{\ell})r_{,\beta}r_{,n}\right\}
\end{aligned} \tag{122}$$

$$\begin{aligned}
P^*_{\alpha\beta\gamma} &= \frac{D(1-\nu)}{4\pi r^2}\{z^2K_0(z)(4r_{,\alpha}r_{,\beta}r_{,\gamma}r_{,n} - \delta_{\alpha\gamma}r_{,\beta}r_{,n}\\
&- r_{,\gamma}r_{,\beta}n_\alpha - \delta_{\beta\gamma}r_{,\alpha}r_{,n} - n_\beta r_{,\alpha}r_{,\gamma})\\
&+ zK_1(z)(32r_{,\alpha}r_{,\beta}r_{,\gamma}r_{,n} + 2\delta_{\gamma\beta}n_\alpha + 2\delta_{\gamma\alpha}n_\beta - 4n_\gamma r_{,\alpha}r_{,\beta} - 6n_\beta r_{,\alpha}r_{,\gamma}\\
&- 6n_\alpha r_{,\beta}r_{,\gamma} - 6\delta_{\beta\gamma}r_{,\alpha}r_{,n} - 6\delta_{\alpha\gamma}r_{,\beta}r_{,n} - 4\delta_{\alpha\beta}r_{,\gamma}r_{,n})\\
&+ 2A_1(z)(48r_{,\alpha}r_{,\beta}r_{,\gamma}r_{,n} - 8\delta_{\alpha\gamma}r_{,\beta}r_{,n} - 8r_{,\gamma}r_{,\beta}n_\alpha - 8r_{,\alpha}r_{,\beta}n_\gamma\\
&- 8\delta_{\alpha\beta}r_{,\gamma}r_{,n} - 8\delta_{\beta\gamma}r_{,\alpha}r_{,n} - 8r_{,\alpha}r_{,\gamma}n_\beta + 2\delta_{\alpha\gamma}n_\beta + 2\delta_{\beta\gamma}n_\alpha + 2\delta_{\alpha\beta}n_\gamma)\}\\
&+ \frac{D(1-\nu)}{4\pi\ell}\left\{\frac{1}{\ell(1-\nu)}\left[\frac{\alpha_4}{1-\nu}(\frac{r}{\ell})K_1(\frac{r}{\ell}) - 2K_0(\frac{r}{\ell})\right]\right.\\
&\times [\nu^2\delta_{\alpha\beta}n_\gamma + \nu(1-\nu)(\delta_{\alpha\beta}r_{,\gamma}r_{,n} + n_\gamma r_{,\alpha}r_{,\beta}) + (1-\nu)^2 r_{,\alpha}r_{,\beta}r_{,\gamma}r_{,n}]\\
&+ \frac{\alpha_4}{\ell}K_0(\frac{r}{\ell})[6r_{,\alpha}r_{,\beta}r_{,\gamma}r_{,n} - n_\gamma r_{,\alpha}r_{,\beta} - n_\beta r_{,\alpha}r_{,\gamma}\\
&- n_\alpha r_{,\beta}r_{,\gamma} - \delta_{\gamma\beta}r_{,\alpha}r_{,n} - \delta_{\alpha\gamma}r_{,\beta}r_{,n} - \delta_{\alpha\beta}r_{,\gamma}r_{,n}]\\
&- \frac{2\nu}{(1-\nu)^2}\frac{\alpha_4}{\ell}K_0(\frac{r}{\ell})n_\gamma\delta_{\alpha\beta}\\
&+ \left[\frac{\alpha_4}{r}K_1(\frac{r}{\ell}) - \frac{4}{\ell\lambda^2r^2}A_1(\frac{r}{\ell})\right]\\
&\times [\delta_{\gamma\beta}n_\alpha + \delta_{\alpha\gamma}n_\beta + \delta_{\alpha\beta}n_\gamma - 4n_\gamma r_{,\alpha}r_{,\beta} - 4n_\beta r_{,\alpha}r_{,\gamma}\\
&- 4n_\alpha r_{,\beta}r_{,\gamma} - 4\delta_{\gamma\beta}r_{,\alpha}r_{,n} - 4\delta_{\alpha\gamma}r_{,\beta}r_{,n} - 4\delta_{\alpha\beta}r_{,\gamma}r_{,n} + 24r_{,\alpha}r_{,\beta}r_{,\gamma}r_{,n}]\\
&+ \frac{2}{r}K_1(\frac{r}{\ell})[(1-\nu)(n_\beta r_{,\alpha}r_{,\gamma} + n_\alpha r_{,\beta}r_{,\gamma} + n_\gamma r_{,\alpha}r_{,\beta}\\
&+ \delta_{\gamma\beta}r_{,\alpha}r_{,n} + \delta_{\alpha\gamma}r_{,\beta}r_{,n} + \delta_{\alpha\beta}r_{,\gamma}r_{,n} - 8r_{,\alpha}r_{,\beta}r_{,\gamma}r_{,n})\\
&+ \left. 2\nu(\delta_{\alpha\beta}n_\gamma - n_\gamma r_{,\alpha}r_{,\beta} - \delta_{\alpha\beta}r_{,\gamma}r_{,n}]\right\}\\
P^*_{3\beta\alpha} &= \frac{D(1-\nu)\lambda^2}{4\pi r}\{zK_1(z)[2r_{,\alpha}r_{,\beta}r_{,n} - n_\beta r_{,\alpha} - \delta_{\alpha\beta}r_{,n}]
\end{aligned}$$

$$
\begin{aligned}
& + \; 2A_1(z)[4r_{,\alpha}r_{,\beta}r_{,n} - \delta_{\alpha\beta}r_{,n} - n_\beta r_{,\alpha} - r_{,\beta}n_\alpha]\} \\
& + \; \frac{D(1-\nu)}{8\pi\ell}\left\{ \left[\frac{4}{r\ell}A_1(\frac{r}{\ell}) - \lambda^2 K_1(\frac{r}{\ell})\left(\alpha_4 - \frac{1-\nu}{\alpha_3}\right)\right]\right. \\
& \times \; [\delta_{\alpha\beta}r_{,n} + n_\beta r_{,\alpha} + r_{,\beta}n_\alpha - 4r_{,\alpha}r_{,\beta}r_{,n}] \\
& + \; \lambda^2\left[\left(\alpha_4 - \frac{1-\nu}{\alpha_3}\right)(\frac{r}{\ell})K_0(\frac{r}{\ell}) - 2\alpha_5(1-\nu)K_1(\frac{r}{\ell})\right] \\
& \times \; \left.\left[\frac{\nu}{1-\nu}n_\alpha r_{,\beta} + r_{,\alpha}r_{,\beta}r_{,n}\right]\right\}
\end{aligned}
$$

$$
\begin{aligned}
P^*_{\alpha\beta 3} \; = & \; \frac{D(1-\nu)\lambda^2}{4\pi r}\{zK_1(z)[n_\beta r_{,\alpha} + n_\alpha r_{,\beta} - 2r_{,\alpha}r_{,\beta}r_{,n}] \\
& + \; 2A_1(z)[n_\beta r_{,\alpha} + n_\alpha r_{,\beta} + \delta_{\alpha\beta}r_{,n} - 4r_{,\alpha}r_{,\beta}r_{,n}]\} \\
& + \; \frac{D(1-\nu)}{8\pi\ell^3}\left\{\left[4(\frac{\ell}{r})A_1(\frac{r}{\ell}) - 2(1-\alpha_1\ell^2)K_1(\frac{r}{\ell})\right]\right. \\
& \times \; (4r_{,\alpha}r_{,\beta}r_{,n} - n_\beta r_{,\alpha} - n_\alpha r_{,\beta} - \delta_{\alpha\beta}r_{,n}) \\
& + \; \left[4(2-\alpha_1\ell^2)K_1(\frac{r}{\ell}) - 2(1-\alpha_1\ell^2)(\frac{r}{\ell})K_0(\frac{r}{\ell})\right] \\
& \times \; \left.\left[\frac{\nu}{1-\nu}\delta_{\alpha\beta}r_{,n} + r_{,\alpha}r_{,\beta}r_{,n}\right]\right\} \\
P^*_{3\beta 3} \; = & \; \frac{D(1-\nu)\lambda^4}{4\pi}[B_1(z)n_\beta - A_1(z)r_{,\beta}r_{,n}] \\
& + \; \frac{D(1-\nu)\lambda^2}{8\pi\ell^2}\left\{(\alpha_5 - \alpha_1\ell^2)\left[K_0(\frac{r}{\ell})n_\beta - (\frac{r}{\ell})K_1(\frac{r}{\ell})r_{,\beta}r_{,n}\right]\right. \\
& + \; \left. 2\alpha_5 K_0(\frac{r}{\ell})r_{,\beta}r_{,n} - 2\alpha_5(\frac{\ell}{r})K_1(\frac{r}{\ell})(n_\beta - 2r_{,\beta}r_{,n})\right\}
\end{aligned}
\tag{123}
$$

$$
\begin{aligned}
W^*_{\alpha\beta} \; = & \; \frac{\ell(1-\nu)}{4\pi\alpha_3}\left\{(\frac{r}{\ell})K_0(\frac{r}{\ell})\left[\frac{\nu}{1-\nu}\delta_{\alpha\beta}r_{,n} + r_{,\alpha}r_{,\beta}r_{,n}\right]\right. \\
& + \; \left.(\frac{\ell}{r})\left[2A_1(\frac{r}{\ell}) + (\frac{r}{\ell})K_1(\frac{r}{\ell})\right](4r_{,\alpha}r_{,\beta}r_{,n} - \delta_{\alpha\beta}r_{,n} - n_\beta r_{,\alpha} - n_\alpha r_{,\beta})\right\} \\
W^*_{3\beta} \; = & \; \frac{1}{4\pi\alpha_3}\left\{(\frac{r}{\ell})K_1(\frac{r}{\ell})r_{,\beta}r_{,n} - K_0(\frac{r}{\ell})n_\beta\right\}
\end{aligned}
\tag{124}
$$

Case 3 ($\kappa < 1$)

$$
\begin{aligned}
U^*_{\alpha\beta\gamma} \; = & \; \frac{-1}{2\pi r}\{zK_1(z)[2r_{,\alpha}r_{,\beta}r_{,\gamma} - \delta_{\alpha\gamma}r_{,\beta} - \delta_{\gamma\beta}r_{,\alpha}] \\
& + \; 2A_1(z)[4r_{,\alpha}r_{,\beta}r_{,\gamma} - \delta_{\alpha\gamma}r_{,\beta} - \delta_{\gamma\beta}r_{,\alpha} - \delta_{\alpha\beta}r_{,\gamma}]\} \\
& - \; \frac{\ell^2}{8\lambda^2\sin 2\psi}\left\{ -[\Psi_5 + \alpha_2\Psi_3]\left[2r_{,\alpha}r_{,\beta}r_{,\gamma} + \frac{2\nu}{1-\nu}\delta_{\alpha\beta}r_{,\gamma}\right]\right.
\end{aligned}
$$

$$
\begin{aligned}
& + \left[\frac{4}{r^2}(\Psi_3 + \alpha_2\Psi_1) - \frac{2}{r}(\Psi_4 + \alpha_2\Psi_2)\right] \\
& \times \left[4r_{,\alpha}r_{,\beta}r_{,\gamma} - \delta_{\alpha\gamma}r_{,\beta} - \delta_{\gamma\beta}r_{,\alpha} - \delta_{\alpha\beta}r_{,\gamma}]\right\} \\
U^*_{3\beta\alpha} &= \frac{\lambda^2}{2\pi}[B_1(z)\delta_{\alpha\beta} - A_1(z)r_{,\alpha}r_{,\beta}] \\
& + \frac{\ell^2}{8\sin 2\psi}\left\{\frac{1}{r}[\Psi_3 + \alpha_1\Psi_1](\delta_{\alpha\beta} - 2r_{,\alpha}r_{,\beta}) + [\Psi_4 + \alpha_1\Psi_2]r_{,\alpha}r_{,\beta}\right\} \\
U^*_{\alpha\beta 3} &= \frac{-\ell^2}{8\alpha_3\sin 2\psi}\left\{\Psi_2\left[(1-\nu)r_{,\alpha}r_{,\beta} + \nu\delta_{\alpha\beta}\right] + \frac{1-\nu}{r}\Psi_1[\delta_{\alpha\beta} - 2r_{,\alpha}r_{,\beta}]\right\} \\
U^*_{3\beta 3} &= \frac{-\ell^2}{8\alpha_3\sin 2\psi}\Psi_3 r_{,\beta} \qquad (125)
\end{aligned}
$$

$$
\begin{aligned}
U^*_{\alpha\beta 3,n} &= \frac{\ell^2}{8\alpha_3\sin 2\psi}\left\{\Psi_3[\nu\delta_{\alpha\beta}r_{,n} + (1-\nu)r_{,\alpha}r_{,\beta}r_{,n}] + \frac{1-\nu}{r}\left[\Psi_2 - \frac{2}{r}\Psi_1\right]\right. \\
& \times \left.(4r_{,\alpha}r_{,\beta}r_{,n} - \delta_{\alpha\beta}r_{,n} - n_\beta r_{,\alpha} - n_\alpha r_{,\beta})\right\} \\
U^*_{3\beta 3,n} &= \frac{-\ell^2}{8\alpha_3\sin 2\psi}\left\{\frac{\Psi_3}{r}[n_\beta - 2r_{,\beta}r_{,n}] + \Psi_4 r_{,\beta}r_{,n}\right\} \qquad (126)
\end{aligned}
$$

$$
\begin{aligned}
P^*_{\alpha\beta\gamma} &= \frac{D(1-\nu)}{4\pi r^2}\{z^2K_0(z)(4r_{,\alpha}r_{,\beta}r_{,\gamma}r_{,n} - \delta_{\alpha\gamma}r_{,\beta}r_{,n} \\
& - r_{,\gamma}r_{,\beta}n_\alpha - \delta_{\beta\gamma}r_{,\alpha}r_{,n} - n_\beta r_{,\alpha}r_{,\gamma}) \\
& + zK_1(z)(32r_{,\alpha}r_{,\beta}r_{,\gamma}r_{,n} + 2\delta_{\gamma\beta}n_\alpha + 2\delta_{\gamma\alpha}n_\beta - 4n_\gamma r_{,\alpha}r_{,\beta} - 6n_\beta r_{,\alpha}r_{,\gamma} \\
& - 6n_\alpha r_{,\beta}r_{,\gamma} - 6\delta_{\beta\gamma}r_{,\alpha}r_{,n} - 6\delta_{\alpha\gamma}r_{,\beta}r_{,n} - 4\delta_{\alpha\beta}r_{,\gamma}r_{,n}) \\
& + 2A_1(z)(48r_{,\alpha}r_{,\beta}r_{,\gamma}r_{,n} - 8\delta_{\alpha\gamma}r_{,\beta}r_{,n} - 8r_{,\gamma}r_{,\beta}n_\alpha - 8r_{,\alpha}r_{,\beta}n_\gamma \\
& - 8\delta_{\alpha\beta}r_{,\gamma}r_{,n} - 8\delta_{\beta\gamma}r_{,\alpha}r_{,n} - 8r_{,\alpha}r_{,\gamma}n_\beta + 2\delta_{\alpha\gamma}n_\beta + 2\delta_{\beta\gamma}n_\alpha + 2\delta_{\alpha\beta}n_\gamma)\} \\
& + \frac{D(1-\nu)\ell^2}{16\lambda^2\sin 2\psi}\left\{[\Psi_6 + \alpha_2\Psi_4]\left[4r_{,\alpha}r_{,\beta}r_{,\gamma}r_{,n}\right.\right. \\
& + \left.\frac{4\nu}{1-\nu}\left(\delta_{\alpha\beta}r_{,\gamma}r_{,n} + n_\gamma r_{,\alpha}r_{,\beta} + \frac{\nu}{1-\nu}\delta_{\alpha\beta}n_\gamma\right)\right] \\
& + \frac{1}{r}[\Psi_5 + \alpha_2\Psi_3]\left[4n_\gamma r_{,\alpha}r_{,\beta} + 4n_\beta r_{,\alpha}r_{,\gamma} + 4n_\alpha r_{,\beta}r_{,\gamma} - 32r_{,\alpha}r_{,\beta}r_{,\gamma}r_{,n}\right. \\
& + 4\delta_{\beta\gamma}r_{,\alpha}r_{,n} + 4\delta_{\alpha\gamma}r_{,\beta}r_{,n} + 4\delta_{\alpha\beta}r_{,\gamma}r_{,n} \\
& + \left.\frac{4\nu}{1-\nu}(2\delta_{\alpha\beta}n_\gamma - 2n_\gamma r_{,\alpha}r_{,\beta} - 2\delta_{\alpha\beta}r_{,\gamma}r_{,n})\right]
\end{aligned}
$$

$$
\begin{aligned}
&+ \quad \frac{4}{r^2}\left[\frac{2}{r}(\Psi_3+\alpha_2\Psi_1)-(\Psi_4+\alpha_2\Psi_2)\right] \\
&\times \quad (24r_{,\alpha}r_{,\beta}r_{,\gamma}r_{,n}-4\delta_{\alpha\gamma}r_{,\beta}r_{,n}-4r_{,\gamma}r_{,\beta}n_\alpha-4r_{,\alpha}r_{,\beta}n_\gamma-4\delta_{\alpha\beta}r_{,\gamma}r_{,n} \\
&- \quad 4\delta_{\beta\gamma}r_{,\alpha}r_{,n}-4r_{,\alpha}r_{,\gamma}n_\beta+\delta_{\alpha\gamma}n_\beta+\delta_{\beta\gamma}n_\alpha+\delta_{\alpha\beta}n_\gamma)\Big\}
\end{aligned}
$$

$$
\begin{aligned}
P^*_{3\beta\alpha} &= \frac{D(1-\nu)\lambda^2}{4\pi r}\{zK_1(z)[2r_{,\alpha}r_{,\beta}r_{,n}-n_\beta r_{,\alpha}-\delta_{\alpha\beta}r_{,n}] \\
&+ 2A_1(z)[4r_{,\alpha}r_{,\beta}r_{,n}-\delta_{\alpha\beta}r_{,n}-n_\beta r_{,\alpha}-r_{,\beta}n_\alpha]\} \\
&+ \frac{D(1-\nu)\ell^2}{16\sin 2\psi}\left\{\left(\frac{4}{r^2}[\Psi_3+\alpha_1\Psi_1]-\frac{2}{r}[\Psi_4+\alpha_1\Psi_2]\right)\right. \\
&\times [4r_{,\alpha}r_{,\beta}r_{,n}-\delta_{\alpha\beta}r_{,n}-n_\beta r_{,\alpha}-r_{,\beta}n_\alpha] \\
&- \left.[\Psi_5+\alpha_1\Psi_3]\left(2r_{,\alpha}r_{,\beta}r_{,n}+\frac{2\nu}{1-\nu}r_{,\beta}n_\alpha\right)\right\} \\
P^*_{\alpha\beta 3} &= \frac{D(1-\nu)\lambda^2}{4\pi r}\{zK_1(z)[n_\beta r_{,\alpha}+n_\alpha r_{,\beta}-2r_{,\alpha}r_{,\beta}r_{,n}] \\
&+ 2A_1(z)[n_\beta r_{,\alpha}+n_\alpha r_{,\beta}+\delta_{\alpha\beta}r_{,n}-4r_{,\alpha}r_{,\beta}r_{,n}]\} \\
&+ \frac{D(1-\nu)\ell^2}{8\sin 2\psi}\left\{[\Psi_5+\alpha_1\Psi_3]\left[r_{,\alpha}r_{,\beta}r_{,n}+\frac{\nu}{1-\nu}\delta_{\alpha\beta}r_{,n}\right]\right. \\
&+ \left[\frac{2}{r^2}(\Psi_3+\alpha_1\Psi_1)-\frac{1}{r}(\Psi_4+\alpha_1\Psi_2)\right] \\
&\times \left.[n_\beta r_{,\alpha}+n_\alpha r_{,\beta}+\delta_{\alpha\beta}r_{,n}-4r_{,\alpha}r_{,\beta}r_{,n}]\right\} \\
P^*_{3\beta 3} &= \frac{D(1-\nu)\lambda^4}{4\pi}[B_1(z)n_\beta-A_1(z)r_{,\beta}r_{,n}] \\
&+ \frac{D(1-\nu)\lambda^2\ell^2}{16\sin 2\psi}\left\{\frac{1}{r}[\alpha_5\Psi_3+\alpha_1\Psi_1](n_\beta-2r_{,\beta}r_{,n})\right. \\
&+ \left.[\alpha_5\Psi_4+\alpha_1\Psi_2]r_{,\beta}r_{,n}\right\}
\end{aligned}
\tag{127}
$$

$$
\begin{aligned}
W^*_{\alpha\beta} &= \frac{-\ell^2}{8\alpha_3\sin 2\psi}\left\{r_{,n}\Psi_1[\nu\delta_{\alpha\beta}+(1-\nu)r_{,\alpha}r_{,\beta}]\right. \\
&- \left.\frac{1-\nu}{r}\left[\frac{2\Psi_{-1}}{r}-\Psi_0\right](4r_{,\alpha}r_{,\beta}r_{,n}-\delta_{\alpha\beta}r_{,n}-n_\beta r_{,\alpha}-n_\alpha r_{,\beta})\right\} \\
W^*_{3\beta} &= \frac{\ell^2}{8\alpha_3\sin 2\psi}\left\{\Psi_2 r_{,\beta}r_{,n}+\frac{\Psi_1}{r}[n_\beta-2r_{,\beta}r_{,n}]\right\}
\end{aligned}
\tag{128}
$$

Chapter 2

Boundary element analysis of thick Reissner plates in bending

A. El-Zafrany

School of Mechanical Engineering, Cranfield University, Cranfield, Bedford, MK43 0AL, UK

Email: a.el-zafrany@cranfield.ac.uk

Abstract

This chapter presents a simplified approach for the derivation of boundary integral equations and fundamental solutions for thick Reissner plates in bending. Explicit expressions for fundamental solution parameters, suitable for plates with arbitrary shapes are derived using Hankel integral transforms. Simplified expressions for loading kernel functions are presented here for the first time, and reductions of domain loading integral terms are described for concentrated forces and moments, and for uniformly- and linearly-distributed loading. Corner and singularity problems are also discussed, and analytical integral expressions for singular kernel functions are listed. Several numerical examples with different loading and boundary conditions have been included and the boundary element results agree very well with corresponding analytical solutions.

Notation

D	plate flexural rigidity
E	Young's modulus of the plate material
e_x, e_y, e_z	arbitrary constant parameters
F	Concentrated shear force normal to the plate
f^*	fundamental solution as defined by eqn (88)
$\boldsymbol{H}$, $\boldsymbol{H}^*$	differential operator matrices
h	plate thickness
$(\hat{i},\hat{j}) \equiv (\hat{i}_1,\hat{i}_2)$	unit vectors in the x and y directions, respectively

i, j, k	indices with values equal to 1, 2 and 3
$I_n(z)$, $K_n(z)$	modified Bessel functions as defined in Appendix A.
l, m, n	directional cosines of the outward normal to the plate surface
$(l, m) \equiv (l_1, l_2)$	directional cosines of the outward normal to the boundary of the plate midplane
(l_i, m_i)	directional cosines of the normal at a source point (x_i, y_i)
M_x, M_y, M_{xy}	internal moments per unit length, as defined by eqn (40)
M_x^*, M_y^*, M_{xy}^*	weighting-function parameters, as defined by eqns (55)
M_n, M_{nt}	boundary moments per unit length as defined by eqn (61)
$\hat{n} \equiv \hat{n}_1$	a unit vector in the normal direction to Γ, $\hat{n} = l\hat{i} + m\hat{j}$
$q(x, y)$	total distributed shear force intensity at (x,y)
Q_x, Q_y	internal shear forces per unit length, as defined by eqn (25)
Q_n	boundary shear force per unit length
Q_x^*, Q_y^*	weighting-function parameters as defined by eqns (56)
R_o	half the length of the eth constant boundary element
r	distance between field and source points
S_x, S_y, S_z	surface tractions in the x, y, and z directions, respectively
$\boldsymbol{T}$	matrix of kernel functions, as defined by eqn (83)
T_x, T_y	bending moments in the x and y directions, respectively
$\hat{t} \equiv \hat{n}_2$	a unit vector in the tangential direction to Γ, $\hat{t} = -m\hat{i} + l\hat{j}$
$\boldsymbol{U}$	matrix of kernel functions, as defined by eqn (82)
$\boldsymbol{U}^*$	matrix of kernel functions, as defined by eqn (80)
u, v, w	displacement components in the x, y, and z directions, respectively, $\equiv (u_1, u_2, u_3)$
(x, y, z)	Cartesian coordinates of a point inside the plate, $\equiv (x_1, x_2, x_3)$
$(x, y) \equiv (x_1, x_2)$	Cartesian coordinates of a point in the midplane of the plate
(x_i, y_i)	Cartesian coordinates of the ith source point
z	refers occasionally to the third Cartesian axis, but it is mainly used as the parameter λr in Bessel functions
z_o	non-dimensional parameter = λR_o
α, β, γ	indices with values limited to 1, 2
Γ	the boundary of the domain Ω of the plate midplane
δ_{ij}	the Kronecker delta
$\delta(x - x_i, y - y_i)$	two-dimensional Dirac delta function
ϵ_{ij}	strain tensor, $\epsilon_x \equiv \epsilon_{11}$, $\epsilon_y \equiv \epsilon_{22}$, $\gamma_{xy} \equiv 2\epsilon_{12}$, etc.

θ_x, θ_y, θ_n, θ_t	average slope angles as defined by eqns (34), (35), (60)
θ_x^*, θ_y^*, θ_n^*, θ_t^*	weighting-function parameters
λ	thickness parameter $\equiv \sqrt{10}/h$
μ	shear modulus
ν	Poisson's ratio of the plate material
τ_{ij}	stress tensor, $\sigma_x \equiv \tau_{11}$, $\sigma_y \equiv \tau_{22}$, $\sigma_z \equiv \tau_{33}$, $\tau_{xy} \equiv \tau_{12}$, $\tau_{yz} \equiv \tau_{23}$, $\tau_{zx} \equiv \tau_{31}$
Ω	the domain of the plate midplane
$\dfrac{\partial}{\partial n}$	$\equiv \dfrac{\partial}{\partial n_1} = l\dfrac{\partial}{\partial x} + m\dfrac{\partial}{\partial y}$
$\dfrac{\partial}{\partial t}$	$\equiv \dfrac{\partial}{\partial n_2} = -m\dfrac{\partial}{\partial x} + l\dfrac{\partial}{\partial y}$

1 Introduction

With the enormous development of the boundary element method in the last two decades, it is natural to find some work in the literature dealing with plate bending problems. Indirect boundary integral equation solutions of Kirchhoff plate-bending problems were presented by Altiero and Sikarskie[1], and Tottenham.[2] Direct boundary element formulations for thin plates were demonstrated by several authors.[3-5] Recently, El-Zafrany *et al.*[6] introduced a modified Kirchhoff theory in which the effect of the transverse stress σ_z is considered, with boundary element formulations based upon three degrees-of-freedom per node. The first paper on the boundary element analysis of thick Reissner plates was introduced by Vander Weeën[7], who employed the Hörmander method for the derivation of the fundamental solution. Antes[8] introduced a regular approach based upon a modification of Trefftz method. Karam and Telles[9], and Long *et al.*[10] presented additional developments based upon Weeën's fundamental solution. Recently, El-Zafrany *et al.*[11] introduced a modified fundamental solution suitable for plates with arbitrary shapes, which was derived by using Hankel integral transforms and strain functions. They [12] later modified the fundamental solution such that the parts of kernel functions representing the effect of transverse stresses, as obtained by means of Reissner's theory, were separated, hence allowing the development of an efficient computer program for thin and thick plates.

In this chapter, derivations of boundary element theory for thick Reissner plates is reviewed, using a weighted-residual approach and integral transforms. Numerical treatment of boundary integral equations is discussed with emphasis on corner problems with isoparametric elements, and singular integrals with constant boundary elements. Reduction of loading domain integrals for uniformly- and linearly-distributed loading, and concentrated shear forces and bending moments are demonstrated.

2 Review of governing equations

A plate is a structural element with two flat surfaces, and it is usually defined in terms of a midplane and thickness. The thickness is much smaller than other dimensions, and is measured in the direction normal to the midplane, which divides the thickness into two equal halves. Plates may be classified into thin and thick plates according to the value of the ratio of the thickness to a span length, and whether transverse stresses have an effect on plate behaviour.

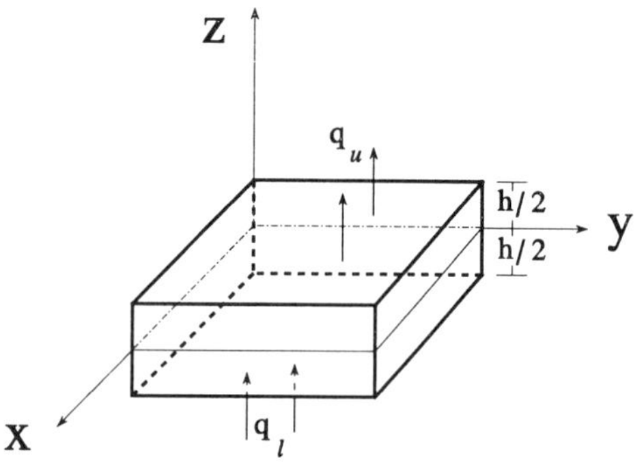

Figure 1. Plate reference axes.

2.1 Basic assumptions

Consider a plate defined in terms of a midplane in the Cartesian x-y plane, and a uniform thickness h measured in the z-direction, as shown in Figure 1. The upper surface of the plate $(z = h/2)$ is considered to be subjected to a shear force per unit area $q_u(x,y)$ and the lower surface $(z = -h/2)$ is subjected to a shear force per unit area $q_l(x,y)$.

The plate theory presented in this section is based upon the following assumptions and approximations:

(i) The material of the plate is homogeneous, isotropic, and linearly elastic. The stress-strain relationships are, therefore, governed by the generalized Hooke's law, i.e.

$$\tau_{ij} = 2\mu\left[\epsilon_{ij} + \frac{\nu}{1-2\nu}\left(\epsilon_x + \epsilon_y + \epsilon_z\right)\delta_{ij}\right] \tag{1}$$

(ii) Displacement gradients are small enough, such that Cauchy's strain-displacement relationships are acceptable, i.e.

$$\epsilon_{ij} = \frac{1}{2}\left(\frac{\partial u_j}{\partial x_i} + \frac{\partial u_i}{\partial x_j}\right) \tag{2}$$

(iii) The midplane of the plate remains unstrained after bending, i.e. it is a neutral plane in bending.

(iv) The lateral deflection w, the displacement component in the z-direction, is independent of z, i.e. $w \approx w(x,y)$.

2.2 Transverse stress modelling

Considering the equation of surface tractions at the upper surface of the plate, then it is clear that:

$$S_x = l\,\sigma_x + m\,\tau_{yx} + n\,\tau_{zx} = 0$$

$$S_y = l\,\tau_{xy} + m\,\sigma_y + n\,\tau_{zy} = 0$$

$$S_z = l\,\tau_{xz} + m\,\tau_{yz} + n\,\sigma_z = q_u$$

With $l = 0,\ m = 0,\ n = 1$, it can be deduced that:

$$\tau_{xz} = \tau_{yz} = 0, \qquad \sigma_z = q_u \tag{3}$$

Similarly, at the lower surface of the plate:

$$\tau_{xz} = \tau_{yz} = 0, \quad \sigma_z = -q_l \tag{4}$$

Hence, it can also be proved that:

$$\gamma_{xz} = \gamma_{yz} = 0 \quad \text{at } z = \pm h/2 \tag{5}$$

Using a three-point Lagrangian interpolation, the distributions of γ_{xz} and γ_{yz} along the thickness of the plate may be obtained in terms of their values at $z_1 = -h/2$, $z_2 = 0$, $z_3 = +h/2$, leading to the following expressions:

$$\gamma_{xz} = \gamma^{o}_{xz}(x,y)\left[1 - \frac{4z^2}{h^2}\right] \tag{6}$$

$$\gamma_{yz} = \gamma^{o}_{yz}(x,y)\left[1 - \frac{4z^2}{h^2}\right] \tag{7}$$

where γ^{o}_{xz}, γ^{o}_{yz} are the values of transverse shear strains at $z = 0$.

The transverse shear stresses can be defined over the thickness, from eqn (1), as follows:

$$\tau_{xz} = \mu\,\gamma_{xz} = \mu\,\gamma^{o}_{xz}\left[1 - \frac{4z^2}{h^2}\right] \tag{8}$$

$$\tau_{yz} = \mu\,\gamma_{yz} = \mu\,\gamma^{o}_{yz}\left[1 - \frac{4z^2}{h^2}\right] \tag{9}$$

and internal shear forces per unit length are defined such that:

$$Q_x = \int_{-h/2}^{h/2} \tau_{xz} \, \mathrm{d}z \equiv \frac{2}{3} \mu \gamma^{\mathrm{o}}_{xz} h \tag{10}$$

$$Q_y = \int_{-h/2}^{h/2} \tau_{yz} \, \mathrm{d}z \equiv \frac{2}{3} \mu \gamma^{\mathrm{o}}_{yz} h \tag{11}$$

Hence, eqns (8) and (9) can be rewritten as follows:

$$\tau_{xz} = \frac{3 Q_x}{2 h} \left[1 - \frac{4 z^2}{h^2} \right] \tag{12}$$

$$\tau_{yz} = \frac{3 Q_y}{2 h} \left[1 - \frac{4 z^2}{h^2} \right] \tag{13}$$

which agree with the equations given by Reissner.[13]

Consider the equation of infinitesimal equilibrium in the z-direction, without domain loading, i.e.

$$\frac{\partial \tau_{xz}}{\partial x} + \frac{\partial \tau_{yz}}{\partial y} + \frac{\partial \sigma_z}{\partial z} = 0 \tag{14}$$

Integrating the previous equation with respect to z, over the full thickness of the plate, it can be shown that:

$$q = - \left(\frac{\partial Q_x}{\partial x} + \frac{\partial Q_y}{\partial y} \right) \tag{15}$$

where $$q = q_u + q_l \tag{16}$$

Substituting from eqns (12), (13) into (14), then:

$$\frac{\partial \sigma_z}{\partial z} = - \frac{3}{2 h} \left(\frac{\partial Q_x}{\partial x} + \frac{\partial Q_y}{\partial y} \right) \left(1 - \frac{4 z^2}{h^2} \right) \equiv \frac{3 q}{2 h} \left[1 - \frac{4 z^2}{h^2} \right] \tag{17}$$

Integrating the previous equation, and using the boundary conditions, it can be proved that:

$$\sigma_z = \frac{q_u - q_l}{2} + \frac{q}{2} \left(\frac{3 z}{h} - \frac{4 z^3}{h^3} \right) \tag{18}$$

2.3 Displacement components

It can be deduced from eqns (2), (12), and (13) that:

$$\gamma_{xz} = \frac{3\,Q_x}{2\,\mu\,h}\left[1 - \frac{4\,z^2}{h^2}\right] \equiv \frac{\partial u}{\partial z} + \frac{\partial w}{\partial x} \tag{19}$$

$$\gamma_{yz} = \frac{3\,Q_y}{2\,\mu\,h}\left[1 - \frac{4\,z^2}{h^2}\right] \equiv \frac{\partial v}{\partial z} + \frac{\partial w}{\partial y} \tag{20}$$

i.e.

$$\frac{\partial u}{\partial z} = -\frac{\partial w}{\partial x} + \frac{3\,Q_x}{2\,\mu\,h}\left[1 - \frac{4\,z^2}{h^2}\right] \tag{21a}$$

and

$$\frac{\partial v}{\partial z} = -\frac{\partial w}{\partial y} + \frac{3\,Q_y}{2\,\mu\,h}\left[1 - \frac{4\,z^2}{h^2}\right] \tag{21b}$$

Integrating the previous equations with respect to z, it can be proved that:

$$u(x,y,z) = u^{\circ}(x,y) - z\frac{\partial w}{\partial x} + \frac{3\,Q_x}{2\,\mu\,h}\left[z - \frac{4\,z^3}{3\,h^2}\right] \tag{22a}$$

$$v(x,y,z) = v^{\circ}(x,y) - z\frac{\partial w}{\partial y} + \frac{3\,Q_y}{2\,\mu\,h}\left[z - \frac{4\,z^3}{3\,h^2}\right] \tag{22b}$$

where u°, v° are the values of displacement components in the midplane of the plate, and they are generated by in-plane loadings. For plate bending problems, i.e. in the absence of any in-plane loading: $u^{\circ} = v^{\circ} \approx 0$, and eqns (22) can be simplified as follows:

$$u(x,y,z) = -z\frac{\partial w}{\partial x} + \frac{3\,Q_x}{2\,\mu\,h}\left[z - \frac{4\,z^3}{3\,h^2}\right] \tag{23}$$

$$v(x,y,z) = -z\frac{\partial w}{\partial y} + \frac{3\,Q_y}{2\,\mu\,h}\left[z - \frac{4\,z^3}{3\,h^2}\right] \tag{24}$$

To simplify the thick plate problem, transverse shear stresses and strains averaged over the thickness may be employed without violating equilibrium or strain energy contributions. Defining average transverse shear stresses $\bar{\tau}_{xz}$, $\bar{\tau}_{yz}$ so as to maintain internal equilibrium, then:

$$(Q_x, Q_y) = \int_{-h/2}^{h/2} (\bar{\tau}_{xz}, \bar{\tau}_{yz})\, dz \equiv \int_{-h/2}^{h/2} (\tau_{xz}, \tau_{yz})\, dz \tag{25}$$

i.e.

$$\bar{\tau}_{xz} = \frac{Q_x}{h} \equiv \frac{2}{3}\mu\gamma^{o}_{xz}, \quad \bar{\tau}_{yz} = \frac{Q_y}{h} \equiv \frac{2}{3}\mu\gamma^{o}_{yz} \tag{26}$$

Defining average transverse shear strains $\bar{\gamma}_{xz}, \bar{\gamma}_{yz}$ so as to maintain strain energy contributions, then:

$$\int_{-h/2}^{h/2} \bar{\tau}_{xz}\bar{\gamma}_{xz}\, dz \equiv \int_{-h/2}^{h/2} \tau_{xz}\gamma_{xz}\, dz \tag{27}$$

and

$$\int_{-h/2}^{h/2} \bar{\tau}_{yz}\bar{\gamma}_{yz}\, dz \equiv \int_{-h/2}^{h/2} \tau_{yz}\gamma_{yz}\, dz \tag{28}$$

Considering first eqn (27), then it can be deduced that:

$$h\bar{\tau}_{xz}\bar{\gamma}_{xz} = \frac{9\bar{\tau}^2_{xz}}{4\mu} \int_{-h/2}^{h/2} \left(1 - \frac{4z^2}{h^2}\right)^2 dz$$

i.e.

$$\bar{\gamma}_{xz} = \bar{\tau}_{xz} / \frac{5}{6}\mu = \frac{6Q_x}{5\mu h} \tag{29a}$$

Similarly, it can be deduced from eqn (28) that:

$$\bar{\gamma}_{yz} = \bar{\tau}_{yz} / \frac{5}{6}\mu = \frac{6Q_y}{5\mu h} \tag{29b}$$

If averaged displacement components are defined according to strain-displacement equations of averaged strains, then:

$$\bar{\gamma}_{xz} = \frac{\partial \bar{u}}{\partial z} + \frac{\partial w}{\partial x} \tag{30}$$

$$\bar{\gamma}_{yz} = \frac{\partial \bar{v}}{\partial z} + \frac{\partial w}{\partial y} \tag{31}$$

and it can be deduced that:

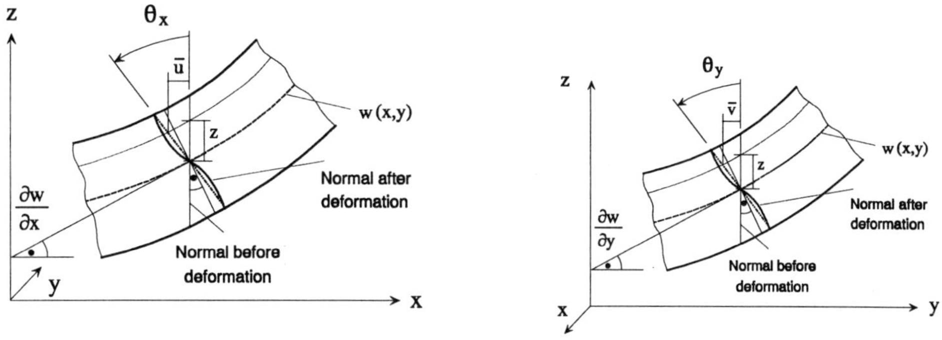

Figure 2. Deformed *x-z* and *y-z* sections of a thick plate.

$$\bar{u}(x,y,z) = -z\left(\frac{\partial w}{\partial x} - \bar{\gamma}_{xz}\right) \equiv z\,\theta_x \tag{32}$$

$$\bar{v}(x,y,z) = -z\left(\frac{\partial w}{\partial y} - \bar{\gamma}_{yz}\right) \equiv z\,\theta_y \tag{33}$$

where $$\theta_x = -\frac{\partial w}{\partial x} + \bar{\gamma}_{xz} \equiv -\frac{\partial w}{\partial x} + \frac{6\,Q_x}{5\,\mu\,h} \tag{34}$$

and $$\theta_y = -\frac{\partial w}{\partial y} + \bar{\gamma}_{yz} \equiv -\frac{\partial w}{\partial y} + \frac{6\,Q_y}{5\,\mu\,h} \tag{35}$$

which are the corresponding average slope angles.

Equations similar to (32) and (33) can be obtained geometrically, if one assumes that normals to the midplane before deformation remain straight but not necessarily normal after deformation (Mindlin[14]), as demonstrated by Figure 2.

2.4 Stress and strain parameters

Transverse shear strain and stress components are as given by eqns (6), (7), (12), and (13). The *x-y* strain components can be obtained from displacement components defined by eqns (23) and (24), as follows:

$$\epsilon_x = \frac{\partial u}{\partial x} = -z\,\frac{\partial^2 w}{\partial x^2} + \frac{3}{2\,\mu\,h}\left(z - \frac{4\,z^3}{3\,h^2}\right)\frac{\partial Q_x}{\partial x} \tag{36a}$$

$$\epsilon_y = \frac{\partial v}{\partial y} = -z\,\frac{\partial^2 w}{\partial y^2} + \frac{3}{2\,\mu\,h}\left(z - \frac{4\,z^3}{3\,h^2}\right)\frac{\partial Q_y}{\partial y} \tag{36b}$$

$$\gamma_{xy} = \frac{\partial v}{\partial x} + \frac{\partial u}{\partial y} = -2z\frac{\partial^2 w}{\partial x\,\partial y} + \frac{3}{2\mu h}\left(z - \frac{4z^3}{3h^2}\right)\left(\frac{\partial Q_x}{\partial y} + \frac{\partial Q_y}{\partial x}\right) \tag{36c}$$

Since the transverse stress σ_z is already defined, by means of eqn (18), it is advantageous to separate its term in stress-strain equations. Hence, the generalized Hooke's equations for x-y components can be written as follows:

$$\epsilon_x = \frac{1}{E}\left(\sigma_x - \nu\,\sigma_y\right) - \frac{\nu}{E}\sigma_z \tag{37a}$$

$$\epsilon_y = \frac{1}{E}\left(\sigma_y - \nu\,\sigma_x\right) - \frac{\nu}{E}\sigma_z \tag{37b}$$

$$\gamma_{xy} = \frac{\tau_{xy}}{\mu} \tag{37c}$$

from which it can be deduced that:

$$\sigma_x = \frac{E}{(1-\nu^2)}\left(\epsilon_x + \nu\,\epsilon_y\right) + \frac{\nu\,\sigma_z}{1-\nu} \tag{38a}$$

$$\sigma_y = \frac{E}{(1-\nu^2)}\left(\nu\,\epsilon_x + \epsilon_y\right) + \frac{\nu\,\sigma_z}{1-\nu} \tag{38b}$$

$$\tau_{xy} = \frac{E}{2(1+\nu)}\gamma_{xy} \tag{38c}$$

Substituting from eqns (36a)-(36c) into (38a)-(38c), it can be proved that:

$$\sigma_x = -\frac{zE}{(1-\nu^2)}\left(\frac{\partial^2 w}{\partial x^2} + \nu\frac{\partial^2 w}{\partial y^2}\right) + \frac{\nu\,(q_l - q_u)}{2(1-\nu)} + \frac{f(z)}{1-\nu}\left(\frac{\partial Q_x}{\partial x} + \nu\frac{\partial Q_y}{\partial y} + \nu\frac{q}{2}\right) \tag{39a}$$

$$\sigma_y = -\frac{zE}{(1-\nu^2)}\left(\nu\frac{\partial^2 w}{\partial x^2} + \frac{\partial^2 w}{\partial y^2}\right) + \frac{\nu\,(q_l - q_u)}{2(1-\nu)} + \frac{f(z)}{1-\nu}\left(\nu\frac{\partial Q_x}{\partial x} + \frac{\partial Q_y}{\partial y} + \nu\frac{q}{2}\right) \tag{39b}$$

$$\tau_{xy} = -\frac{z(1-\nu)E}{(1-\nu^2)}\frac{\partial^2 w}{\partial x\,\partial y} + \frac{f(z)}{2}\left(\frac{\partial Q_x}{\partial y} + \frac{\partial Q_y}{\partial x}\right) \tag{39c}$$

where $f(z) = 3(z/h) - 4(z/h)^3$

Internal bending moments per unit length are defined as follows:

$$\left\{M_x, M_y, M_{xy}\right\} = \int_{-h/2}^{h/2} z\left\{\sigma_x, \sigma_y, \tau_{xy}\right\}\mathrm{d}z \tag{40}$$

Hence, by integrating over the plate thickness, the bending moments per unit length can be obtained as follows:

$$M_x = -D\left(\frac{\partial^2 w}{\partial x^2} + \nu\frac{\partial^2 w}{\partial y^2}\right) + \frac{h^2}{5(1-\nu)}\left(\frac{\partial Q_x}{\partial x} + \nu\frac{\partial Q_y}{\partial y} + \nu\frac{q}{2}\right) \tag{41a}$$

$$M_y = -D\left(\nu\frac{\partial^2 w}{\partial x^2} + \frac{\partial^2 w}{\partial y^2}\right) + \frac{h^2}{5(1-\nu)}\left(\nu\frac{\partial Q_x}{\partial x} + \frac{\partial Q_y}{\partial y} + \nu\frac{q}{2}\right) \tag{41b}$$

$$M_{xy} = -(1-\nu)D\frac{\partial^2 w}{\partial x\,\partial y} + \frac{h^2}{10}\left(\frac{\partial Q_x}{\partial y} + \frac{\partial Q_y}{\partial x}\right) \tag{41c}$$

Using average slope angles, as defined by eqns (34) and (35), it can be shown that:

$$Q_x = \frac{5\mu h}{6}\left(\frac{\partial w}{\partial x} + \theta_x\right) \equiv \frac{(1-\nu)}{2}D\lambda^2\left(\theta_x + \frac{\partial w}{\partial x}\right) \tag{42a}$$

$$Q_y = \frac{5\mu h}{6}\left(\frac{\partial w}{\partial y} + \theta_y\right) \equiv \frac{(1-\nu)}{2}D\lambda^2\left(\theta_y + \frac{\partial w}{\partial y}\right) \tag{42b}$$

and substituting from eqns (34), (35), (42a) and (42b) into (41a)-(41c), the moment equations can be simplified as follows:

$$M_x = D\left(\frac{\partial\theta_x}{\partial x} + \nu\frac{\partial\theta_y}{\partial y}\right) + \frac{\nu q}{(1-\nu)\lambda^2} \tag{43a}$$

$$M_y = D\left(\nu\frac{\partial\theta_x}{\partial x} + \frac{\partial\theta_y}{\partial y}\right) + \frac{\nu q}{(1-\nu)\lambda^2} \tag{43b}$$

$$M_{xy} = \frac{(1-\nu)D}{2}\left(\frac{\partial\theta_x}{\partial y} + \frac{\partial\theta_y}{\partial x}\right) \tag{43c}$$

2.5 Equations of equilibrium over the thickness

In the absence of body forces, the infinitesimal equations of equilibrium at any point inside the plate are expressed as follows:

$$\frac{\partial\sigma_x}{\partial x} + \frac{\partial\tau_{yx}}{\partial y} + \frac{\partial\tau_{zx}}{\partial z} = 0 \tag{44a}$$

$$\frac{\partial\tau_{xy}}{\partial x} + \frac{\partial\sigma_y}{\partial y} + \frac{\partial\tau_{zy}}{\partial z} = 0 \tag{44b}$$

$$\frac{\partial\tau_{xz}}{\partial x} + \frac{\partial\tau_{yz}}{\partial y} + \frac{\partial\sigma_z}{\partial z} = 0 \tag{44c}$$

The equations of equilibrium over the thickness are based upon relationships between components of internal moments and shear forces (per unit length), and can be obtained from infinitesimal equations of equilibrium. Multiplying eqn (44a) by z, and integrating the resulting equation over the plate thickness then:

$$\int_{-h/2}^{h/2} z\left(\frac{\partial\sigma_x}{\partial x} + \frac{\partial\tau_{yx}}{\partial y} + \frac{\partial\tau_{zx}}{\partial z}\right)\mathrm{d}z = 0 \tag{45}$$

Notice that:

$$\int_{-h/2}^{h/2} z\frac{\partial\sigma_x}{\partial x}\,\mathrm{d}z = \frac{\partial M_x}{\partial x}, \qquad \int_{-h/2}^{h/2} z\frac{\partial\tau_{xy}}{\partial y}\,\mathrm{d}z = \frac{\partial M_{xy}}{\partial y}$$

and

$$\int_{-h/2}^{h/2} z\frac{\partial\tau_{zx}}{\partial z}\,\mathrm{d}z = z\,\tau_{xz}\Big|_{-h/2}^{h/2} - \int_{-h/2}^{h/2}\tau_{xz}\,\mathrm{d}z \equiv -Q_x$$

hence, eqn (45) can be simplified as follows:

$$\frac{\partial M_x}{\partial x} + \frac{\partial M_{xy}}{\partial y} - Q_x = 0 \tag{46a}$$

Similarly, by multiplying eqn (44b) by z and integrating over the plate thickness it can be proved that:

$$\frac{\partial M_{xy}}{\partial x} + \frac{\partial M_y}{\partial y} - Q_y = 0 \tag{46b}$$

Integrating eqn (44c) over the plate thickness, it can also be shown that:

$$\frac{\partial Q_x}{\partial x} + \frac{\partial Q_y}{\partial y} + q = 0 \tag{46c}$$

where q is as defined by eqn (16). Substituting from eqns (41a)-(41c) into (46a) and (46b), the following equations can be obtained for shear forces per unit length:

$$Q_x = -D\frac{\partial}{\partial x}(\nabla^2 w) + \frac{h^2}{10}\left[\nabla^2 Q_x - \frac{1}{(1-\nu)}\frac{\partial q}{\partial x}\right] \tag{47a}$$

$$Q_y = -D\frac{\partial}{\partial y}(\nabla^2 w) + \frac{h^2}{10}\left[\nabla^2 Q_y - \frac{1}{(1-\nu)}\frac{\partial q}{\partial y}\right] \tag{47b}$$

Substituting into eqn (46c), then the partial differential equation of lateral deflection of thick plates can be expressed as follows:

$$D\nabla^4 w = q - \frac{(2-\nu)h^2}{10(1-\nu)}\nabla^2 q \tag{48}$$

which is a biharmonic differential equation, and it is similar to that deduced from Kirchhoff theory for thin plates in cases with: $\nabla^2 q = 0$. Notice also that by substituting from eqns (42a,b) and (43a,b,c) into (46a), (46b) and (46c), simultaneous partial differential equations in terms of lateral deflection and slope angles can be derived, and expressed as follows:

$$\nabla^2\theta_x + \left(\frac{1+\nu}{1-\nu}\right)\frac{\partial}{\partial x}\left(\frac{\partial\theta_x}{\partial x} + \frac{\partial\theta_y}{\partial y}\right) - \lambda^2\left(\theta_x + \frac{\partial w}{\partial x}\right) = -\frac{2\nu}{D\lambda^2(1-\nu)^2}\frac{\partial q}{\partial x} \tag{49a}$$

$$\nabla^2\theta_y + \left(\frac{1+\nu}{1-\nu}\right)\frac{\partial}{\partial y}\left(\frac{\partial\theta_x}{\partial x} + \frac{\partial\theta_y}{\partial y}\right) - \lambda^2\left(\theta_y + \frac{\partial w}{\partial y}\right) = -\frac{2\nu}{D\lambda^2(1-\nu)^2}\frac{\partial q}{\partial y} \tag{49b}$$

$$\frac{\partial\theta_x}{\partial x} + \frac{\partial\theta_y}{\partial y} + \nabla^2 w = -\frac{2q}{(1-\nu)D\lambda^2} \tag{49c}$$

3 Derivation of Boundary Integral Equations

Considering approximate solutions θ_x, θ_y and w, which satisfy the given boundary conditions, a weighted-residual expression can be obtained from the equations of equilibrium over the thickness (eqns 46), as follows:

$$\iint_\Omega \left[\theta_x^* \left(\frac{\partial M_x}{\partial x} + \frac{\partial M_{xy}}{\partial y} - Q_x \right) + \theta_y^* \left(\frac{\partial M_{xy}}{\partial x} + \frac{\partial M_y}{\partial y} - Q_y \right) \right.$$

$$\left. + w^* \left(\frac{\partial Q_x}{\partial x} + \frac{\partial Q_y}{\partial y} + q \right) \right] \mathrm{d}x \, \mathrm{d}y = 0 \tag{50}$$

where θ_x^*, θ_y^* and w^* are given weighting functions. Using the following integration by parts theorems (El-Zafrany[15]):

$$\iint_\Omega f \frac{\partial g}{\partial x} \, \mathrm{d}x \, \mathrm{d}y = \oint_\Gamma f g \, l \, \mathrm{d}\Gamma - \iint_\Omega \frac{\partial f}{\partial x} g \, \mathrm{d}x \, \mathrm{d}y \tag{51}$$

$$\iint_\Omega f \frac{\partial g}{\partial y} \, \mathrm{d}y \, \mathrm{d}x = \oint_\Gamma f g \, m \, \mathrm{d}\Gamma - \iint_\Omega \frac{\partial f}{\partial y} g \, \mathrm{d}y \, \mathrm{d}x \tag{52}$$

where f, g are functions of x, y, then eqn (50) can be modified to

$$\oint_\Gamma \theta_x^* (l M_x + m M_{xy}) \, \mathrm{d}\Gamma - \iint_\Omega \left[M_x \frac{\partial \theta_x^*}{\partial x} + M_{xy} \frac{\partial \theta_x^*}{\partial y} + Q_x \theta_x^* \right] \mathrm{d}x \mathrm{d}y$$

$$+ \oint_\Gamma \theta_y^* (l M_{xy} + m M_y) \, \mathrm{d}\Gamma - \iint_\Omega \left[M_{xy} \frac{\partial \theta_y^*}{\partial x} + M_y \frac{\partial \theta_y^*}{\partial y} + Q_y \theta_y^* \right] \mathrm{d}x \mathrm{d}y$$

$$+ \oint_\Gamma w^* (l Q_x + m Q_y) \, \mathrm{d}\Gamma - \iint_\Omega \left[Q_x \frac{\partial w^*}{\partial x} + Q_y \frac{\partial w^*}{\partial y} - q w^* \right] \mathrm{d}x \mathrm{d}y = 0$$

Rearranging the resulting terms, the previous expression can be reduced to the following form:

$$\oint_{\Gamma} (\theta_x^* t_x + \theta_y^* t_y + w^* t_z) \mathrm{d}\Gamma + \iint_{\Omega} q w^* \mathrm{d}x \mathrm{d}y$$

$$- \iint_{\Omega} \left[Q_x \left(\theta_x^* + \frac{\partial w^*}{\partial x} \right) + Q_y \left(\theta_y^* + \frac{\partial w^*}{\partial y} \right) \right] \mathrm{d}x \mathrm{d}y$$

$$- \iint_{\Omega} \left[M_x \frac{\partial \theta_x^*}{\partial x} + M_y \frac{\partial \theta_y^*}{\partial y} + M_{xy} \left(\frac{\partial \theta_x^*}{\partial y} + \frac{\partial \theta_y^*}{\partial x} \right) \right] \mathrm{d}x \mathrm{d}y = 0 \tag{53}$$

where $$t_x = l M_x + m M_{xy}, \; t_y = l M_{xy} + m M_y, \; t_z = l Q_x + m Q_y \tag{54}$$

Defining the following functions:

$$M_x^* = D \left(\frac{\partial \theta_x^*}{\partial x} + \nu \frac{\partial \theta_y^*}{\partial y} \right) \tag{55a}$$

$$M_y^* = D \left(\nu \frac{\partial \theta_x^*}{\partial x} + \frac{\partial \theta_y^*}{\partial y} \right) \tag{55b}$$

$$M_{xy}^* = \frac{(1 - \nu)}{2} D \left(\frac{\partial \theta_x^*}{\partial y} + \frac{\partial \theta_y^*}{\partial x} \right) \tag{55c}$$

$$Q_x^* = \frac{(1 - \nu)}{2} D \lambda^2 \left(\theta_x^* + \frac{\partial w^*}{\partial x} \right) \tag{56a}$$

$$Q_y^* = \frac{(1 - \nu)}{2} D \lambda^2 \left(\theta_y^* + \frac{\partial w^*}{\partial y} \right) \tag{56b}$$

then, it can be shown that:

$$M_x \frac{\partial \theta_x^*}{\partial x} + M_{xy} \left(\frac{\partial \theta_x^*}{\partial y} + \frac{\partial \theta_y^*}{\partial x} \right) + M_y \frac{\partial \theta_y^*}{\partial y} \equiv$$

$$M_x^* \frac{\partial \theta_x}{\partial x} + M_{xy}^* \left(\frac{\partial \theta_x}{\partial y} + \frac{\partial \theta_y}{\partial x} \right) + M_y^* \frac{\partial \theta_y}{\partial y} + \frac{\nu q}{(1 - \nu) \lambda^2} \left(\frac{\partial \theta_x^*}{\partial x} + \frac{\partial \theta_y^*}{\partial y} \right) \tag{57}$$

and $$Q_x\left(\theta_x^* + \frac{\partial w^*}{\partial x}\right) \equiv Q_x^*\left(\theta_x + \frac{\partial w}{\partial x}\right) \tag{58a}$$

$$Q_y\left(\theta_y^* + \frac{\partial w^*}{\partial y}\right) \equiv Q_y^*\left(\theta_y + \frac{\partial w}{\partial y}\right) \tag{58b}$$

Substituting from eqns (57) and (58) into (53) and integrating by parts once more, an inverse weighted-residual expression can be obtained in the following form:

$$\oint_\Gamma (t_x\theta_x^* + t_y\theta_y^* + t_z w^*)\,\mathrm{d}\Gamma - \oint_\Gamma (t_x^*\theta_x + t_y^*\theta_y + t_z^* w)\,\mathrm{d}\Gamma$$

$$+ \iint_\Omega q\left[w^* - \frac{\nu}{(1-\nu)\lambda^2}\left(\frac{\partial\theta_x^*}{\partial x} + \frac{\partial\theta_y^*}{\partial y}\right)\right]\mathrm{d}x\,\mathrm{d}y$$

$$+ \iint_\Omega \left[\theta_x\left(\frac{\partial M_x^*}{\partial x} + \frac{\partial M_{xy}^*}{\partial y} - Q_x^*\right) + \theta_y\left(\frac{\partial M_{xy}^*}{\partial x} + \frac{\partial M_y^*}{\partial y} - Q_y^*\right)\right.$$

$$\left. + w\left(\frac{\partial Q_x^*}{\partial x} + \frac{\partial Q_y^*}{\partial y}\right)\right]\mathrm{d}x\,\mathrm{d}y = 0 \tag{59}$$

where t_x^*, t_y^* and t_z^* are defined in terms of M_x^*, M_y^*, M_{xy}^*, Q_x^*, Q_y^* by equations similar to (54). To simplify the implementation of boundary conditions, normal and tangential components will be employed, and they are defined as follows:

$$\begin{bmatrix}\theta_n\\ \theta_t\\ w\end{bmatrix} = \begin{bmatrix} l & m & 0\\ -m & l & 0\\ 0 & 0 & 1\end{bmatrix}\begin{bmatrix}\theta_x\\ \theta_y\\ w\end{bmatrix} \tag{60}$$

$$\begin{bmatrix}M_n\\ M_{nt}\\ Q_n\end{bmatrix} = \begin{bmatrix} l & m & 0\\ -m & l & 0\\ 0 & 0 & 1\end{bmatrix}\begin{bmatrix}t_x\\ t_y\\ t_z\end{bmatrix} \tag{61}$$

Notice also that similar equations can be used for corresponding starred parameters, and it can be proved from eqns (55), (56), (60), and (61) that:

$$M_n^* = D\left(\frac{\partial \theta_n^*}{\partial n} + \nu \frac{\partial \theta_t^*}{\partial t}\right) \tag{62a}$$

$$M_{nt}^* = \frac{(1-\nu)}{2} D\left(\frac{\partial \theta_n^*}{\partial t} + \frac{\partial \theta_t^*}{\partial n}\right) \tag{62b}$$

$$Q_n^* = \frac{(1-\nu)}{2} D\lambda^2\left(\theta_n^* + \frac{\partial w^*}{\partial n}\right) \tag{63}$$

which can be represented in the following matrix form:

$$\begin{bmatrix} M_n^* \\ M_{nt}^* \\ Q_n^* \end{bmatrix} = D\begin{bmatrix} \dfrac{\partial}{\partial n} & \nu\dfrac{\partial}{\partial t} & 0 \\ \dfrac{(1-\nu)}{2}\dfrac{\partial}{\partial t} & \dfrac{(1-\nu)}{2}\dfrac{\partial}{\partial n} & 0 \\ \dfrac{(1-\nu)}{2}\lambda^2 & 0 & \dfrac{(1-\nu)}{2}\lambda^2\dfrac{\partial}{\partial n} \end{bmatrix}\begin{bmatrix} \theta_n^* \\ \theta_t^* \\ w^* \end{bmatrix} \tag{64}$$

Using eqns (60) and (61), it can also be shown that:

$$M_n\theta_n^* + M_{nt}\theta_t^* \equiv t_x\theta_x^* + t_y\theta_y^*, \quad M_n^*\theta_n + M_{nt}^*\theta_t \equiv t_x^*\theta_x + t_y^*\theta_y \tag{65}$$

Hence, eqn (59) can be rewritten as follows:

$$\oint_\Gamma (M_n\theta_n^* + M_{nt}\theta_t^* + Q_n w^*)\mathrm{d}\Gamma - \oint_\Gamma (M_n^*\theta_n + M_{nt}^*\theta_t + Q_n^* w)\mathrm{d}\Gamma$$

$$+ \iint_\Omega q\left[w^* - \frac{\nu}{(1-\nu)\lambda^2}\left(\frac{\partial \theta_x^*}{\partial x} + \frac{\partial \theta_y^*}{\partial y}\right)\right]\mathrm{d}x\,\mathrm{d}y$$

$$+ \iint_\Omega \left[\theta_x\left(\frac{\partial M_x^*}{\partial x} + \frac{\partial M_{xy}^*}{\partial y} - Q_x^*\right) + \theta_y\left(\frac{\partial M_{xy}^*}{\partial x} + \frac{\partial M_y^*}{\partial y} - Q_y^*\right)\right.$$

$$\left. + w\left(\frac{\partial Q_x^*}{\partial x} + \frac{\partial Q_y^*}{\partial y}\right)\right]\mathrm{d}x\,\mathrm{d}y = 0 \tag{66}$$

Equation (66) represents a domain integral equation, which can be reduced to a boundary integral equation, with respect to a source point (x_i, y_i), if the weighting functions are defined such that:

$$\frac{\partial M_x^*}{\partial x} + \frac{\partial M_{xy}^*}{\partial y} - Q_x^* = -e_x \delta(x - x_i, y - y_i) \tag{67a}$$

$$\frac{\partial M_{xy}^*}{\partial x} + \frac{\partial M_y^*}{\partial y} - Q_y^* = -e_y \delta(x - x_i, y - y_i) \tag{67b}$$

$$\frac{\partial Q_x^*}{\partial x} + \frac{\partial Q_y^*}{\partial y} = -e_z \delta(x - x_i, y - y_i) \tag{67c}$$

where e_x, e_y and e_z are arbitrary constant parameters, and $\delta(x - x_i,\ y - y_i)$ is a two-dimensional Dirac delta function.

Substituting from eqns (67) into (66) and using properties of the Dirac delta functions,[15] a boundary integral equation can be obtained and expressed in the following form:

$$c_i[e_x\theta_x(x_i, y_i) + e_y\theta_y(x_i, y_i) + e_z w(x_i, y_i)] + \oint_\Gamma (M_n^*\theta_n + M_{nt}^*\theta_t + Q_n^* w)\mathrm{d}\Gamma$$

$$= \oint_\Gamma (M_n\theta_n^* + M_{nt}\theta_t^* + Q_n w^*)\mathrm{d}\Gamma + \iint_\Omega q\left[w^* - \frac{\nu}{(1-\nu)\lambda^2}\left(\frac{\partial\theta_x^*}{\partial x} + \frac{\partial\theta_y^*}{\partial y}\right)\right]\mathrm{d}x\,\mathrm{d}y \tag{68}$$

where
$$c_i \equiv \iint_\Omega \delta(x - x_i, y - y_i)\,\mathrm{d}x\,\mathrm{d}y \tag{69}$$

Notice that eqns (67) represent three simultaneous partial differential equations, and it can be deduced from their solution, as will be explained in section 5, that:

$$\begin{bmatrix} \theta_n^* \\ \theta_t^* \\ w^* \end{bmatrix} = \boldsymbol{U}(x - x_i, y - y_i)\begin{bmatrix} e_x \\ e_y \\ e_z \end{bmatrix} \tag{70}$$

and

$$\begin{bmatrix} M_n^* \\ M_{nt}^* \\ Q_n^* \end{bmatrix} = \boldsymbol{T}(x - x_i, y - y_i) \begin{bmatrix} e_x \\ e_y \\ e_z \end{bmatrix} \tag{71}$$

and hence eqn (68) can be split into the following three boundary integral equations:

$$c_i \theta_x(x_i, y_i) + \oint_\Gamma (T_{11}\theta_n + T_{21}\theta_t + T_{31} w)\,\mathrm{d}\Gamma = \oint_\Gamma (U_{11} M_n + U_{21} M_{nt} + U_{31} Q_n)\,\mathrm{d}\Gamma + \iint_\Omega L_1\, q\, \mathrm{d}x\mathrm{d}y \tag{72a}$$

$$c_i \theta_y(x_i, y_i) + \oint_\Gamma (T_{12}\theta_n + T_{22}\theta_t + T_{32} w)\,\mathrm{d}\Gamma = \oint_\Gamma (U_{12} M_n + U_{22} M_{nt} + U_{32} Q_n)\,\mathrm{d}\Gamma + \iint_\Omega L_2\, q\, \mathrm{d}x\mathrm{d}y \tag{72b}$$

$$c_i w(x_i, y_i) + \oint_\Gamma (T_{13}\theta_n + T_{23}\theta_t + T_{33} w)\,\mathrm{d}\Gamma = \oint_\Gamma (U_{13} M_n + U_{23} M_{nt} + U_{33} Q_n)\,\mathrm{d}\Gamma + \iint_\Omega L_3\, q\, \mathrm{d}x\mathrm{d}y \tag{72c}$$

where L_1, L_2, L_3 are defined such that:

$$w^* - \frac{\nu}{(1 - \nu)\lambda^2} \left(\frac{\partial \theta_x^*}{\partial x} + \frac{\partial \theta_y^*}{\partial y} \right) \equiv L_1 e_x + L_2 e_y + L_3 e_z \tag{73}$$

4 Fundamental Solution

If the starred parameters in eqns (67) are expressed in terms of θ_x^*, θ_y^* and w^*, as defined by eqns (55), (56), then eqns (67) can be rewritten as three simultaneous partial differential equations in terms of θ_x^*, θ_y^*, w^*, and they can be expressed in the following matrix form:

$$\boldsymbol{H}\begin{bmatrix}\theta_x^*\\ \theta_y^*\\ w^*\end{bmatrix} = \delta(x-x_i, y-y_i)\begin{bmatrix}e_x\\ e_y\\ e_z\end{bmatrix} \tag{74}$$

where $\boldsymbol{H}$ is a matrix of differential operators, defined as follows:

$$H_{\alpha\beta} = \frac{D}{2}\left[(1-\nu)(\nabla^2-\lambda^2)\delta_{\alpha\beta} + (1+\nu)\frac{\partial^2}{\partial x_\alpha \partial x_\beta}\right]$$

$$H_{3\alpha} = (1-\nu)\lambda^2\frac{\partial}{\partial x_\alpha} = -H_{\alpha 3}$$

$$H_{33} = (1-\nu)\lambda^2\nabla^2$$

with $\alpha = 1,2, \quad \beta = 1,2, \quad (x_1, x_2) \equiv (x, y).$

A solution based upon Hörmander's method[16] was introduced by Weeën[7], but a simpler alternative approach will be given here. Strain functions can be employed similar to that based upon the Galerkin vector for two- and three-dimensional elasticity problems, so that eqn (74) can be decoupled.

Defining the strain functions f_x^*, f_y^* and f_z^* such that:

$$\begin{bmatrix}\theta_x^*\\ \theta_y^*\\ w^*\end{bmatrix} = \boldsymbol{H}^*\begin{bmatrix}f_x^*\\ f_y^*\\ f_z^*\end{bmatrix} \tag{75}$$

where
$$H_{\alpha\beta}^* = 2\delta_{\alpha\beta}\nabla^4 - \frac{\partial^2}{\partial x_\alpha \partial x_\beta}\left[(1+\nu)\nabla^2 + (1-\nu)\lambda^2\right]$$

$$H_{3\alpha}^* = -(1-\nu)\frac{\partial}{\partial x_\alpha}(\nabla^2-\lambda^2) = -H_{\alpha 3}^*$$

$$H_{33}^* = (\nabla^2-\lambda^2)\left[2\nabla^2 - (1-\nu)\lambda^2\right]/\lambda^2,$$

then it can be shown that eqn (74) can be reduced to the following form:

$$D(1-\nu)(\nabla^2-\lambda^2)\nabla^4 \begin{bmatrix} f_x^* \\ f_y^* \\ f_z^* \end{bmatrix} = -\delta(x-x_i, y-y_i) \begin{bmatrix} e_x \\ e_y \\ e_z \end{bmatrix} \tag{76}$$

and it is clear that f_x^*, f_y^* and f_z^* can be defined in terms of one function f^* such that:

$$f_x^* = e_x f^*, \quad f_y^* = e_y f^*, \quad f_z^* = e_z f^* \tag{77}$$

where
$$D(1-\nu)(\nabla^2-\lambda^2)\nabla^4 f^* = -\delta(x-x_i, y-y_i) \tag{78}$$

Hence, eqn (75) can be modified as follows:

$$\begin{bmatrix} \theta_x^* \\ \theta_y^* \\ w^* \end{bmatrix} = \boldsymbol{U}^* \begin{bmatrix} e_x \\ e_y \\ e_z \end{bmatrix} \tag{79}$$

where
$$\boldsymbol{U}^* = \boldsymbol{H}^* f^* \tag{80}$$

and it can be deduced from eqns (60), (70), and (80) that:

$$\begin{bmatrix} \theta_n^* \\ \theta_t^* \\ w^* \end{bmatrix} = \begin{bmatrix} l & m & 0 \\ -m & l & 0 \\ 0 & 0 & 1 \end{bmatrix} \boldsymbol{U}^* \begin{bmatrix} e_x \\ e_y \\ e_z \end{bmatrix} \equiv \boldsymbol{U} \begin{bmatrix} e_x \\ e_y \\ e_z \end{bmatrix} \tag{81}$$

where

$$\boldsymbol{U} = \begin{bmatrix} l & m & 0 \\ -m & l & 0 \\ 0 & 0 & 1 \end{bmatrix} \boldsymbol{U}^* \tag{82}$$

Using eqns (64) and (81) it can also be shown that:

$$T = D \begin{bmatrix} \dfrac{\partial}{\partial n} & \nu \dfrac{\partial}{\partial t} & 0 \\ \dfrac{(1-\nu)}{2} \dfrac{\partial}{\partial t} & \dfrac{(1-\nu)}{2} \dfrac{\partial}{\partial n} & 0 \\ \dfrac{(1-\nu)}{2} \lambda^2 & 0 & \dfrac{(1-\nu)}{2} \lambda^2 \dfrac{\partial}{\partial n} \end{bmatrix} U \tag{83}$$

The matrices of kernel functions, U, T are dependent on the function f^*, which will be derived here by means of integral transforms.

Due to the expected symmetry of f^* with respect to (x_i, y_i), a double Fourier transform applied to eqn (78) will reduce it to the following algebraic equation:

$$D(1-\nu)\rho^4(\rho^2+\lambda^2)\bar{f}^*(\rho) = \frac{1}{2\pi} \tag{84}$$

where $\bar{f}^*(\rho)$ is the Hankel transform of $f^*(r)$ defined as follows[17]:

$$\bar{f}^*(\rho) = \int_0^\infty r f^* \mathrm{J}_0(\rho r)\mathrm{d}r \tag{85}$$

and J_0 is a Bessel function,[18] $r = \sqrt{(x-x_i)^2+(y-y_i)^2}$.

Hence, it can be deduced from eqn (84) that

$$\bar{f}^*(\rho) = \frac{1}{2\pi D(1-\nu)\lambda^4}\left[\frac{\lambda^2}{\rho^4} - \frac{1}{\rho^2} + \frac{1}{\rho^2+\lambda^2}\right] \tag{86}$$

and $f^*(r)$ can be obtained by means of the inverse Hankel transform, defined as follows:

$$f^*(r) = \int_0^\infty \rho \bar{f}^* \mathrm{J}_0(\rho r)\,\mathrm{d}\rho \tag{87}$$

i.e.

$$f^* = \frac{1}{2\pi D(1-\nu)\lambda^4}\left[\frac{z^2}{4}(\log z - 1) + \log z + \mathrm{K}_0(z)\right] \tag{88}$$

where $z = \lambda r$, and $\mathrm{K}_0(z)$ is a modified Bessel function.[18]

5 Derivation of Kernel Functions

Using the following differential theorems (El-Zafrany[15]), with m, n, l being integer parameters:

$$\frac{\partial}{\partial x_\alpha}\left(\frac{1}{r^n}\frac{\partial r}{\partial x_\beta}\right) = \frac{1}{r^{n+1}}\left[\delta_{\alpha\beta} - (n+1)\left(\frac{\partial r}{\partial x_\alpha}\right)\left(\frac{\partial r}{\partial x_\beta}\right)\right] \tag{89}$$

$$\begin{aligned}\frac{\partial}{\partial x_\gamma}\left[\frac{1}{r^l}\left(\frac{\partial r}{\partial x_\alpha}\right)^m\left(\frac{\partial r}{\partial x_\beta}\right)^n\right] &= \frac{1}{r^{l+1}}\left[\left(\frac{\partial r}{\partial x_\alpha}\right)^{m-1}\left(\frac{\partial r}{\partial x_\beta}\right)^{n-1}\right] \\ &\times\left[m\,\delta_{\alpha\gamma}\frac{\partial r}{\partial x_\beta} + n\,\delta_{\beta\gamma}\frac{\partial r}{\partial x_\alpha} - (m+n+l)\frac{\partial r}{\partial x_\alpha}\frac{\partial r}{\partial x_\beta}\frac{\partial r}{\partial x_\gamma}\right]\end{aligned} \tag{90}$$

together with the properties of Bessel functions listed in Appendix A, the elements of the matrix $\boldsymbol{U}^*$ can be obtained as follows:

$$\begin{aligned}U^*_{\alpha\beta} = -\frac{1}{\pi D(1-\nu)}&\left\{\frac{\partial r}{\partial x_\alpha}\frac{\partial r}{\partial x_\beta}\left[A(z) + \frac{(1-\nu)}{4}\right]\right. \\ &\left. - \delta_{\alpha\beta}\left[B(z) - \frac{(1-\nu)}{4}\left(\log z - \frac{1}{2}\right)\right]\right\}\end{aligned} \tag{91a}$$

$$U^*_{\alpha 3} = -\frac{r}{4\pi D}\frac{\partial r}{\partial x_\alpha}\left(\log z - \frac{1}{2}\right) \tag{91b}$$

$$U^*_{3\beta} = \frac{r}{4\pi D}\frac{\partial r}{\partial x_\beta}\left(\log z - \frac{1}{2}\right) \equiv -U^*_{\beta 3} \tag{91c}$$

$$U^*_{33} = -\frac{1}{\pi D(1-\nu)\lambda^2}\left[\log z - \frac{(1-\nu)}{8}z^2(\log z - 1)\right] \tag{91d}$$

where $\alpha = 1,2, \quad \beta = 1,2, \quad z = \lambda r$

$$A(z) = K_o(z) + \frac{2}{z}K_1(z) - \frac{2}{z^2} \tag{92}$$

$$B(z) = \frac{1}{2}\left[A(z) + K_o(z)\right] \tag{93}$$

Using eqn (82), it can be shown that:

$$U_{\alpha\beta} = -\frac{1}{\pi D(1-\nu)}\left\{\frac{\partial r}{\partial n_\alpha}\frac{\partial r}{\partial x_\beta}\left[A(z)+\frac{(1-\nu)}{4}\right]\right.$$

$$\left.-\left(\hat{n}_\alpha\cdot\hat{i}_\beta\right)\left[B(z)-\frac{(1-\nu)}{4}\left(\log z-\frac{1}{2}\right)\right]\right\} \tag{94a}$$

$$U_{\alpha 3} = -\frac{r}{4\pi D}\frac{\partial r}{\partial n_\alpha}\left(\log z-\frac{1}{2}\right) \tag{94b}$$

$$U_{3\beta} = \frac{r}{4\pi D}\frac{\partial r}{\partial x_\beta}\left(\log z-\frac{1}{2}\right) \equiv U^*_{3\beta} \tag{94c}$$

$$U_{33} = -\frac{1}{\pi D(1-\nu)\lambda^2}\left[\log z-\frac{(1-\nu)}{8}z^2(\log z-1)\right] \equiv U^*_{33} \tag{94d}$$

Similarly, it can be proved from eqns (83) and (94) that:

$$T_{1\beta} = -\frac{\lambda}{\pi(1-\nu)}\left\{(1+\nu)\frac{\partial r}{\partial x_\beta}C(z)+g\frac{\partial r}{\partial x_\beta}F(z)\right.$$

$$\left.+\left[l_\beta\frac{\partial r}{\partial n}+\nu\left(\hat{t}\cdot\hat{i}_\beta\right)\frac{\partial r}{\partial t}\right]G(z)\right\} \tag{95a}$$

$$T_{2\beta} = -\frac{\lambda}{2\pi}\left\{\left[\left(\hat{t}\cdot\hat{i}_\beta\right)\frac{\partial r}{\partial n}+l_\beta\frac{\partial r}{\partial t}\right]G(z)+2\frac{\partial r}{\partial n}\frac{\partial r}{\partial t}\frac{\partial r}{\partial x_\beta}F(z)\right\} \tag{95b}$$

$$T_{3\beta} = -\frac{\lambda^2}{2\pi}\left\{\frac{\partial r}{\partial n}\frac{\partial r}{\partial x_\beta}A(z)-l_\beta B(z)\right\} \tag{95c}$$

$$T_{13} = -\frac{1}{4\pi}\left[(1+\nu)\left(\log z-\frac{1}{2}\right)+g\right] \tag{95d}$$

$$T_{23} = -\frac{(1-\nu)}{4\pi}\frac{\partial r}{\partial n}\frac{\partial r}{\partial t} \tag{95e}$$

$$T_{33} = -\frac{1}{2\pi r}\frac{\partial r}{\partial n} \tag{95f}$$

where

$$C(z) = \frac{A(z)}{z} + \frac{(1-\nu)}{4z} \tag{96}$$

$$F(z) = -\frac{4A(z)}{z} - \left[K_1(z) - \frac{1}{z}\right] - \frac{(3-\nu)}{2z} \tag{97}$$

$$G(z) = \frac{2A(z)}{z} + \left[K_1(z) - \frac{1}{z}\right] + \frac{(3-\nu)}{2z} \tag{98}$$

$$g = \left(\frac{\partial r}{\partial n}\right)^2 + \nu\left(\frac{\partial r}{\partial t}\right)^2 \tag{99}$$

It can also be proved from eqns (73) and (79) that:

$$L_j = U_{3j}^* - \frac{\nu}{(1-\nu)\lambda^2}\left[\frac{\partial U_{1j}^*}{\partial x} + \frac{\partial U_{2j}^*}{\partial y}\right] \tag{100}$$

6 Reduction of Domain Loading Terms

Domain loading terms are the terms appearing in the boundary integral equations (eqns 72) as:

$$\iint_{\Omega} L_j q \, \mathrm{d}x \, \mathrm{d}y$$

and will be simplified for some types of loading as explained in this section.

6.1 Introductory theorems

(a) *Defining* $$\phi^* = \frac{r^2}{8\pi D}\left(\log z - 1\right) \tag{101}$$

then $$\nabla^2 \phi^* = \frac{\log z}{2\pi D} \tag{102}$$

where $z = \lambda r$

This theorem can be proved by direct differentiation, using:

$$\nabla^2 f(r) = \frac{\mathrm{d}^2 f}{\mathrm{d}r^2} + \frac{1}{r}\frac{\mathrm{d}f}{\mathrm{d}r} \tag{103}$$

(b) *Defining*

$$\psi^* = \frac{r^4}{128\pi D}\left(\log z - \frac{3}{2}\right) \tag{104}$$

then

$$\nabla^2 \psi^* = \phi^* \tag{105}$$

which can also be proved by direct differentiation.

(c) *Theorem:*

$$\phi^* = -(1-\nu)(\nabla^2 - \lambda^2) f^* \tag{106}$$

where f^* is as defined by eqn (88).

Proof: From different fundamental solution theorems, it is clear that:

$$(\nabla^2 - \lambda^2)\,\mathrm{K_o}(z) = -2\pi\,\delta(x - x_i, y - y_i) \tag{107}$$

and

$$\nabla^2(\log z) = 2\pi\,\delta(x - x_i, y - y_i) \tag{108}$$

i.e. it can be deduced that:

$$(\nabla^2 - \lambda^2)\left[\mathrm{K_o}(z) + \log z\right] = -\lambda^2 \log z \tag{109}$$

Rewriting eqn (88) as:

$$f^* = \frac{\phi^*}{(1-\nu)\lambda^2} + \frac{1}{2\pi D\lambda^4(1-\nu)}\left[\mathrm{K_o}(z) + \log z\right]$$

then it can be shown that:

$$(\nabla^2 - \lambda^2) f^* = \frac{\nabla^2\phi^*}{\lambda^2(1-\nu)} - \frac{\phi^*}{(1-\nu)} - \frac{\log z}{2\pi D\lambda^2(1-\nu)}$$

$$\equiv -\frac{\phi^*}{(1-\nu)}$$

The following useful results can be proved from eqns (80) and (106):

(i)

$$U^*_{3\alpha} = -U^*_{\alpha 3} = \frac{\partial \phi^*}{\partial x_\alpha} \tag{110}$$

(ii) $$U_{33}^{*} = \left[1 - \frac{2}{(1-\nu)\lambda^2} \nabla^2 \right] \phi^{*} \tag{111}$$

(iii) $$\frac{\partial U_{1\alpha}^{*}}{\partial x} + \frac{\partial U_{2\alpha}^{*}}{\partial y} = - \frac{\partial}{\partial x_\alpha} (\nabla^2 \phi^{*}) \tag{112}$$

(iv) $$\frac{\partial U_{13}^{*}}{\partial x} + \frac{\partial U_{23}^{*}}{\partial y} = - \nabla^2 \phi^{*} \tag{113}$$

6.2 Simplification of domain loading kernel functions

Using the previous theorems, then the domain loading functions defined by eqn (100) can be simplified as follows:

$$L_\alpha = \left[1 + \frac{\nu}{(1-\nu)\lambda^2} \nabla^2 \right] \frac{\partial \phi^{*}}{\partial x_\alpha} \tag{114}$$

$$L_3 = \left[1 - \frac{(2-\nu)}{(1-\nu)\lambda^2} \nabla^2 \right] \phi^{*} \tag{115}$$

Notice also that:

$$\frac{\partial \phi^{*}}{\partial x_\alpha} = \frac{r}{4 \pi D} \left(\log z - \frac{1}{2} \right) \frac{\partial r}{\partial x_\alpha} \tag{116}$$

and

$$\frac{\partial}{\partial x_\alpha} (\nabla^2 \phi^{*}) = \frac{1}{2 \pi D r} \frac{\partial r}{\partial x_\alpha} \tag{117}$$

6.3 Case of concentrated forces and moments

6.3.1 Representation in terms of domain loading intensity

Concentrated forces and moments may be represented in terms of distributions over the domain, using Dirac delta functions, as explained next.

(a) Case of a concentrated force

Consider the case where a concentrated force F in the z direction is acting on a plate at midplane point (x_l, y_l). Defining a small circle of radius ϵ centred at (x_l, y_l), as shown in Figure 3, then an equivalent loading intensity can be defined as follows:

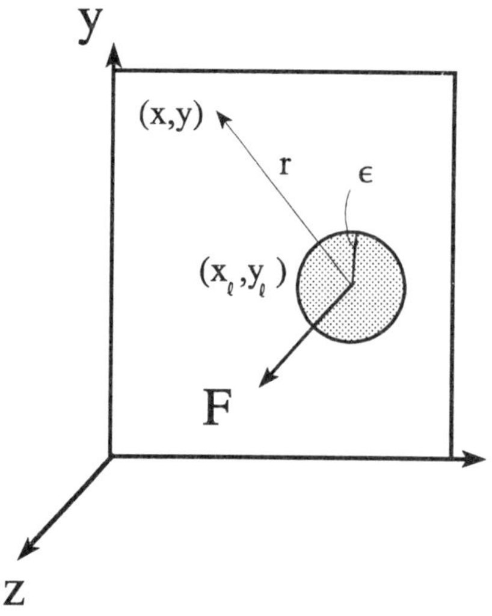

Figure 3. Concentrated force acting on a plate.

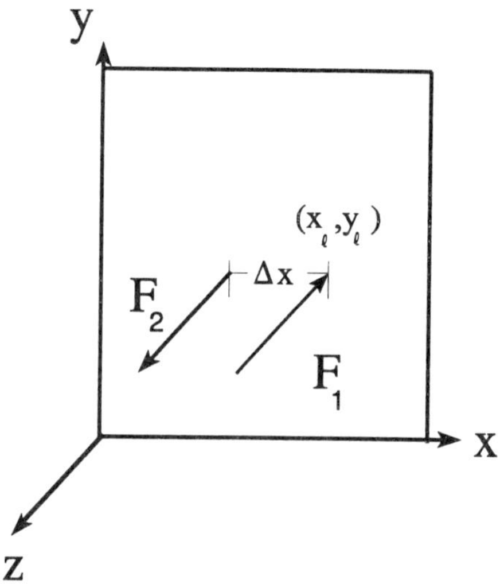

Figure 4. Concentrated couple acting on a plate.

$$q(x,y) = 0 \qquad \text{for } r > 0$$

$$q(x,y) = \lim_{\epsilon \to 0} \frac{F}{\pi \epsilon^2} \qquad \text{for } r \to 0$$

where $$r = \sqrt{(x - x_l)^2 + (y - y)^2}$$

Hence, it can be deduced from the definition of two-dimensional Dirac delta functions[15] that:

$$q(x,y) = F\,\delta(x - x_l, y - y_l) \tag{118}$$

(b) Case of a concentrated moment

A concentrated moment T_y in the y direction, acting at (x_l, y_l), can be represented by a couple of two forces, as shown in Figure 4, where:

$$F_2 = -F_1 = \lim_{\Delta x \to 0} \frac{T_y}{\Delta x}$$

Using the previous analysis, the loading intensity equivalent to every force can be expressed as follows:

$$q(F_1) = F_1\,\delta(x - x_l, y - y_l) \equiv -\lim_{\Delta x \to 0} \frac{T_y\,\delta(x - x_l, y - y_l)}{\Delta x}$$

$$q(F_2) = F_2\,\delta(x - x_l + \Delta x, y - y_l) \equiv \lim_{\Delta x \to 0} \frac{T_y\,\delta(x + \Delta x - x_l, y - y_l)}{\Delta x}$$

Hence, the resultant loading intensity is:

$$\begin{aligned} q(x,y) &\equiv q(F_1) + q(F_2) \\ &= T_y \lim_{\Delta x \to 0} \frac{\delta(x + \Delta x - x_l, y - y_l) - \delta(x - x_l, y - y_l)}{\Delta x} \end{aligned}$$

i.e. $$q(x,y) = T_y \frac{\partial}{\partial x}\delta(x - x_l, y - y_l) \tag{119}$$

Similarly, if there is a concentrated moment T_x in the x direction, acting at (x_l, y_l), it can be represented by a distributed loading with intensity:

$$q(x,y) = -T_x \frac{\partial}{\partial y}\delta(x - x_l, y - y_l) \tag{120}$$

6.3.2 Reduction of domain integrals

From previous analysis, if there are a concentrated shear force F and bending moments T_x and T_y acting at a point (x_l, y_l) inside the domain Ω, a distribution of an equivalent domain loading intensity can be obtained as follows:

$$q \equiv \left(-T_x\frac{\partial}{\partial y} + T_y\frac{\partial}{\partial x} + F\right)\delta(x - x_l, y - y_l) \tag{121}$$

Using the following properties of Dirac delta function[15]:

$$\iint_\Omega f(x,y)\,\delta(x - x_l, y - y_l)\,\mathrm{d}x\mathrm{d}y = f(x_l, y_l) \tag{122}$$

$$\iint_\Omega f(x,y)\frac{\partial}{\partial x_\beta}\delta(x - x_l, y - y_l)\,\mathrm{d}x\mathrm{d}y = -\frac{\partial f}{\partial x_\beta} \text{ at } (x_l, y_l) \tag{123}$$

then it can be proved that:

$$\iint_{\Omega} L_j \, q \, \mathrm{d}x \, \mathrm{d}y \equiv \iint_{\Omega} L_j \left(-T_x \frac{\partial}{\partial y} + T_y \frac{\partial}{\partial x} + F \right) \delta(x - x_l, y - y_l) \, \mathrm{d}x \, \mathrm{d}y$$

$$= \left(T_x \frac{\partial}{\partial y} - T_y \frac{\partial}{\partial x} + F \right) L_j \qquad \text{at } x = x_l, \; y = y_l \tag{124}$$

6.4 Case of uniformly or linearly-distributed loading

Substituting from eqn (105) into (114) and (115), the domain loading functions can be redefined as follows:

$$L_\alpha = \frac{\partial}{\partial x_\alpha} \nabla^2 \left[\psi^* + \frac{\nu \, \phi^*}{(1 - \nu) \, \lambda^2} \right] \tag{125}$$

$$L_3 = \nabla^2 \left[\psi^* - \frac{(2 - \nu) \, \phi^*}{(1 - \nu) \, \lambda^2} \right] \tag{126}$$

Using integration by parts theorems as given by eqns (51) and (52), it can be proved, for cases with $q(x,y)$ being a linear function of (x,y), that:

$$\iint_{\Omega} L_\alpha \, q \, \mathrm{d}x \, \mathrm{d}y = \oint_{\Gamma} V_\alpha \, q \, \mathrm{d}\Gamma - \oint_{\Gamma} S_\alpha \frac{\partial q}{\partial x_\alpha} \mathrm{d}\Gamma \tag{127}$$

$$\iint_{\Omega} L_3 \, q \, \mathrm{d}x \, \mathrm{d}y = \oint_{\Gamma} V_3 \, q \, \mathrm{d}\Gamma - \oint_{\Gamma} S_3 \frac{\partial q}{\partial n} \mathrm{d}\Gamma \tag{128}$$

where

$$V_\alpha = l_\alpha \left[1 + \frac{\nu}{(1 - \nu) \, \lambda^2} \nabla^2 \right] \phi^* \tag{129}$$

$$V_3 = \frac{\partial}{\partial n} \left[\psi^* - \frac{(2 - \nu) \, \phi^*}{(1 - \nu) \, \lambda^2} \right] \tag{130}$$

$$S_\alpha = \frac{\partial}{\partial n} \left[\psi^* + \frac{\nu \, \phi^*}{(1 - \nu) \, \lambda^2} \right] \tag{131}$$

$$S_3 = \psi^* - \frac{(2 - \nu) \, \phi^*}{(1 - \nu) \, \lambda^2} \tag{132}$$

7 Numerical Treatment of Boundary Integral Equations

The boundary integral equations (72) are discretized in terms of boundary integrals over constant or isoparametric boundary elements, and the standard procedure of the BEM[15] can be adopted. Some additional problems will be discussed here.

7.1 Corner problem for isoparametric elements

Due to the discontinuity of the normal directional cosines (l,m) at boundary corners, and the singularities in some of the kernel functions, additional terms known as *jump functions*[15] may appear when a boundary integral crosses a corner point, where a source point is defined. A simplified approach is presented here, by modifying the boundary integral equations (eqns (72)) to the following:

$$c_{11}\theta_x(x_i,y_i) + c_{12}\theta_y(x_i,y_i) + c_{13}w(x_i,y_i) + \oint_{\grave{\Gamma}}(T_{11}\theta_n + T_{21}\theta_t + T_{31}w)\,d\Gamma$$
$$= \oint_{\Gamma}(U_{11}M_n + U_{21}M_{nt} + U_{31}Q_n)\,d\Gamma + \iint_{\Omega} L_1 q\,d\Gamma \qquad (133a)$$

$$c_{21}\theta_x(x_i,y_i) + c_{22}\theta_y(x_i,y_i) + c_{23}w(x_i,y_i) + \oint_{\grave{\Gamma}}(T_{12}\theta_n + T_{22}\theta_t + T_{32}w)\,d\Gamma$$
$$= \oint_{\Gamma}(U_{12}M_n + U_{22}M_{nt} + U_{32}Q_n)\,d\Gamma + \iint_{\Omega} L_2 q\,d\Gamma \qquad (133b)$$

$$c_{31}\theta_x(x_i,y_i) + c_{32}\theta_y(x_i,y_i) + c_{33}w(x_i,y_i) + \oint_{\grave{\Gamma}}(T_{13}\theta_n + T_{23}\theta_t + T_{33}w)\,d\Gamma$$
$$= \oint_{\Gamma}(U_{13}M_n + U_{23}M_{nt} + U_{33}Q_n)\,d\Gamma + \iint_{\Omega} L_3 q\,d\Gamma \qquad (133c)$$

where $\grave{\Gamma}$ represents the boundary without any corner, i.e. the contour integrals are carried out without corner effects been considered.

The coefficients $c_{\alpha\beta}$, which contain the corner effects, may be evaluated mathematically. A practical approach is to consider that eqns (133) should satisfy rigid translation and rotation conditions, as follows:

(i) Applying a rigid translation in the z-direction, w = constant, $\theta_x = \theta_y = 0$, all forces = 0, then it can be deduced from eqns (133) that:

$$c_{j3} = -\oint_{\grave{\Gamma}} T_{3j}\,d\Gamma \qquad (134)$$

(ii) Applying a rigid rotation, such that: $\theta_x = \alpha =$ constant, $w_i = 0$, $\theta_y = 0$, all forces = 0, then $w(x,y) = -\alpha(x - x_i)$, and it can be shown that:

$$c_{j1} = -\oint_{\Gamma}\left[(lT_{1j} - mT_{2j}) - (x - x_i)T_{3j}\right]\mathrm{d}\Gamma \tag{135}$$

(iii) Similarly applying a rigid rotation such that: $\theta_y = \beta$, $w_i = 0$, $\theta_x = 0$, all forces = 0, then $w(x,y) = -\beta(y - y_i)$ and:

$$c_{j2} = -\oint_{\Gamma}\left[(mT_{1j} + lT_{2j}) - (y - y_i)T_{3j}\right]\mathrm{d}\Gamma \tag{136}$$

The $c_{\alpha\beta}$ coefficients as defined by the previous equations have been found to automatically correct numerical errors in the evaluated singular integrals of $T_{\alpha\beta}$ expressions, if they are estimated at every source point. Other parameters may have only a logarithmic singularity, which can be dealt with using a special quadrature technique as discussed in Reference 15. Further details associated with the use of isoparametric elements for the analysis of thick Reissner plates in bending, including the analysis of singular integrals, are given by El-Zafrany & Fadhil.[19]

7.2 Singular integrals for the case of constant elements

Consider the case of straight line constant boundary elements. When a source point (x_i, y_i) is the midpoint of the eth constant element, then most of the kernel functions will have infinite values within the integration region of that element. Using integral theorems given in Appendix B, analytical expressions of the integrals of the kernel functions over the *eth* element can be evaluated as follows:

$$\int_e U_{11}\mathrm{d}\Gamma = -\frac{l_i F_1}{2\pi D(1-\nu)} \tag{137a}$$

$$\int_e U_{12}\mathrm{d}\Gamma = -\frac{m_i F_1}{2\pi D(1-\nu)} \tag{137b}$$

$$\int_e U_{21}\mathrm{d}\Gamma = -\frac{m_i F_2}{2\pi D(1-\nu)} \tag{137c}$$

$$\int_e U_{22}\mathrm{d}\Gamma = \frac{l_i F_2}{2\pi D(1-\nu)} \tag{137d}$$

$$\int_e U_{33}\,d\Gamma = -\frac{1}{2\pi D}\left[\frac{4R_o(\log z_o - 1)}{(1-\nu)\lambda^2} - \frac{R_o^3}{6}\log z_o + \frac{2}{9}R_o^3\right] \tag{137e}$$

$$\int_e U_{\alpha 3}\,d\Gamma = \int_e U_{3\beta}\,d\Gamma = 0 \tag{137f}$$

$$\int_e T_{\alpha\beta}\,d\Gamma = \int_e T_{23}\,d\Gamma = \int_e T_{33}\,d\Gamma = 0 \tag{138a}$$

$$\int_e T_{13}\,d\Gamma = -\frac{R_o}{2\pi}\left[(1+\nu)\log z_o - \frac{3+\nu}{2}\right] \tag{138b}$$

$$\int_e T_{31}\,d\Gamma = -\frac{l_i\lambda}{\pi}\left[K_1(z_o) - \frac{1}{z_o}\right] \tag{138c}$$

$$\int_e T_{32}\,d\Gamma = -\frac{m_i\lambda}{\pi}\left[K_1(z_o) - \frac{1}{z_o}\right] \tag{138d}$$

$$\int_e V_1\,d\Gamma = \frac{l_i F_3}{2\pi D} \tag{139a}$$

$$\int_e V_2\,d\Gamma = \frac{m_i F_3}{2\pi D} \tag{139b}$$

$$\int_e V_3\,d\Gamma = 0 \tag{139c}$$

$$\int_e S_\alpha\,d\Gamma = 0 \tag{140a}$$

$$\int_e S_3\,d\Gamma = \frac{1}{4\pi D}\left[\frac{R_o^5}{80}\left(\log z_o - \frac{17}{10}\right) - \frac{(2-\nu)}{\lambda^2(1-\nu)}\frac{R_o^3}{3}\left(\log z_o - \frac{4}{3}\right)\right] \tag{140b}$$

where $$F_1 = 4\left[K_1(z_o) - \frac{1}{z_o}\right] + (1-\nu) R_o \left(\log z_o - \frac{3}{2}\right) \tag{141}$$

$$F_2 = 4\left[K_1(z_o) - \frac{1}{z_o}\right] - (1-\nu) R_o \left(\log z_o - \frac{1}{2}\right) + 4\int_0^{z_o} K_o(z)\,dz \tag{142}$$

$$F_3 = \frac{R_o^3}{6}\left(\log z_o - \frac{4}{3}\right) + \frac{2\nu R_o}{(1-\nu)}\left(\log z_o - 1\right) \tag{143}$$

R_o = half the element length, $z_o = \lambda R_o$, (l_i, m_i) are the directional cosines of the outward normal to the element at (x_i, y_i), and the integration of K_o is given in Appendix A.3.

8. Evaluation of stress components

8.1 Linearized bending stresses

For practical engineering applications, bending stresses may be approximated in terms of averaged linearized values $(\bar{\sigma}_x, \bar{\sigma}_y, \bar{\tau}_{xy})$, which satisfy equilibrium equations, i.e.

$$\left\{M_x, M_y, M_{xy}\right\} = \int_{-h/2}^{h/2} z\left\{\sigma_x, \sigma_y, \tau_{xy}\right\} dz \equiv \int_{-h/2}^{h/2} z\left\{\bar{\sigma}_x, \bar{\sigma}_y, \bar{\tau}_{xy}\right\} dz \tag{144}$$

Hence it can be deduced that:

$$\left\{\bar{\sigma}_x, \bar{\sigma}_y, \bar{\tau}_{xy}\right\} = \frac{12z}{h^3}\left\{M_x, M_y, M_{xy}\right\} \tag{145}$$

To estimate the error in the maximum values of stress components, which are at plate surfaces, the actual values are obtained from eqns (39) at $z = \pm h/2$, and it can be proved that:

$$\left(\bar{\sigma}_x - \sigma_x\right)_{\text{at } z = \pm h/2} = \frac{1}{10(1-\nu)}\left\{\pm\left[(2-\nu)\frac{\partial Q_x}{\partial x} + \nu\frac{\partial Q_y}{\partial y}\right] + 5(q_u - q_l)\right\} \tag{146a}$$

$$\left(\bar{\sigma}_y - \sigma_y\right)_{\text{at } z = \pm h/2} = \frac{1}{10(1-\nu)}\left\{\pm\left[\nu\frac{\partial Q_x}{\partial x} + (2-\nu)\frac{\partial Q_y}{\partial y}\right] + 5(q_u - q_l)\right\} \tag{146b}$$

$$\left(\bar{\tau}_{xy} - \tau_{xy}\right)_{\text{at } z = \pm h/2} = \pm \frac{1}{10}\left[\frac{\partial Q_x}{\partial y} + \frac{\partial Q_y}{\partial x}\right] \tag{146c}$$

The evaluation of stress components given in this section is based on averaged linearized values.

8.2 Stress components at boundary points

For constant elements, the results from boundary element analysis are first assumed to have constant distributions over each element. Taking the averages at every connecting point between elements, a linear distribution can, therefore, be considered over each element, based on the averaged values. Hence, constant elements can be dealt with, in this section, in a way similar to isoparametric elements with two end nodes. Isoparametric elements are described by means of the following equations, which represent parametric equations for the n-node element in terms of the parameter ξ:

$$x(\xi) = \sum_{j=1}^{n} x_j N_j(\xi), \qquad y(\xi) = \sum_{j=1}^{n} y_j N_j(\xi) \tag{147}$$

where

$$N_j(\xi) \equiv L_j^n(\xi) = \prod_{\substack{r=1,\\ r\neq j}}^{n} \frac{(n-1)\,\xi - (r-1)}{j-r} \tag{148}$$

which represents a one-dimensional Lagrangian shape function, and j represents the local number of the jth point on the element. Hence, it can be deduced that:

$$\frac{dx}{d\xi} = \sum_{j=1}^{n} x_j \frac{dN_j}{d\xi}, \qquad \frac{dy}{d\xi} = \sum_{j=1}^{n} y_j \frac{dN_j}{d\xi} \tag{149}$$

where

$$\frac{dN_j}{d\xi} = \sum_{\substack{r=1,\\ r\neq j}}^{n} \left[\left(\frac{n-1}{j-r}\right) \prod_{\substack{s=1,\\ s\neq r, s\neq j}}^{n} \left(\frac{(n-1)\xi - (s-1)}{j-s}\right)\right] \tag{150}$$

and it can also be shown that:

$$\frac{d\Gamma}{d\xi} = \sqrt{\left(\frac{dx}{d\xi}\right)^2 + \left(\frac{dy}{d\xi}\right)^2} \equiv J \tag{151}$$

Results at boundary element points are: θ_n, θ_t, w, M_n, M_{nt}, Q_n, and isoparametric equations are employed to find their values at any point on the element, i.e.

$$\theta_n(\xi) = \sum_{j=1}^{n} (\theta_n)_j N_j(\xi), \quad \ldots, \quad Q_n(\xi) = \sum_{j=1}^{n} (Q_n)_j N_j(\xi) \tag{152}$$

and it can also be deduced that:

$$\frac{\partial \theta_n}{\partial t} = \frac{1}{J} \sum_{j=1}^{n} (\theta_n)_j \frac{\partial N_j}{\partial \xi}, \quad \frac{\partial \theta_t}{\partial t} = \frac{1}{J} \sum_{j=1}^{n} (\theta_t)_j \frac{\partial N_j}{\partial \xi}, \quad \frac{\partial w}{\partial t} = \frac{1}{J} \sum_{j=1}^{n} w_j \frac{\partial N_j}{\partial \xi} \tag{153}$$

Hence, the following equation can be used to find the normal derivative of θ_n:

$$\frac{\partial \theta_n}{\partial n} = \frac{M_n}{D} - \nu \frac{\partial \theta_t}{\partial t} - \frac{\nu q}{(1-\nu) D \lambda^2} \tag{154}$$

and the moment M_t can be obtained from:

$$M_t = D \left(\nu \frac{\partial \theta_n}{\partial n} + \frac{\partial \theta_t}{\partial t} \right) + \frac{\nu q}{(1-\nu) \lambda^2} \tag{155}$$

The corresponding bending stress components, with respect to normal and tangential directions, can be found as follows:

$$\left\{ \bar{\sigma}_n, \bar{\sigma}_t, \bar{\tau}_{nt} \right\} = \frac{6}{h^2} \left\{ M_n, M_t, M_{nt} \right\} \tag{156}$$

The stress components with respect to x and y axes are obtained from the following relationships:

$$\sigma_x = l^2 \sigma_n + m^2 \sigma_t - 2 l m \tau_{nt} \tag{157a}$$

$$\sigma_y = m^2 \sigma_n + l^2 \sigma_t + 2 l m \tau_{nt} \tag{157b}$$

$$\tau_{xy} = lm(\sigma_n - \sigma_t) + (l^2 - m^2) \tau_{nt} \tag{157c}$$

The normal derivative of w can be obtained from:

$$\frac{\partial w}{\partial n} = \frac{Q_n}{\frac{1}{2} D \lambda^2 (1-\nu)} - \theta_n \tag{158}$$

and its derivatives with respect to x and y are:

$$\frac{\partial w}{\partial x} = l \frac{\partial w}{\partial n} - m \frac{\partial w}{\partial t}, \quad \frac{\partial w}{\partial y} = m \frac{\partial w}{\partial n} + l \frac{\partial w}{\partial t} \tag{159}$$

Similarly $\theta_x = l\,\theta_n - m\,\theta_t, \quad \theta_y = m\,\theta_n + l\,\theta_t$ (160)

Hence, Q_x, Q_y can be calculated from previous values by means of eqns (42), and the corresponding transverse shear stresses τ_{xz}, τ_{yz} can, therefore, be estimated using eqns (12) and (13).

8.3 Results at internal points

The lateral deflection and the rotation components can be obtained at an internal point (x_i, y_i) simply by considering it as the source point in the boundary integral equations. For points inside the domain $c_i = 1$, and eqns (72) can be reduced to:

$$\theta_x(x_i, y_i) = \iint_\Omega L_1\, q\, \mathrm{d}x\mathrm{d}y - \oint_\Gamma (T_{11}\theta_n + T_{21}\theta_t + T_{31}w)\,\mathrm{d}\Gamma + \oint_\Gamma (U_{11}M_n + U_{21}M_{nt} + U_{31}Q_n)\,\mathrm{d}\Gamma \tag{161a}$$

$$\theta_y(x_i, y_i) = \iint_\Omega L_2\, q\, \mathrm{d}x\mathrm{d}y - \oint_\Gamma (T_{12}\theta_n + T_{22}\theta_t + T_{32}w)\,\mathrm{d}\Gamma + \oint_\Gamma (U_{12}M_n + U_{22}M_{nt} + U_{32}Q_n)\,\mathrm{d}\Gamma \tag{161b}$$

$$w(x_i, y_i) = \iint_\Omega L_3\, q\, \mathrm{d}x\mathrm{d}y - \oint_\Gamma (T_{13}\theta_n + T_{23}\theta_t + T_{33}w)\,\mathrm{d}\Gamma + \oint_\Gamma (U_{13}M_n + U_{23}M_{nt} + U_{33}Q_n)\,\mathrm{d}\Gamma \tag{161c}$$

Using first principles,[15] their derivatives with respect to x and y can be obtained from the following boundary integral equations:

$$\frac{\partial \theta_x}{\partial x_\gamma} = -\iint_\Omega \frac{\partial L_1}{\partial x_\gamma}\, q\, \mathrm{d}x\mathrm{d}y + \oint_\Gamma \left(\frac{\partial T_{11}}{\partial x_\gamma}\theta_n + \frac{\partial T_{21}}{\partial x_\gamma}\theta_t + \frac{\partial T_{31}}{\partial x_\gamma} w \right)\mathrm{d}\Gamma - \oint_\Gamma \left(\frac{\partial U_{11}}{\partial x_\gamma} M_n + \frac{\partial U_{21}}{\partial x_\gamma} M_{nt} + \frac{\partial U_{31}}{\partial x_\gamma} Q_n \right)\mathrm{d}\Gamma \tag{162a}$$

$$\frac{\partial \theta_y}{\partial x_\gamma} = -\iint_\Omega \frac{\partial L_2}{\partial x_\gamma} q \, \mathrm{d}x\mathrm{d}y + \oint_\Gamma \left(\frac{\partial T_{12}}{\partial x_\gamma} \theta_n + \frac{\partial T_{22}}{\partial x_\gamma} \theta_t + \frac{\partial T_{32}}{\partial x_\gamma} w \right) \mathrm{d}\Gamma$$
$$- \oint_\Gamma \left(\frac{\partial U_{12}}{\partial x_\gamma} M_n + \frac{\partial U_{22}}{\partial x_\gamma} M_{nt} + \frac{\partial U_{32}}{\partial x_\gamma} Q_n \right) \mathrm{d}\Gamma \qquad (162b)$$

$$\frac{\partial w}{\partial x_\gamma} = -\iint_\Omega \frac{\partial L_3}{\partial x_\gamma} q \, \mathrm{d}x\mathrm{d}y + \oint_\Gamma \left(\frac{\partial T_{13}}{\partial x_\gamma} \theta_n + \frac{\partial T_{23}}{\partial x_\gamma} \theta_t + \frac{\partial T_{33}}{\partial x_\gamma} w \right) \mathrm{d}\Gamma$$
$$- \oint_\Gamma \left(\frac{\partial U_{13}}{\partial x_\gamma} M_n + \frac{\partial U_{23}}{\partial x_\gamma} M_{nt} + \frac{\partial U_{33}}{\partial x_\gamma} Q_n \right) \mathrm{d}\Gamma \qquad (162c)$$

and internal shear forces and moments can be obtained, using eqns (42) and (43) and previous values. Hence, the bending stress components can be estimated from equations similar to (156), and transverse stresses from eqns (12) and (13).

An alternative approach is to use the finite difference method, where the following approximations are considered for the numerical evaluation of the derivatives of a function $f(x, y)$ at (x_i, y_i):

$$\left.\frac{\partial f}{\partial x}\right|_{\text{at } x_i,\, y_i} \approx \frac{f(x_i + \Delta x, y_i) - f(x_i - \Delta x, y_i)}{2\,\Delta x} \qquad (163)$$

$$\left.\frac{\partial f}{\partial y}\right|_{\text{at } x_i,\, y_i} \approx \frac{f(x_i, y_i + \Delta y) - f(x_i, y_i - \Delta y)}{2\,\Delta y} \qquad (164)$$

Hence, the derivatives of θ_x, θ_y, w can be obtained from previous approximations, using values estimated from boundary integral equations (eqns 161), at four source points:

$$(x_i + \Delta x, y_i),\ (x_i - \Delta x, y_i),\ (x_i, y_i + \Delta y),\ (x_i, y_i - \Delta y)$$

9 Numerical examples

Boundary element computer programs based upon the derivations presented in this chapter have been coded in FORTRAN and tested on a PC. Some numerical examples were analyzed with the developed software and the results were compared with the corresponding analytical solutions given by Weeën[7] and Debbih,[20] and are summarized in this section.

9.1 Simply-supported circular plate

This case represents a solid thick circular plate, with outer diameter = 2 m, and thickness = 0.25 m. The plate material has Young's modulus = 2.05×10^{11} N/m^2, and Poisson's ratio = 0.3. Boundary element meshes with 30 constant elements equally spaced on the boundary were employed in the analysis, and the following boundary conditions were applied at all boundary nodes:

$$w = 0, \qquad M_n = M_{nt} = 0$$

The plate was subjected first to a uniformly-distributed loading with intensity $q = 4.0 \times 10^5$ N/m^2, then the loading was replaced with a concentrated force $F = 4.0 \times 10^5$ N, acting at the centre of the plate. Internal nodes on the diameter parallel to the x axis were selected, and the radial distributions of the deflection w, slope θ_r, and the moment M_r were plotted against analytical solutions based upon Reissner's plate-bending theory, as shown in Figures 5, 6, and 7, respectively. It is clear from those figures that the boundary element results are very close to the Reissner's theory analytical solution.

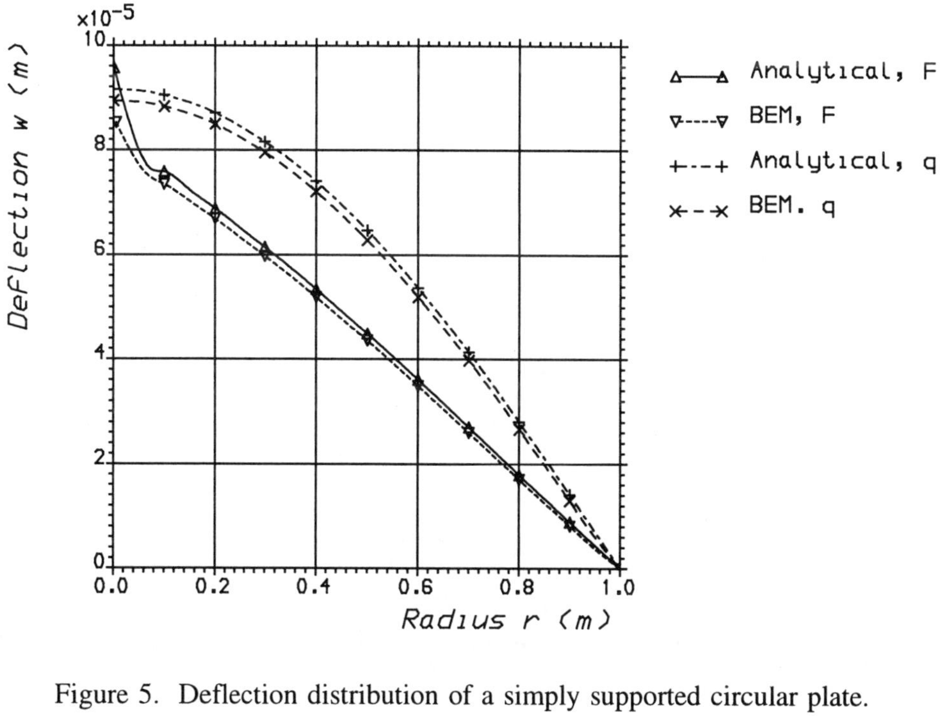

Figure 5. Deflection distribution of a simply supported circular plate.

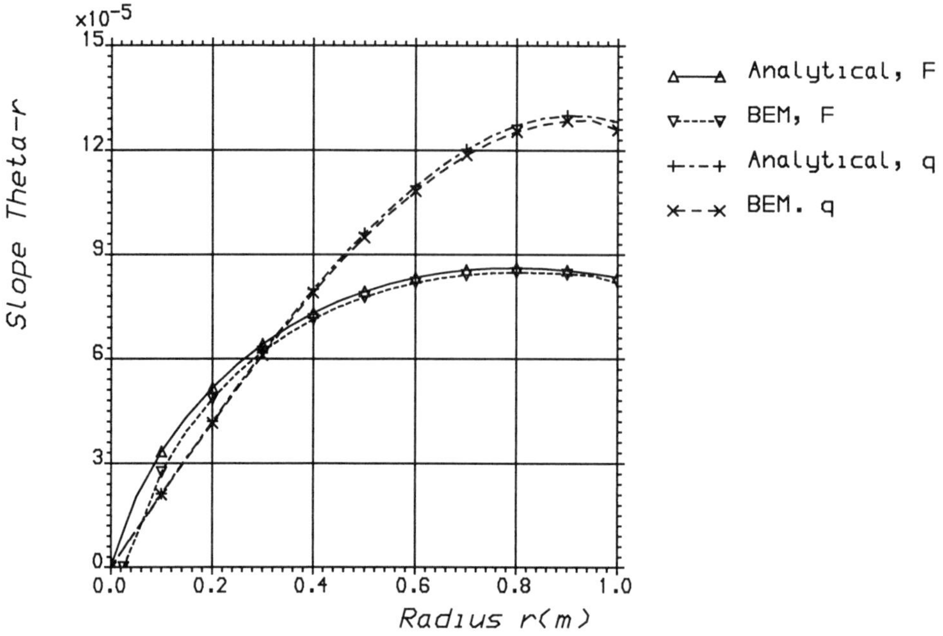

Figure 6. Slope distribution of a simply-supported circular plate.

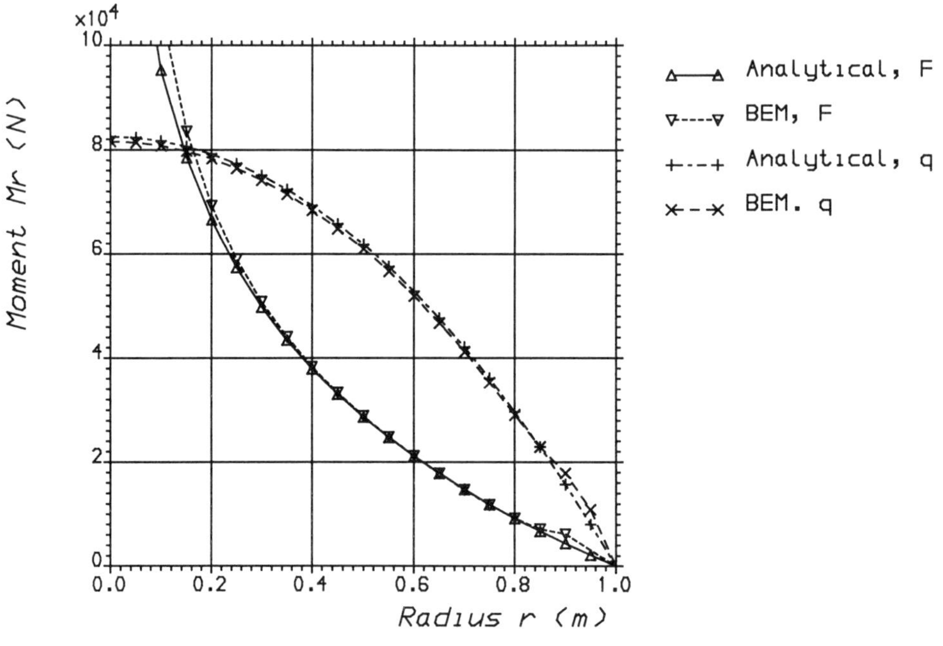

Figure 7. Moment distribution of a simply-supported circular plate.

9.2 Clamped circular plate

This case has the same dimensions, material properties, and boundary element meshes, as in the previous case, but the values of loading parameters are: $q = 1.4 \times 10^6$ N/m^2, and $F = 1.0 \times 10^6$ N, and the following boundary conditions were applied at all boundary nodes:

$$w = 0, \qquad \theta_n = \theta_t = 0$$

The radial distributions of the deflection w, slope θ_r, and the moment M_r were also plotted against the corresponding analytical solutions of Reissner's plate-bending theory, as shown in Figures 8, 9, and 10, respectively. It can be seen from those figures that the boundary element results have a good agreement with corresponding analytical solutions.

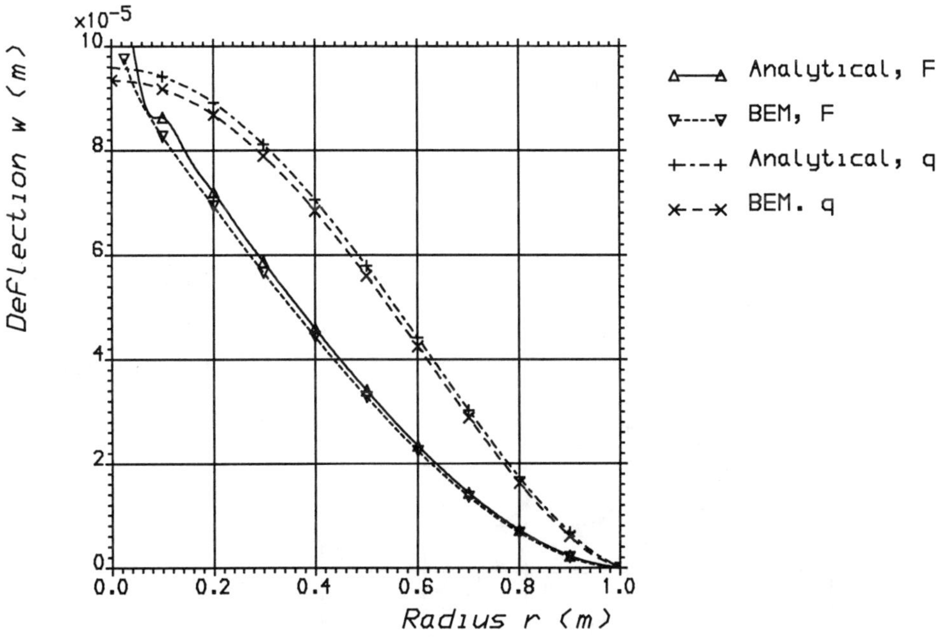

Figure 8. Deflection distribution of a clamped circular plate.

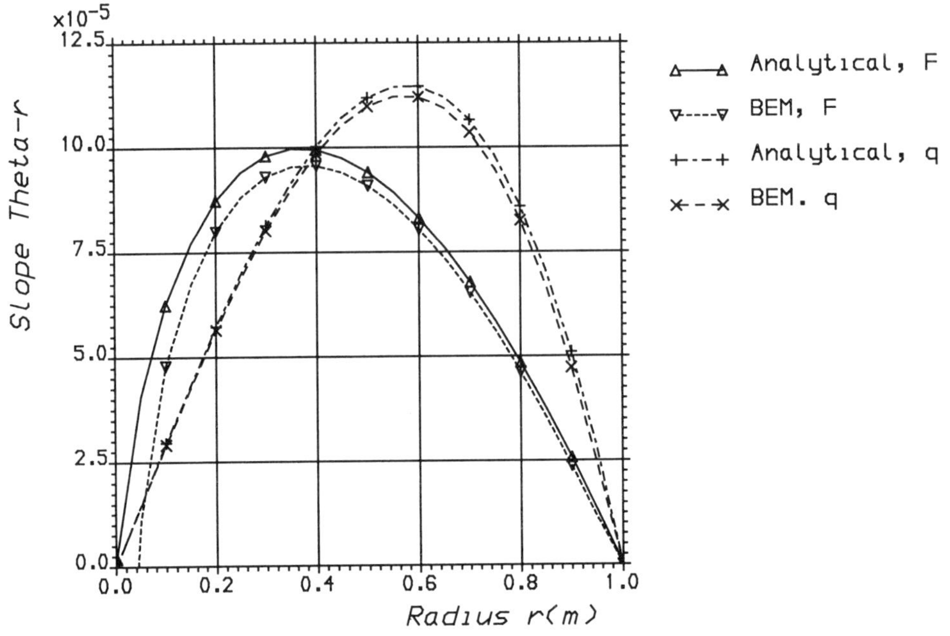

Figure 9. Slope distribution of a clamped circular plate.

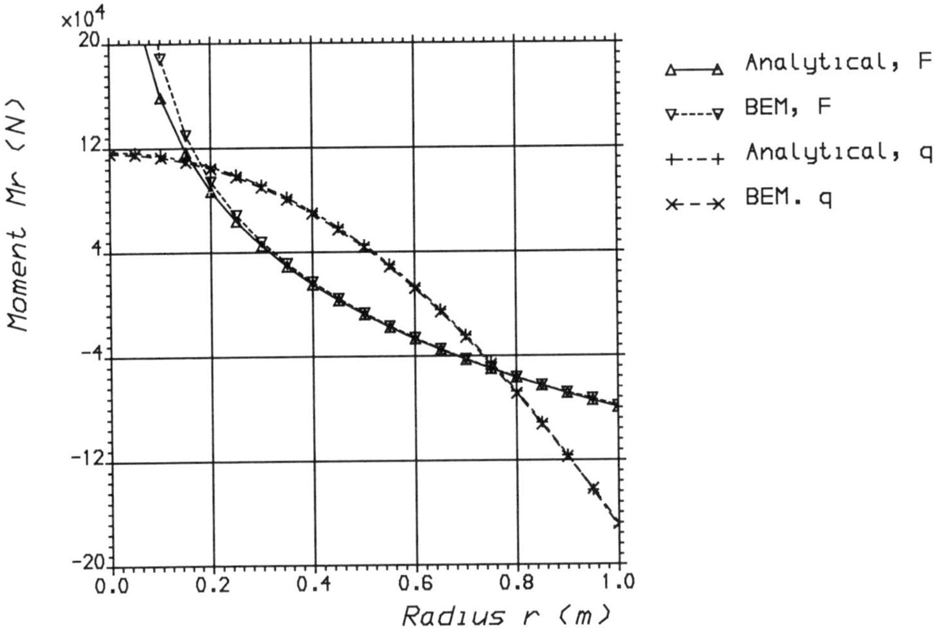

Figure 10. Moment distribution of a clamped circular plate.

9.3 Simply-supported rectangular plate

To test the validity of the boundary element derivations for thin plates, a thin square plate of thickness 0.5 m, and side length = 20 m was analysed. The plate centre is the origin of the Cartesian coordinates, and its intersecting sides are parallel to the x and y axes. The material properties are similar to the previous cases, and two types of loading were also tested with this case: a uniformly-distributed loading with intensity $q = 3.5 \times 10^5$ N/m^2, and a concentrated force $F = 4.8 \times 10^7$ N acting at the plate centre. Boundary element meshes with 24 constant elements equally spaced on the boundary were employed in the analysis, together with boundary conditions similar to those mentioned in Section 9.1. The distributions of deflection w, slope θ_x, and moment M_x along the central line, which is parallel to the x axis, were plotted versus corresponding analytical solutions based on Kirchhoff's plate-bending theory, as shown in Figures 11, 12, and 13, respectively. Those figures indicate that a good agreement between boundary element results and corresponding analytical solutions has been achieved, irrespective of the presence of corners.

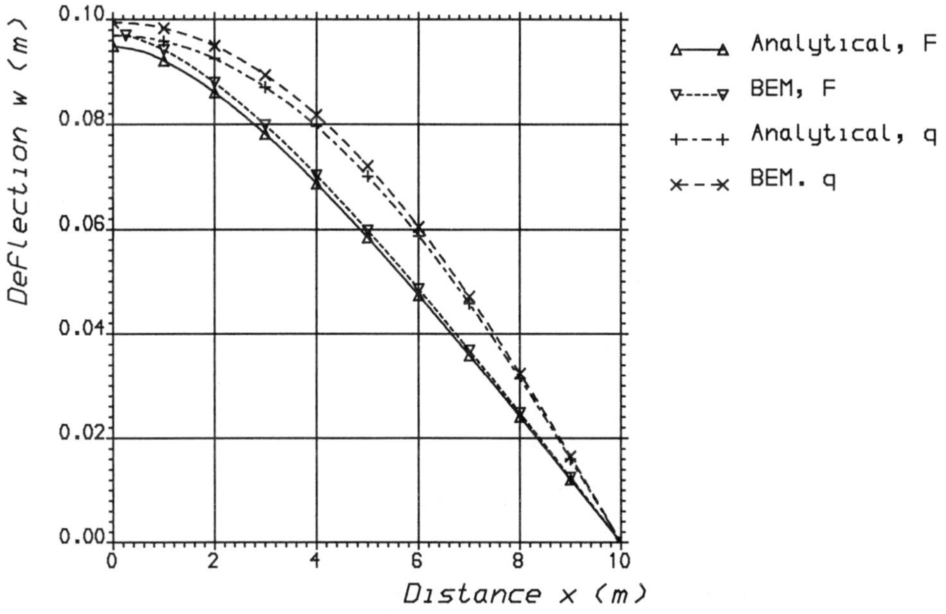

Figure 11. Deflection distribution of a simply supported square plate.

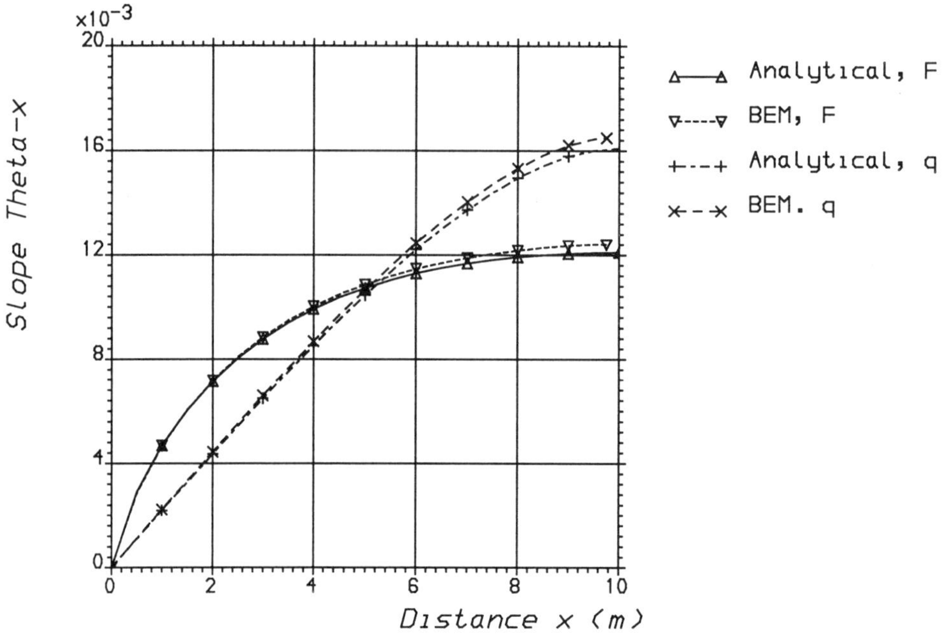

Figure 12. Slope distribution of a simply supported square plate.

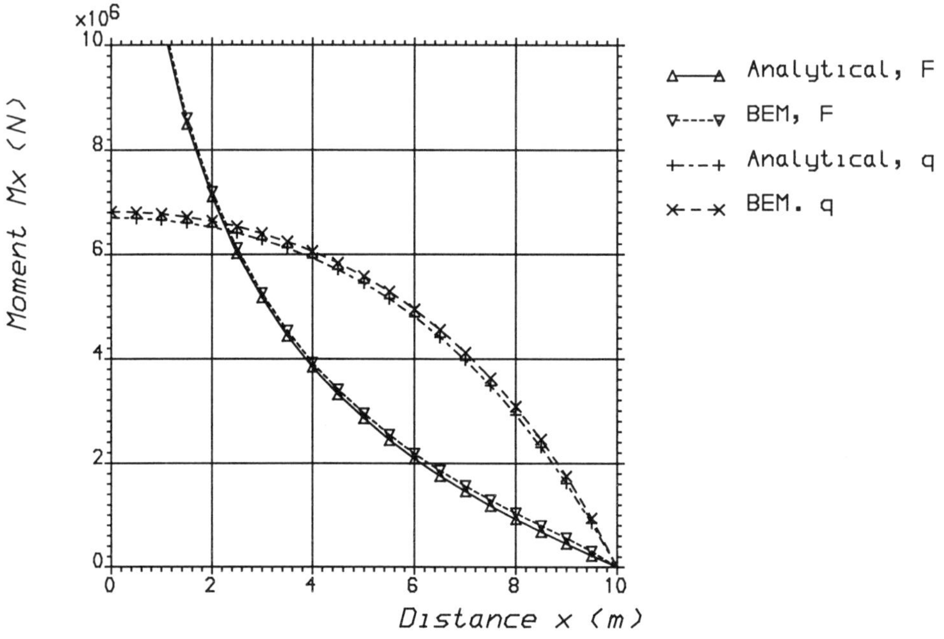

Figure 13. Moment distribution of a simply supported square plate.

10 Concluding remarks

It is clear from the previous examples that the boundary element derivations for plates with arbitrary shapes and boundary conditions, as presented in this chapter, are correct and lead to accurate results for thin and thick plates. The simplified equations for domain loading terms, and analytical singular integral expressions have functioned properly even with constant boundary elements. Internal moments estimated from finite differences of deflection and slope values are also very accurate.

References

[1] Altiero, N. J. & Sikarskie, D. L. A boundary integral method applied to plates of arbitrary plan form, *Computers & Structures*, **9**, 163-168, 1978.

[2] Tottenham, H. The boundary element method for plates and shells, in *Development in Boundary Element Methods - I*, eds. P. K. Banerjee & R. Butterfield, Applied Science Publishers, London, 1979.

[3] Bezine, G. & Gamby, D. A. A new integral equation formulation for plate bending problems, in *Recent Advances in Boundary Element Methods*, ed. C. A. Brebbia, Pentech Press, London, 1978.

[4] Stern, M. A general boundary integral formulation for the numerical solution of plate bending problems, *International Journal of Solids & Structures*, **15**, 769-782, 1979.

[5] Hartmann, F. & Zotemantel, R. The direct boundary element method in plate bending, *International Journal for Numerical Methods in Engineering*, **23**, 2049-2069, 1986.

[6] El-Zafrany, A., Debbih, M. & Fadhil, S. A modified Kirchhoff theory for boundary element bending analysis of thin plates. *International Journal of Solids & Structures,* **31**, 2885-2899,1994.

[7] Vander Weeën, F. Application of the boundary integral equation method to Reissner's plate model, *International Journal for Numerical Methods in Engineering*, **18**, 1-10, 1982.

[8] Antes, H. On a regular boundary integral equation and a modified Trefftz method in Reissner's plate theory, *Engineering Analysis*, **1**, 149-153, 1984.

[9] Karam, V. J. & Telles, J. C. F. On boundary elements for Reissner's plate theory, *Engineering Analysis*, **5**, 21-27, 1988.

[10] Long, S. Y., Brebbia, C. A. & Telles, J. C. F. Boundary element bending analysis of moderately thick plates, *Engineering Analysis*, **5**, 64-74, 1988.

[11] El-Zafrany, A., Debbih, M. & Fadhil, S. Boundary element analysis of thick Reissner plates in bending. *Engineering Analysis,* **14**, 159-169, 1995.

[12] El-Zafrany, A., Fadhil, S. & Debbih, M. An efficient approach for boundary element bending analysis of thin and thick plates. *Computers & Structures,* **56**, 565-576, 1995.

[13] Reissner, E. The effect of transverse shear deformation on the bending of elastic plates, *ASME Journal of Applied Mechanics*, **12**, A69-A77, 1945.

[14] Mindlin, R. D. Influence of rotary inertia and shear on flexural motions of isotropic elastic plates. *ASME Journal of Applied Mechanics,* **18**, 31-38, 1951.

[15] El-Zafrany, *A. Techniques of the Boundary Element Method.* Ellis Horwood, Chichester, England, 1993.

[16] Hörmander, H. *Linear Partial Differential Operators.* Springer Verlag, Berlin, 1963.

[17] Ditkin. V. A. & Prudnikov, A. P. *Integral Transforms and Operational Calculus.* Pergamon Press, Oxford, 1965.

[18] Abramowitz, M. & Stegun, I. A. (eds.) *Handbook of Mathematical Functions.* Dover Publications, New York, 1965.

[19] El-Zafrany, A. & Fadhil, S. The use of isoparametric boundary elements for the analysis of thick Reissner plates in bending, in *Boundary Elements XIX*, eds. M. Marchetti, C. A. Brebbia & M. H. Aliabadi, Computational Mechanics Publications, Southampton, 1997.

[20] Debbih, M. Boundary element stress analysis of thin and thick plates. PhD Thesis, Cranfield Institute of Technology, Bedford, England, 1989.

Appendix A: Modified Bessel functions

A.1 Definitions

The modified Bessel function $K_o(z)$ is a particular singular integral to the following differential equation:

$$(\nabla^2 - \lambda^2) f(r) = \delta(x - x_i, y - y_i)$$

i.e.

$$f(r) = -\mathrm{K}_0(z)/(2\pi)$$

where

$$z = \lambda r, \quad r = \sqrt{(x - x_i)^2 + (y - y_i)^2}$$

and

$$\nabla^2 \equiv \frac{\mathrm{d}^2}{\mathrm{d}r^2} + \frac{1}{r}\frac{\mathrm{d}}{\mathrm{d}r}$$

Generally speaking, the solution of the following ordinary differential equation:

$$z^2 \frac{\mathrm{d}^2 f}{\mathrm{d}z^2} + z \frac{\mathrm{d}f}{\mathrm{d}z} - (z^2 + n^2) f = 0$$

can be expressed as follows[18]:

$$f(z) = c_1 \mathrm{I}_n(z) + c_2 \mathrm{K}_n(z)$$

where c_1, c_2 are integration constants and, I_n, K_n are the nth order modified Bessel functions of the first and second kind, respectively. The infinite series of I_n and K_n for $n = 0, 1$ can be expressed as follows:

$$\mathrm{I}_0(z) = 1 + \frac{z^2/4}{(1!)^2} + \frac{(z^2/4)^2}{(2!)^2} + \frac{(z^2/4)^3}{(3!)^2} + \cdots$$

$$\equiv \sum_{m=0}^{\infty} \frac{(z^2/4)^m}{(m!)^2}$$

$$\mathrm{I}_1(z) = \frac{z}{2}\left[1 + \frac{z^2/4}{2\,(1!)^2} + \frac{(z^2/4)^2}{3\,(2!)^2} + \frac{(z^2/4)^3}{4\,(3!)^2} + \cdots\right]$$

$$\equiv \sum_{m=0}^{\infty} \frac{(z/2)^{(2m+1)}}{(m+1)\,(m!)^2}$$

$$\mathrm{K}_0(z) = -\left[\log\left(\frac{z}{2}\right) + \gamma\right]\mathrm{I}_0(z) + \sum_{m=1}^{\infty} \frac{(z^2/4)^m}{(m!)^2}\,\phi(m)$$

$$\mathrm{K}_1(z) = \left[\log\left(\frac{z}{2}\right) + \gamma\right]\mathrm{I}_1(z) + \frac{1}{z} - \frac{z}{4}$$

$$- \sum_{m=1}^{\infty} \frac{(z/2)^{(2m+1)}}{(m+1)\,(m!)^2}\left[\phi(m) + \frac{1}{2\,(m+1)}\right]$$

where $\gamma = 0.5772\ 15664\ 90153\ 28606\ ...$, which is Euler's constant, and

$$\phi(m) = \sum_{j=1}^{m} \frac{1}{j}$$

A.2 Derivatives of Bessel functions

$$\frac{\mathrm{d}\mathrm{K}_o(z)}{\mathrm{d}z} = -\ \mathrm{K}_1(z)$$

$$\frac{\mathrm{d}^2\mathrm{K}_o(z)}{\mathrm{d}z^2} = \mathrm{K}_o(z) + \frac{1}{z}\mathrm{K}_1(z) \equiv B_1(z)$$

Notice also that, with $z = \lambda\, r$, then:

$$\frac{\mathrm{d}\mathrm{K}_o(z)}{\mathrm{d}r} = -\ \lambda\, \mathrm{K}_1(z)$$

$$\frac{\mathrm{d}^2\mathrm{K}_o(z)}{\mathrm{d}r^2} = \lambda^2 B_1(z)$$

Defining $\quad A_1(z) = \mathrm{K}_o(z) + \dfrac{2\,\mathrm{K}_1(z)}{z}$

then it can be deduced that:

$$\frac{\mathrm{d}}{\mathrm{d}z}\left(\frac{\mathrm{K}_1}{\mathrm{z}}\right) = -\ \frac{A_1(z)}{z}$$

$$\frac{\mathrm{d}}{\mathrm{d}z}\left[B_1(z)\right] = -\mathrm{K}_1(z) - \frac{A_1(z)}{z}$$

$$\frac{\mathrm{d}}{\mathrm{d}z}\left[A_1(z)\right] = -\mathrm{K}_1(z) - \frac{2\,A_1(z)}{z}$$

Using $\quad \dfrac{\partial f(z)}{\partial n_j} = \dfrac{\mathrm{d}f}{\mathrm{d}z}\dfrac{\partial z}{\partial n_j} \equiv \lambda\,\dfrac{\mathrm{d}f}{\mathrm{d}z}\dfrac{\partial r}{\partial n_j}$

the following relationships can be proved:

$$\frac{\partial}{\partial n_1} \mathrm{K}_o(z) = -\lambda \frac{\partial r}{\partial n_1} \mathrm{K}_1(z)$$

$$\frac{\partial^2}{\partial n_1 \partial n_2} \mathrm{K}_o(z) = \lambda^2 \left[-\left(\hat{n}_1 \cdot \hat{n}_2\right) \frac{\mathrm{K}_1(z)}{z} + \frac{\partial r}{\partial n_1} \frac{\partial r}{\partial n_2} A_1(z) \right]$$

$$\frac{\partial^3}{\partial n_1 \partial n_2 \partial n_3} \mathrm{K}_o(z) = \lambda^3 \left\{ \left[\left(\hat{n}_2 \cdot \hat{n}_3\right) \frac{\partial r}{\partial n_1} + \left(\hat{n}_3 \cdot \hat{n}_1\right) \frac{\partial r}{\partial n_2} + \left(\hat{n}_1 \cdot \hat{n}_2\right) \frac{\partial r}{\partial n_3} \right.\right.$$

$$\left.\left. - 4 \frac{\partial r}{\partial n_1} \frac{\partial r}{\partial n_2} \frac{\partial r}{\partial n_3} \right] \frac{A_1(z)}{z} - \frac{\partial r}{\partial n_1} \frac{\partial r}{\partial n_2} \frac{\partial r}{\partial n_3} \mathrm{K}_1(z) \right\}$$

where $\hat{n}_1$, $\hat{n}_2$, $\hat{n}_3$ are unit vectors in the directions of lines n_1, n_2, n_3 (representing x, y, normal, or tangential directions), and with $\hat{n}_k = l_k \hat{i} + m_k \hat{j}$, it can be shown that:

$$\frac{\partial r}{\partial n_k} = \left(\nabla r \cdot \hat{n}_k\right) = \frac{l_k\left(x - x_i\right) + m_k\left(y - y_i\right)}{r}$$

A.3 Useful integral theorems

$$\int_0^{z_o} \mathrm{K}_o(z)\,\mathrm{d}z = -z_o \left[\log\left(\frac{z_o}{2}\right) + \gamma - 1 \right]$$

$$+ 2 \sum_{m=1}^{\infty} \frac{(z_o/2)^{(2m+1)}}{(2m+1)(m!)^2} \left[\phi(m) + \frac{1}{(2m+1)} - \log\left(\frac{z_o}{2}\right) - \gamma \right]$$

$$\int_0^{z_o} \left[\frac{\mathrm{K}_1(z)}{z} - \frac{1}{z^2} \right] \mathrm{d}z = -\mathrm{K}_1(z_o) + \frac{1}{z_o} - \int_0^{z_o} \mathrm{K}_o(z)\,\mathrm{d}z$$

$$\int_0^{z_o} B(z)\,\mathrm{d}z = -\mathrm{K}_1(z_o) + \frac{1}{z_o}$$

Appendix B Integration over constant elements

Consider a constant element (element e), which is a straight line ab, and its midpoint c is the source point (x_i, y_i), as shown in Figure 14. The field point (x,y) is any point on the element, and r represents the distance between the source point and the field point. The directional cosines of the normal to the element at any field point have the same values: (l_i, m_i), and in this appendix, $f(r)$ represents any function of r which does not contain directional cosines.

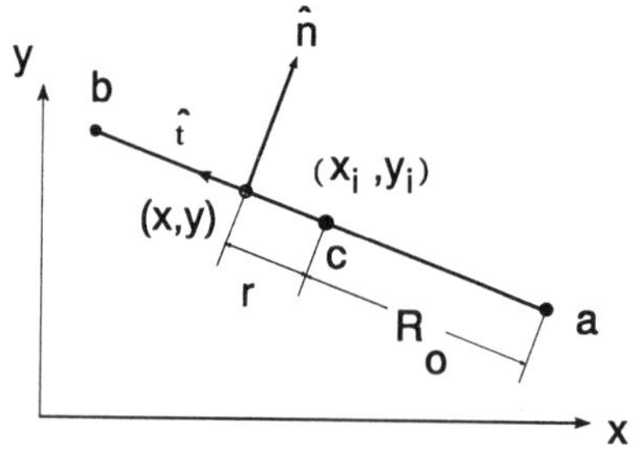

Figure 14. Straight line element, with a midside source point.

The integral domain of the element consists of two equal parts each with length R_0, the first part is from point a to c, and the second part is from point c to d. Since the value of r is zero or positive, the following useful results can be concluded, where the lower sign is valid on the first part, and the upper sign is valid on the second part:

$$\mathrm{d}\Gamma = \pm\, \mathrm{d}r, \quad \frac{\partial r}{\partial x} = \mp\, m_i, \quad \frac{\partial r}{\partial y} = \pm\, l_i$$

Hence, it can be deduced that:

$$\frac{\partial r}{\partial n} = 0, \quad \frac{\partial r}{\partial t} = \pm\, 1$$

and the following theorems can be proved for integrals over the element e:

$$\int_e f(r)\, \mathrm{d}\Gamma = 2 \int_0^{R_0} f(r)\, \mathrm{d}r$$

$$\int_e f(r) \frac{\partial r}{\partial x_\alpha}\, \mathrm{d}\Gamma = 0$$

$$\int_e f(r) \frac{\partial r}{\partial x_\alpha} \frac{\partial r}{\partial x_\beta}\, \mathrm{d}\Gamma = 2 \int_0^{R_0} f(r) \frac{\partial r}{\partial x_\alpha} \frac{\partial r}{\partial x_\beta}\, \mathrm{d}r$$

With $z = \lambda\, r$, it can also be shown that:

$$\int_{e} f(z)\, \mathrm{d}\Gamma = \frac{2}{\lambda} \int_{0}^{z_o} f(z)\, \mathrm{d}z$$

where $z_o = \lambda\, R_o$. This is useful with Bessel function expressions.

Chapter 3

Elastoplastic analysis of Reissner's plate using the boundary element method

G.O. Ribeiro[a] & W.S.Venturini[b]
[a]*Department of Structural Engineering, Federal University of Minas, Av. de Contorno 842, 30110-060 Belo Horizonte, Brazil*
Email: gabriel@ufmg.br
[b]*São Carlos School of Engineering, University of São Paulo, Av. Carlos Botelho 1465, 13560 S. Carlos, Brazil*
Email: venturin@sc.usp.br

Abstract

This work deals with a formulation of the boundary element method (BEM) applied to plate bending problem using the Reissner's theory, which makes it possible to achieve a more consistent numerical procedure in which the three physical conditions along the plate boundary can be enforced. The boundary element approach using this theory is extended to consider initial stress fields applied over the plate domain, as well as to take into account the occurrence of several kinds of loads, concentrated and distributed either over sub-domains or along internal lines.

The presence of initial stress fields enables one to analyse temperature and shrinkage effects, and to formulate procedures to deal with physical non-linearities. A procedure to deal with elastoplastic behavior is implemented using an incremental and iterative algorithm based on the initial stiffness method, where the true plastic solution is achieved by applying initial stress fields.

1 Introduction

Analysis of plate bending problems using the boundary element method (BEM) has attracted the attention of many researchers during the last years. The first works discussing that use of boundary methods appeared in conjunction with the Kirchhoff's theory, e.g. Bezine,[1] Stern[2] and Altiero & Sikarskie.[3] Those works, as well as several other more recent publications, have pointed out the capability of the method, taking into account its accuracy and confidence, to model plates in bending exhibiting either linear or non-linear behaviours, e.g. Hartmann & Zotemantel[4], Ghosh & Mukherjee[5],

Venturini & Paiva[6], Katsikadelis & Armenakas[7], Tanaka[8], Morjaria & Mukerjee[9], Moshaiov & Vorus[10], Providakis & Beskos[11].

The first work using boundary elements to deal with moderated thick plates was carried out by Weeën[12,13], in 1982, where the required integral equations have been achieved employing the Reissner's theory, e.g. Reissner[14]. Other authors, following the same theory, have more recently given relevant contributions to this BEM plate bending formulation, e.g. Ribeiro & Venturini[15] and Karan & Telles[16]. Although studied less, techniques based on that plate theory have already shown to be efficient and accurate when used for practical purposes.

Using Reissner's hypothesis gives a more consistent BEM formulation for plate bending. For this case, six boundary values are defined along the plate contour, therefore allowing one to prescribe three physical conditions at any boundary node. Besides the classical deflections and rotations, assumed in the classical theory as the only degrees of freedom, for the Reissner's hypotheses the transverse shear strain energy is also taken in account. Thus, by adopting this model, the influence of the thickness and the condition of either releasing or restricting the twist rotation in the tangential vertical plane along the boundary plate can be analysed as well.

In this paper, the integral equations based on the Reissner's theory are extended to take into consideration fields of initial stresses, or initial moments in generalized sense. After properly deriving displacement and effort integral representations, analyses of plate problems including temperature and shrinkage effects are possible, as well as the consideration of non-linear behaviours, particularly elastoplastic plates. In addition, extra integral terms are derived for displacement and effort representation to consider particular domain loadings; for instance, distributed moments and line transversal loads have been taken into account in this work.

Computational implementations were made by adopting boundary elements with linear geometry and quadratic approximation for all boundary values. For simplicity, the algebraic equations required to solve the problem in terms of boundary values are achieved adopting only outside collocation points, avoiding therefore the evaluation of some singular integrals.

The numerical procedure is accomplished by an incremental and iterative algorithm based on the initial stiffness method, for which the plastic solution is achieved by conveniently applying the initial stress fields in the classical explicit sense. Two families of elastoplastic models have been implemented. The first criteria considered in this work are those written in terms of efforts, particularly dependent only on bending and twisting moments. For this case, the von Mises and Tresca elastoplastic models have been implemented into the developed computer code. Other models can now be easily implemented following the same steps. The second family of elastoplastic models implemented together with the developed algorithm is based on the layer technique. For that case, we preferred computing bending and twisting moments by using a gaussian scheme. In this process, the onset of plastic deformation is verified at each Gauss point defined along the plate thickness. Again, von Mises and Tresca criteria have been implemented, but are now written in terms of stresses.

Finally, some examples are shown to illustrate the accuracy reached using the plate bending BEM Reissner's algorithm either to verify elastic answer due to particular loads or to carry out elastoplastic analysis for layered and non-layered cases.

2 Governing equations of Reissner's theory

A flat plate of thickness *h,* referred to a Cartesian system of coordinates with axes x_1 and x_2 lying on its middle surface and axis x_3 perpendicular to that plane, is considered. It is assumed that the plate supports a distributed load q in the x_3 direction, and distributed external moments m_α, both acting on the plate middle plane. For this plate, the equilibrium equations are given by the following expressions:

$$M_{\alpha\beta,\beta} - Q_\alpha + m_\alpha = 0 \tag{1a}$$

$$Q_{\alpha,\alpha} + q = 0 \tag{1b}$$

where $M_{\alpha\beta}$ are bending and twisting moments, Q_α represents shear forces, with the Greek subscripts taken in the range {1, 2}.

In order to be concise, rotations and deflections ϕ_α, w) are assumed as generalized values represented by the notation (u_α, u_3). Note that ϕ_α means rotations in the plane $x_\alpha x_3$ and w is the deflection in x_3 direction. Similarly, generalized boundary tractions, p_i, are adopted to represent boundary bending and twisting moments as well as shear forces. The plate domain is denoted by Ω, while its boundary is represented by Γ.Over Γ, the following boundary conditions may be assumed: $u_i = \bar{u}_i$ on Γ_1 and $p_i = \bar{p}_i$ on Γ_2, where $\Gamma_1 \cup \Gamma_2 = \Gamma$, p_i being the generalized boundary forces (tractions) given by,

$$p_\alpha = M_{\alpha\beta}\, n_\beta \tag{2a}$$

$$p_3 = Q_\alpha\, n_\alpha \tag{2b}$$

where n_α stands for the director cosines of the outward normal.

As usual, the generalized stress × displacement relations can be expressed in the following form:

$$M_{\alpha\beta} = \frac{D(1-\nu)}{2}\left(u_{\alpha,\beta} + u_{\beta,\alpha} + \frac{2\nu}{1-\nu}\, u_{\gamma,\gamma}\, \delta_{\alpha\beta} \right) + \frac{\nu q\, \delta_{\alpha\beta}}{(1-\nu)\,\lambda^2} \tag{3a}$$

$$Q_\alpha = \frac{D\,(1-\nu^2)\,\lambda^2}{2}\left(u_\alpha + u_{3,\alpha} \right) \tag{3b}$$

where $\lambda = \sqrt{10}/h$ is the characteristic parameter of Reissner's model, and $D = Eh^3/(1-\nu^2)$ is the flexural rigidity.

The generalized strains can also be expressed in terms of the generalized displacements, as follows:

$$\chi_{\alpha\beta} = \frac{1}{2}\left(u_{\alpha,\beta} + u_{\beta,\alpha}\right) \tag{4a}$$

$$\psi_{\alpha} = u_{\alpha} + u_{3,\alpha} \tag{4b}$$

where $\chi_{\alpha\beta}$ and ψ_α are bending (curvatures) and shear strains, respectively.

In order to deal with initial strain fields, always required to analyse problems with prescribed temperature and shrinkage loads, as well as to model non-linear behaviours, one must express the total generalized strains as a sum of two parts: $\chi^e_{\alpha\beta}$, the elastic part which is obtained due to the domain loads and the boundary values prescribed along Γ, and $\chi^0_{\alpha\beta}$, which represents generalized strain initial values (temperature or shrinkage).

Thus, the total generalized strains are given by,

$$\chi_{\alpha\beta} = \chi^e_{\alpha\beta} + \chi^0_{\alpha\beta} \tag{5a}$$

$$\psi_{\alpha} = \psi^e \tag{5b}$$

In order to take into account an initial strain field in the BEM formulation it is better to define the corresponding initial moment field, which is easily obtained from eqn (3), as follows,

$$M^0_{\alpha\beta} = \frac{D(1-\nu)}{2}\left(u^0_{\alpha,\beta} + u^0_{\beta,\alpha} + \frac{2\nu}{1-\nu}u^0_{\gamma,\gamma}\delta_{\alpha\beta}\right) \tag{6}$$

Thus, the total moments $M_{\alpha\beta}$ are given by,

$$M_{\alpha\beta} = M^e_{\alpha\beta} - M^0_{\alpha\beta} \tag{7}$$

where $M^e_{\alpha\beta}$ is the elastic part of the total moment.

From the above expressions, one can derive the BEM relations, both integral representations and the corresponding algebraic equations, appropriate to model temperature and shrinkage effects (e.g Ribeiro & Venturini[15]), as well as to simulate the elastoplastic plates, when they are properly arranged, as it will be shown later on.

From this point, the superscript "o" is adopted to define either temperature and shrinkage equivalent moment loads or the plastic moment components.

3 Integral equations

The boundary integral equation for Reissner's plate, where initial moments fields are considered, can be derived using either the well known weighted residual methods or the Betti's reciprocal work theorem. The generalized displacement representation at a point S, named load point, is expressed as follows:

$$C_{ik}(S)\;u_k(S)=\int_{\Gamma}\left[u^*_{ik}(S,T)\;p_k(T)-p^*_{ik}(S,T)\;u_k(T)\right]d\Gamma(T)$$

$$+\int_{\Omega_q} q(t)\left[u^*_{i3}(S,t)-\frac{\nu}{(1-\nu)\lambda^2}\,u^*_{i\alpha,\alpha}(S,T)\right]d\Omega(t)+\int_{\Omega} u^*_{i\alpha,\beta}(S,T)\,M^0_{\alpha\beta}(t)d\Omega(t)$$

$$+\int_{\Omega_m} u^*_{i\alpha}(S,t)m_\alpha(t)d\Omega(t) \qquad (8)$$

where Ω_q and Ω_m are loaded regions over which transversal load q and moment m_α are applied, respectively.

In eqn (8) and throughout this work, capital letters indicate load or field points along the boundary, while domain points are always defined by small letters. This equation holds for internal points with $C_{ij}=\delta_{ij}$ (Kronecker delta). For load points taken along the boundary, $C_{ij}=\delta_{ij}/2$, while $C_{ij}=0$ is achieved when they are defined out of the plate. The kernels $u^*_{ik}(S,T)$ and $p^*_{ik}(S,T)$ are the fundamental solutions for displacements and tractions respectively (e.g. Weeën[12]). Eqn (8) can be applied to the BEM elastoplastic plate bending analysis, where the initial moment field $M^0_{\alpha\beta}$ is taken to represent the plastic increment components.

The bending moment and the shear force integral representations can be derived by replacing, in eqn (3), the displacements given in eqn (8), taking into account that the derivatives of the domain term, whose density is $M^o_{\alpha\beta}$, involves $1/r$ singularities and therefore must be properly performed. This differentiation is performed by following the general concept of singular integral derivatives given by Miklin[17] and used by Telles & Brebbia[18] and Bui[19] for plane analysis. After differentiating those terms and applying properly Hooke's law, one finds the bending and twisting moment integral representations, as well as the shear force integral equation, as follows,

$$M_{\alpha\beta}(s) = \int_{\Gamma} u^*_{\alpha\beta k}(s,T)\, p_k(T)\, d\Gamma(T) - \int_{\Gamma} p^*_{\alpha\beta k}(s,T)\, u_k(T)\, d\Gamma(T)$$

$$+ \int_{\Omega_q} r^*_{\alpha\beta}(s,t)\, q(t)\, d\Omega_q(t) + \frac{\nu}{(1-\nu)\lambda^2} q + \int_{\Omega_m} t^*_{\alpha\beta\gamma}(s,t)\, m_\gamma\, d\Omega_m(t) \tag{9}$$

$$+ \int e^*_{\alpha\beta\gamma\theta}(s,t)\, M^0_{\gamma\theta}(t)\, d\Omega(t) + g_{\alpha\beta\gamma\theta}(s)\, M^0_{\gamma\theta} - M^0_{\alpha\beta}(s)$$

$$Q_\beta(s) = \int_{\Gamma} u^*_{3\beta k}(s,T)\, p_k(T)\, d\Gamma(T) - \int_{\Gamma} p^*_{3\beta k}(s,T) u_k(T) d\Gamma(T)$$

$$+ \int_{\Omega_q} r^*_{3\beta}(s,t)\, q(t)\, d\Omega_q(t) + \int_{\Omega_m} t^*_{3\beta\gamma}(s,t)\, m_\gamma(t)\, d\Omega_m + \int_{\Omega} e^*_{3\beta\gamma\theta}(s,t)\, M^0_{\gamma\theta}(t)\, d\Omega(t) \tag{10}$$

where the fundamental tensors $u^*_{i\beta k}$ and $p^*_{i\beta k}$ are given by Weeen[12] and in other references (e.g. Weeën[13], Ribeiro & Venturini[15] and Karan &Telles[16]).

It is important to observe that the integrals involving the fundamental tensor $e^*_{i\beta\gamma\theta}$ must be performed in the Cauchy's principal value sense. The fundamental values $r^*_{i\beta}$, $t^*_{i\beta\gamma}$, and $e^*_{i\beta\gamma\theta}$, given in eqns (9) and (10), are obtained from the Weeën's fundamental solution, and are expressed by,

$$r^*_{\alpha\beta} = -\frac{(1-\nu)}{8\pi}\left\{\delta_{\alpha\beta}\left[2\frac{(1+\nu)}{(1-\nu)}\ell nz - 1\right] + 2r_{,\alpha} r_{,\beta} - \frac{4\nu}{(1-\nu)\lambda^2 r^2}(2r_{,\alpha} r_{,\beta} - \delta_{\alpha\beta})\right\} \tag{11a}$$

$$r^*_{3\beta} = \frac{1}{2\pi r^2} r_{,\beta} \tag{11b}$$

$$t^*_{\alpha\beta\gamma} = \frac{1}{8\pi r}\Big\{\left[8A + 4zk_1 + 2(1-\nu)\right](\delta_{\alpha\gamma} r_{,\beta} + \delta_{\beta\gamma} r_{,\alpha}) + 2r_{,\gamma}\delta_{\alpha\beta}(4A + 1 + \nu) +$$

$$-2r_{,\alpha} r_{,\beta} r_{,\gamma}\left[16A + 4zk_1 + 2(1-\nu)\right]\Big\} \tag{12a}$$

$$t^*_{3\beta\gamma} = \frac{\lambda^2}{2\pi}\left[B\,\delta_{\beta\gamma} - A\, r_{,\beta}\, r_{,\gamma}\right] \tag{12b}$$

$$e^{*}_{\alpha\beta\gamma\theta} = -\frac{1}{8\pi r^2}\Big\{\big[16A + 4zk_1 + 2(1-\nu) + 4z^2k_0\big]\big(r_{,\theta}\, r_{,\beta}\,\delta_{\alpha\gamma} + r_{,\theta}\, r_{,\alpha}\,\delta_{\beta\gamma}\big)$$

$$+\big[16A + 4zk_1 + 2(1-\nu)\big]\big(2r_{,\beta}r_{,\gamma}\delta_{\alpha\theta} + 2r_{,\alpha}r_{,\beta}\delta_{\gamma\theta} + 2r_{,\gamma}r_{,\theta}\delta_{\alpha\beta} + 2r_{,\alpha}r_{,\gamma}\delta_{\beta\theta}$$

$$+r_{,\alpha}r_{,\theta}\delta_{\beta\gamma} + r_{,\beta}r_{,\theta}\delta_{\alpha\gamma}\big) - (4A + 1 - \nu)\big(2\delta_{\gamma\theta}\delta_{\alpha\beta} + \delta_{\alpha\theta}\delta_{\gamma\beta} + \delta_{\beta\theta}\delta_{\alpha\gamma}\big) \tag{13a}$$

$$-\big(4A + 4zk_1 + 1 - \nu\big)\big(\delta_{\alpha\theta}\delta_{\beta\gamma} + \delta_{\alpha\gamma}\delta_{\beta\theta}\big) - \big[96A + 32zk_1 + 4z^2k_0 + 8(1-\nu)\big]$$

$$\text{x}\,\big(2r_{,\alpha}r_{,\beta}r_{,\gamma}r_{,\theta}\big) + 4\nu\delta_{\alpha\beta}\big(2r_{,\gamma}r_{,\theta} - \delta_{\gamma\theta}\big)\Big\}$$

$$e^{*}_{3\beta\gamma} = -\frac{\lambda^2}{2\pi r}\Big[\big(A + 2zk_1\big)r_{,\theta}\delta_{\beta\gamma} - \big(4A + zk_1\big)r_{,\beta}r_{,\gamma}r_{,\theta} + A\big(\delta_{\beta\theta}r_{,\gamma} + \delta_{\gamma\theta}r_{,\beta}\big)\Big] \quad \text{13b)}$$

The free term $g_{\alpha\beta\gamma\theta}$ is due to singular integral differentiation and is given by,

$$g^{*}_{\alpha\beta\gamma\theta} = \frac{1}{8}\Big[(3-\nu)\big(\delta_{\alpha\gamma}\delta_{\theta\beta} + \delta_{\beta\gamma}\delta_{\theta\alpha}\big) - \big(1 - 3\nu\big)\delta_{\alpha\beta}\delta_{\gamma\theta}\Big] \tag{14}$$

4 Algebraic equations

The algebraic equations of the BEM are achieved by discretizing the boundary of the plate into elements, over which the boundary values are approximated usually by polynomial functions. For this case, we took linear elements to represent geometrically the boundary and quadratic functions to approach all boundary values, either prescribed or unknown tractions and displacements. As unknown initial moments are also applied over the plate, the domain has to be discretized and these densities approximated over the resulting sub-domains. In this case, we decided to use triangular cells over which the initial moments are approached using linear functions. Assuming boundary and domain values approximated as defined above and writing the algebraic version of eqn (8) for an appropriate number of collocation points, the following linear equation system of equations is achieved,

$$\mathbf{CU} + \hat{\mathbf{H}}\mathbf{U} = \mathbf{GP} + \mathbf{B} + \mathbf{EM}^0 \tag{15}$$

where **U** and **P** are the generalized boundary nodal displacement and traction vectors, and $\mathbf{M^0}$ gives the initial moment values at internal and boundary nodes.

The vector **B**, in eqn (15), represents the influence of the domain loads, that can be either the standard distributed load applied on a particular region Ω_q of the plate surface or the distributed moment acting over the specific area Ω_m. As these loads are known, the corresponding integral terms can be easily transformed to the boundary. One particular practical use for this type of load is the case of line loading. The two dimensional integrals are very easily transformed into line integrals assuming regions of narrow width. These load line integrals can be performed by adopting an appropriate number of elements enough to guarantee the accuracy of the integration scheme.

As in this work the collocation points are placed outside the plate domain, the free term **C** is equal to zero. Matrices **G** and $\overline{\mathbf{H}}$, computed by integrating the product of the shape function times the corresponding fundamental solutions, contain traction and displacement influence coefficients. Matrix **E** represents the influence of the initial moment field over the boundary variables and is computed by performing analytically or numerically the integral of the shape function times the fundamental tensor $u^*_{i\alpha,\beta}$ product over all cells.

Thus, eqn (15) can be simplified to give the classical BEM algebraic representation,

$$\mathbf{HU} = \mathbf{GP} + \mathbf{B} + \mathbf{EM}^0 \tag{16}$$

For a well posed problem, with a sufficient number of prescribed displacements, one can appropriately interchange **H** and **G** columns, to have all unknown quantities in a vector **X**, to obtain,

$$\mathbf{A\,X} = \mathbf{F} + \mathbf{E\,M}^0 \tag{17}$$

where the vector **F** contains the effects due to the boundary prescribed values and the domain applied loads, except the initial moments, and **A** is the matrix of the unknown boundary value coefficients.

By solving eqn (17), one achieves,

$$\mathbf{X} = \mathbf{L} + \mathbf{R\,M}^0 \tag{18}$$

where

$$\mathbf{R} = \mathbf{A}^{-1}\,\mathbf{E} \qquad \text{and} \qquad \mathbf{L} = \mathbf{A}^{-1}\,\mathbf{F} \tag{19}$$

The moment and shear integral representations can be written into their algebraic form as well, using a similar procedure,

$$\mathbf{M} = \mathbf{G'P} - \mathbf{H'U} + \mathbf{B'} + (\mathbf{E'} - \mathbf{I})\mathbf{M}^0 \tag{20a}$$

$$\mathbf{Q} = \mathbf{G}'' - \mathbf{H}''\mathbf{U} + \mathbf{B}'' + \mathbf{E}''\mathbf{M}^0 \tag{20b}$$

where the matrices $\mathbf{H}'$, $\mathbf{H}''$, $\mathbf{G}'$ and $\mathbf{G}''$ give the influences of the boundary displacements and tractions in the bending moments and shear forces, respectively, **I** is the identity matrix, vectors $\mathbf{B}'$ and $\mathbf{B}''$ are the influence of the domain loads, while $\mathbf{E}'$ and $\mathbf{E}''$ represent the influences of the initial moment field.

As usual in non-linear boundary element formulation, the matrix effort eqns (20), can be conveniently arranged to give,

$$\mathbf{M} = \mathbf{F}' - \mathbf{A}'\mathbf{X} + (\mathbf{E}' - \mathbf{I})\mathbf{M}^0 \tag{21a}$$

$$\mathbf{Q} = \mathbf{F}'' - \mathbf{A}''\mathbf{X} + \mathbf{E}''\mathbf{M}^0 \tag{21b}$$

The moment algebraic representation, given by eqn (21a), can be simplified by substituting the unknown vector **X** according to eqn (18). Then, writing only the elastic part of the moment vector, what is more convenient for plastic analysis ($\mathbf{M}^e = \mathbf{M} + \mathbf{M}^0$), the following reduced expression can be achieved:

$$\mathbf{M}^e = \mathbf{N} + \mathbf{S}\mathbf{M}^0 \tag{22}$$

where $\mathbf{M}^e$ represents the elastic part of the moment vector **M** and

$$\mathbf{N} = \mathbf{F}' - \mathbf{A}'\mathbf{L} \tag{23a}$$

$$\mathbf{S} = \mathbf{E}' - \mathbf{A}'\mathbf{R} \tag{23b}$$

The vectors **L** and **N** represent the solution of the problem in terms of boundary values and moments, respectively, in the absence of initial moments. The initial moment effects are given by **R** and **S**, respectively on the boundary unknowns and moments.

Following the same steps one can achieve the reduced shear force matrix equation, as follows:

$$\mathbf{Q} = \mathbf{N}' + \mathbf{S}'\mathbf{M}^0 \tag{24}$$

where $\mathbf{N}'$ and $\mathbf{S}'$ are defined similarly to **N** and **S**.

The algebraic equations obtained above contain matrices that give the influence of the initial moment field on displacement and effort representations. Computing those matrices means performing integrals over linear cells that, due to the complexity of the involved kernels, requires special care. Thus, it is useful to describe the way that was chosen to perform those integrals in this work.

From eqns (7), (8) and (9), we can take the domain terms that have to be properly computed after dividing the domain into cells. For any non-singular cells the integrals are performed numerically using a Gauss quadrature scheme for triangular domain cells. For the case where the load point is placed at the cell corner, an analytical procedure must be adopted to deal with the singularities of type $(1/r)$. Taking into account the complexity of the kernel, it is also convenient to divide the singular cell into three sub-domains, as shown in Figure 1; one small region containing the load point, J_1, and other two non-singular sub-cells, J_2, J_3.

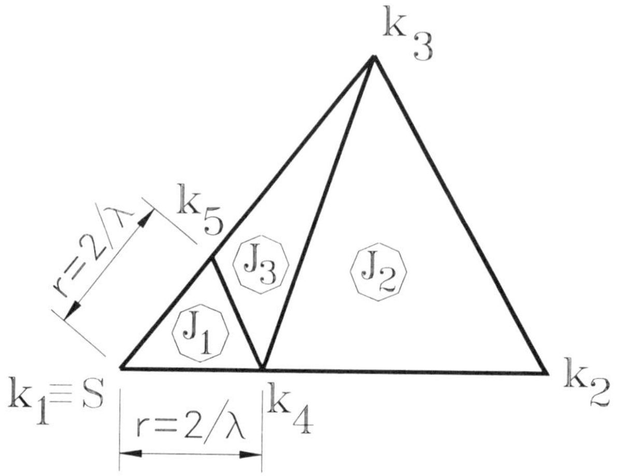

Figure 1. Subdivision of a cell with load point at a particular corner k_1.

The sub-domain J_1, adjacent to the corner load point S, is a region containing field points T, whose distance to S is $r \leq 2/\lambda$. For that region, a semi-analytical integration scheme has been adopted. Taking into account that the integral is to be made over two dimensions, one can perform the integral first analytically along the radial coordinate r, and then numerically over the angular coordinate θ. The integrals over the other two sub-domains J_2 and J_3, indicated in Figure 1, are not singular; they do not contain the load point, therefore they are performed numerically as well.

5 Elastoplastic analysis

Before describing the technique based on the BEM algebraic equations to model non-linear behaviours, in particular elastoplastic models for plate in bending, it is better to summarize the basic relations of the incremental plasticity theory. The following requirements are classically required to model elastoplastic behaviours: (a) definition of an explicit relationship between stress and strain during the elastic range; (b) assumption of a yield criterion to clearly define the elastic material range end and the beginning of the onset of plastic strains; (c) assumption of a stress × strain relationship during the elastoplastic behaviour to give the possibility of evaluating plastic strain components.

It is assumed that during the elastic phase the behavior is governed by the generalized Hooke's law, therefore,

$$M_{\alpha\beta} = C_{\alpha\beta\gamma\theta}\, \chi^{e}_{\alpha\beta} \tag{25}$$

where $C_{\alpha\beta\gamma\theta}$ is the fourth order isotropic elastic tensor.

As usual the yielding criterion can be represented by,

$$F\left(M_{\alpha\beta}, k\right) = f\left(M_{\alpha\beta}\right) - Y(k) = 0 \tag{26}$$

where $Y(k)$ is the yielding moment that depends of a hardening parameter k.

The stress × strain relationship during the yielding is developed under the hypothesis which states that the plastic strain increment $\mathrm{d}\chi^{0}_{\alpha\beta}$ is proportional to the gradient of the plastic potential. Assuming only the case of associate plasticity the strain increment can be written as,

$$\mathrm{d}\chi^{0}_{\alpha\beta} = \mathrm{d}\lambda \frac{\partial f\left(M_{\alpha\beta}\right)}{\partial M_{\alpha\beta}} \tag{27}$$

where $\mathrm{d}\lambda$ is a constant named plastic multiplier.

Taking in account the non-linear nature of the elastoplastic behaviour, an incremental and iterative algorithm must be adopted. This requires that all relations are written into their linear incremental forms at each iteration. The total strain increment, for instance, becomes,

$$\mathrm{d}\chi_{\alpha\beta} = \mathrm{d}\chi^{e}_{\alpha\beta} + \mathrm{d}\chi^{0}_{\alpha\beta} \tag{28}$$

while the elastic moment increment is given by,

$$\mathrm{d}M_{\alpha\beta} = C_{\alpha\beta\gamma\theta}\, \mathrm{d}\chi^{e}_{\alpha\beta} \tag{29}$$

From eqns (28) and (29), one can derive the moment × strain incremental expression as follows:

$$\mathrm{d}\, M_{\alpha\beta} = \left[C_{\alpha\beta\gamma\theta} - \frac{d_{\alpha\beta}\, a_{\mu\rho}\, C_{\mu\rho\gamma\theta}}{a_{\mu\rho}\, d_{\mu\rho} + H'} \right] \mathrm{d}\chi_{\gamma\theta} \tag{30}$$

where,

$$d_{\alpha\beta} = C_{\alpha\beta\gamma\theta}\, a_{\gamma\theta} \tag{31a}$$

$$a_{\gamma\theta} = \frac{\partial f\left(M_{\alpha\beta}\right)}{\partial M_{\alpha\beta}} \tag{31b}$$

H' can be interpreted as the moment × curvature curve slope in the plastic region and $f(M_{\alpha\beta})$ is taken as a uniaxial equivalent moment. Thus,

$$H' = \frac{\mathrm{d}M}{\mathrm{d}\chi_0} = \frac{(EI)_T}{1 - \dfrac{(EI)_T}{EI}} \tag{32}$$

In the above equation, EI is the flexural rigidity modulus in the linear elastic range, $(EI)_T$ is the flexural tangential modulus, obtained from the pure one-dimensional bending test carried out on a plate strip.

Bearing in mind that expressions more convenient for computational purposes are required, the moment increment can be given by its elastic and plastic parts as follows:

$$\mathrm{d}M_{\alpha\beta} = \mathrm{d}M^{e}_{\alpha\beta} - \mathrm{d}M^{0}_{\alpha\beta} \tag{33}$$

where

$$\mathrm{d}\,M^{e}_{\alpha\beta} = C_{\alpha\beta\gamma\theta}\, \mathrm{d}\chi_{\gamma\theta} \tag{34}$$

and

$$\mathrm{d}\,M^{0}_{\alpha\beta} = \frac{d_{\alpha\beta}\, a_{\gamma\theta}}{a_{\gamma\theta}\, d_{\gamma\theta} + H'}\, \mathrm{d}\,M^{e}_{\gamma\theta} \tag{35}$$

The above relations are enough to model plate bending plastic behaviour where the criterion is given in terms of efforts, in particular bending and twisting moments. In order to take into account yielding throughout the plate thickness, an elastoplastic layered model must be adopted. For that case, the state of stresses should be verified at particular points taken along the plate thickness. Thus, expressions similar to the relations (25) ~ (35), but written in terms of stresses and strains are required. The same expressions can be rewritten replacing moment components by stress components and curvatures by strains. In particular, eqns (25) ~ (29) are now given by,

$$\sigma_{\alpha\beta} = C_{\alpha\beta\gamma\theta}\,\varepsilon^{e}_{\alpha\beta} \tag{36}$$

$$F\left(\sigma_{\alpha\beta}, k\right) = f\left(\sigma_{\alpha\beta}\right) - Y(k) = 0 \tag{37}$$

$$\mathrm{d}\varepsilon^{0}_{\alpha\beta} = \mathrm{d}\lambda\,\frac{\partial f\left(\sigma_{\alpha\beta}\right)}{\partial\sigma_{\alpha\beta}} \tag{38}$$

$$\mathrm{d}\,\varepsilon_{\alpha\beta} = \mathrm{d}\,\varepsilon^{\mathrm{e}}_{\alpha\beta} + \mathrm{d}\,\varepsilon^{0}_{\alpha\beta} \tag{39}$$

$$\mathrm{d}\sigma_{\alpha\beta} = C_{\alpha\beta\gamma\theta}\,\mathrm{d}\varepsilon^{e}_{\gamma\theta} \tag{40}$$

In order to take into account that yielding occurs throughout the thickness, it is not necessary to assume a fictitious plate division into layers, as usual made for that problem. Stresses can be computed at selected Gauss points defined along a dimensionless coordinate axis ξ, whose origin is at the plate middle surface. Then, moment components are computed by performing numerically the integral over the thickness. Thus, the coordinate of any point taken through plate thickness can be expressed by,

$$x_3 = \xi\,\frac{h}{2} \tag{41}$$

which gives,

$$\mathrm{d}x_3 = \frac{h}{2}\,\mathrm{d}\xi \tag{42}$$

Thus, bending and twisting moments are now computed according to,

$$M_{\alpha\beta} = \int_{-\frac{h}{2}}^{\frac{h}{2}} \sigma_{\alpha\beta}\, x_3\, dx_3 = \frac{h^2}{4}\int_{-1}^{1} \sigma_{\alpha\beta}\,\xi\, d\xi \tag{43}$$

From an elastic moment increment, one can compute the stress increment at any Gauss point as follows:

$$\mathrm{d}\sigma^{e}_{\varepsilon\beta} = \frac{6}{h^2}\,\mathrm{d}M^{e}_{\alpha\beta}\,\xi \tag{44}$$

Assuming that the elastic increments are computed for all Gauss points along x_3, the elastoplastic analysis can be performed to achieve the actual stresses and the plastic strains, and then the actual moments and plastic curvatures for the plate transverse section can be properly evaluated. After computing all necessary values expressed in moments and curvatures, the elastoplastic analysis can be proceeded following the same steps adopted for the non-layered case.

6 Solution technique

As it has been previously mentioned, the algebraic equations for displacements and efforts can be used to model elastoplastic solids or to consider other non-linear behaviours. For that purpose, those expressions must be written into their incremental form. For instance, the final algebraic representation for displacements, eqn (18), can be taken to express elastic increments due to either an external load increment or the plastic moment field, as follows:

$$\Delta\mathbf{X} = \Delta\mathbf{L} + \mathbf{R}\,\Delta\mathbf{M}^{\mathbf{p}} \tag{45}$$

Similarly, one can write the algebraic effort relations in increments. Thus, the elastic moment increment derived from eqn (22) is given by,

$$\Delta\mathbf{M}^{\mathbf{e}} = \Delta\mathbf{N} + \mathbf{S}\,\Delta\mathbf{M}^{\mathbf{p}} \tag{46}$$

Shear force elastic increment can also be represented in a similar way.

From these expressions, the following steps have to be adopted to achieve elastoplastic plate bending solutions:

1. The problem is solved elastically to obtain all boundary values and elastic efforts, shear forces and particularly bending and twisting moments represented by $\mathbf{M}^{\mathrm{e}}$, which is given by vector $\mathbf{N}$ in eqn (20). Temperature and shrinkage effects can also be given by eqn (20), now using the equivalent initial stress field, $\mathbf{M}^0$, therefore the total elastic moment can be represented by , $\mathbf{M}^{\mathrm{e}} = \mathbf{N} + \mathbf{S}\,\mathbf{M}^0$.
2. According to the defined incremental process, the boundary computed values, $\Delta\mathbf{X}$, and elastic moments increment, $\Delta\mathbf{M}^{\mathbf{e}}$, can be computed. In particular, for an increment i the incremental factor β_i can be used, as follows:

$$\left(\Delta\,\mathbf{L}_i\right) = \beta_i\mathbf{L} \tag{47a}$$

$$\left(\Delta\,\mathbf{M}_i^{\mathbf{e}}\right) = \beta_i\mathbf{M}^{\mathbf{e}} \tag{47b}$$

3 The elastic increments $\Delta\mathbf{L}_i$ and $\Delta\mathbf{M}_i^{\mathbf{e}}$ have to be modified due to the residual moment, $\Delta\,\mathbf{M}^{\mathbf{p}}$, computed at interaction i minus 1, i.e.

$$\Delta \mathbf{L}_i + \mathbf{E}\Delta\mathbf{M}^{\mathrm{p}}_{i-1} \longrightarrow \Delta \mathbf{L}_i \tag{48a}$$

$$\Delta \mathbf{M}^{\mathrm{e}}_i + \mathbf{S}\Delta\mathbf{M}^{\mathrm{p}}_{i-1} \longrightarrow \Delta \mathbf{M}^{\mathrm{e}}_i \tag{48b}$$

4 The elastic moment increment, $\Delta\mathbf{M}^{\mathrm{e}}_i$, is added the actual state of efforts at each point of the plate. From the new value, the criterion is checked. Then, the actual, $\mathbf{M}_i$, and residual, $\Delta \mathbf{M}^{\mathrm{p}}_i$, moments can be computed.

5 The residual moments, $\Delta \mathbf{M}^{\mathrm{p}}_i$ can be accumulated to make easy computing other values that do not depend on the non-linear process.

6 After checking the convergence, it should return to step 3 for a new iteration or to step 2 to add new load increment.

At this point, it is interesting to point out the modifications required to model a plate in bending, assuming that yielding occurs along the plate thickness, therefore the criterion adopted should be given in terms of stresses. This analysis is essentially the same one presented above. After computing the elastic moment increment, $\Delta\mathbf{M}^{\mathrm{e}}_i$, step 3, the Gauss point elastic stress increment, $\Delta\sigma^{\mathrm{e}}_i$, is computed according to eqn (44). The global state of stress is then actualized at all points defined along the thickness. Now, the criterion written in terms of stresses is verified locally, assuming plane stress conditions, giving the actual stress and the residual stress values. From the residual state of stress, $\Delta\sigma^{\mathrm{p}}_i$, residual moments can be computed by performing the integral over the plate thickness, similarly as shown in eqn (43), i.e.

$$\Delta \mathbf{M}^{\mathrm{p}}_i = \frac{h^2}{\mathbf{4}} \int_{-1}^{1} \Delta\sigma^{\mathrm{p}}_i \; \xi \; \mathbf{d}\xi = \frac{h^2}{\mathbf{4}} \sum_{k=1}^{NG} (\Delta\sigma^{\mathrm{p}}_i \; \xi)_k \, \omega_k \tag{49}$$

where NG is the Gauss point number.

After that, the actual moment can be achieved by differences.

7 Examples

Four examples have been selected to illustrate the presented formulation. Linear as well as elastoplastic analyses will be shown to point out the accuracy of the formulation. The two first have been taken to illustrate linear cases where linear external vertical loads and moments are applied over the plate. The two last examples deal with the elastoplastic models discussed in this work. When necessary, the plate stiffness, $Eh^3/[12(1-\nu^2)]$, will be represented by D.

As the first linear example to be discussed, a square simply supported plate subjected to a vertical line load is analysed. Hard conditions, i.e. null rotation parallel to all sides, have been assumed. The boundary has been divided into eight quadratic elements (two elements per side), while the load line was applied along two internal segments, as in Figure 2. The geometric characteristic of the plate is completed by assuming the h/a ratio equal to 0.05. Poisson ration equal to 0.3 was also taken to carry out the numerical analysis.

The numerical values computed for that case show the very good accuracy achieved by the BEM formulation to analyse this kind of loading. Figure 3 gives deflections at the central point, comparing them with the analytical answer, e.g. Timoshenko & Woynosky-Krieger[20]. Bending moments have also been accurately computed. Figure 4 exhibits the comparison of the computed moment M_{11} and M_{22} values with the analytical solution.

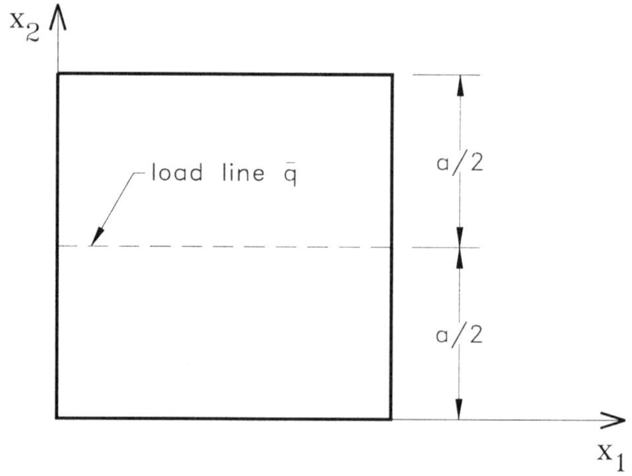

Figure 2. Square plate subjected to a line load. Geometry and discretization.

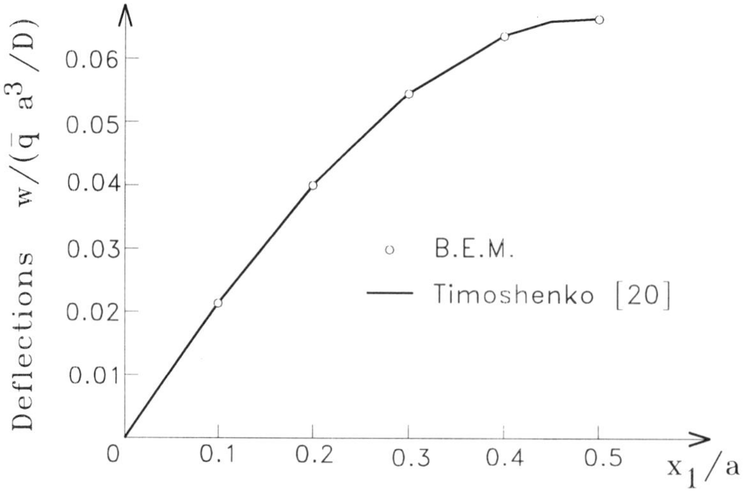

Figure 3. Deflections along the central line.

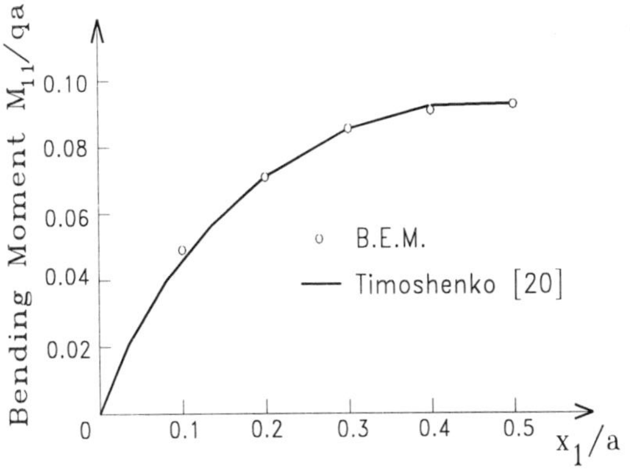

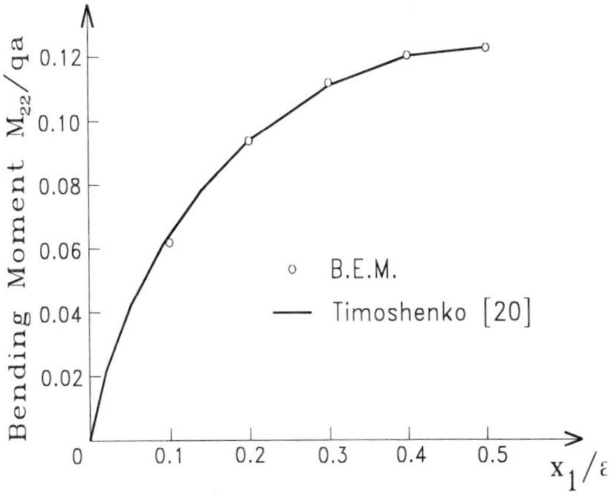

Figure 4. Bending moment along the central line.

The second elastic bending plate analysis deals with a cantilever plate over which two lines of external moments are assumed to act. The plate is rectangular with ratio between its sides equal to 4, as shown in Figure 5. Quadratic elements of the same length ($h = \ell / 8$) have been taken to discretize the plate boundary. The distributed moments along two internal lines are assumed constant, discretized into eight segments. Poisson ratio has been taken equal to 0.0.

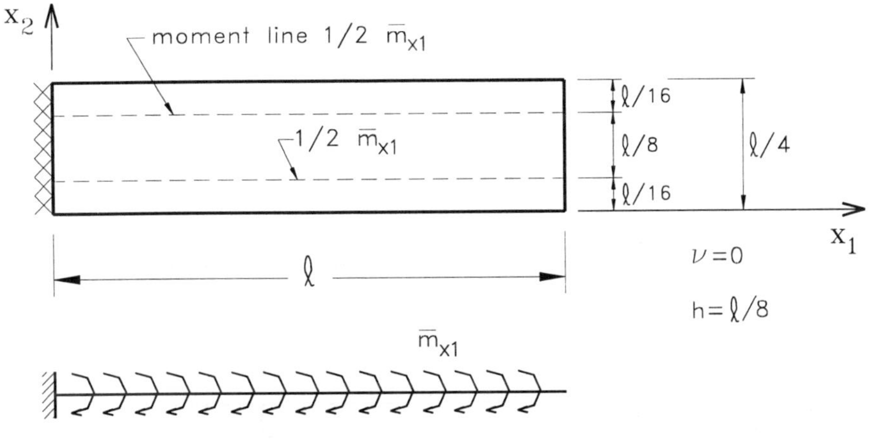

Figure 5. Cantilever plate with distributed moment lines.

The results obtained for this example are compared with the beam theory solution. Table 1 gives the computed generalized displacements, deflections and rotations, along the beam axis, while Table 2 shows the final moment values. As it can be seen for this particular example, the results computed numerically and obtained from the beam's theory are practically the same, therefore pointing out the very good accuracy that the proposed model can reach.

Table 1. Deflections w and rotations ϕ_{x1} along the line $x_2 = \ell/8$.

x_1	$w \dfrac{D}{\overline{m}_{x1}\ell^3}$		$\phi_{x1} \dfrac{D}{\overline{m}_{x1}\ell^2}$	
	BEM	Beam theory	BEM	Beam theory
$\ell/4$	0.0143	0.0143	0.1094	0.1094
$\ell/2$	0.0521	0.0521	0.1875	0.1875
$3\ell/4$	0.1055	0.1055	0.2344	0.2344
ℓ	0.1666	0.1667	0.2494	0.2500

Table 2. Bending moment M_{11} along the line $x_2 = \ell/8$

x_1	$M_{11} \dfrac{1}{\overline{m}_{x1}\ell^3}$	
	BEM	Beam theory
$\ell/4$	2.9997	3.0000
$\ell/2$	2.0002	2.0000
$3\ell/4$	1.0005	1.0000

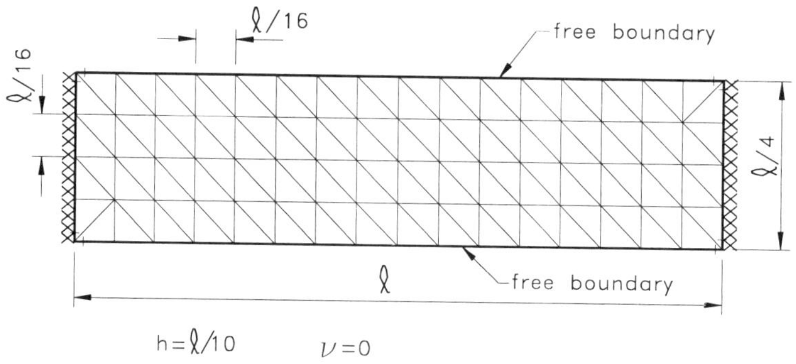

Figure 6. Boundary and domain discretizations. Example 3.

In the first example to deal with elastoplastic behaviour, a 0.8m thick rectangular plate with side lengths equal to $\ell = 8.0$ m and $b = 2.0$ m is analysed. The material exhibits Poisson ratio $\nu = 0.0$, elastic modulus $E = 2.0 \times 10^7 \, \mathrm{kN/m^2}$, while the yielding moment is assumed $M_p = 100 \, \mathrm{kN \times m/m}$. A transversal uniformly distributed load q is applied over the plate area to carry out the numerical analysis. Twenty boundary quadratic elements together with 44 nodes have been adopted to discretize the boundary of the plate, while 128 triangular domain cells and 45 internal points are used to approximate the domain integrals, as shown in Figure 6. Outside collocation points were located at a distance of $0.25 \, \ell_m$ from the boundary, where ℓ_m is the average length of the corresponding node adjacent elements. Traction and displacement discontinuities at the corners were introduced by adopting double nodes. The plate is clamped along the two shortest edges ($b = 2.0$ m), while the other two sides exhibit free deflections and rotations. Initially the plate was analyzed without taking into account elastoplastic effects through the thickness. In particular, two situations regarding the edge constraints are analysed: (a) soft condition: the twist rotations are free and the torsion moments vanish; (b) hard condition: the twist rotations are prevented.

The load has been applied in 17 increments for the soft condition case and in 20 increments when hard conditions have been assumed.

The load × displacement curves for both situations are shown in Figure 7, where displacements are computed for the plate center node. In Figure 7, one can note that after a quite linear behaviour exhibited during the three first increments, the graphic becomes a curved line until the increment number nine, when yield lines are formed along and close the fixed edges. During the next part, from the tenth increment to the fifteenth, for the soft condition case, and from the tenth to the eighteenth increment, for the hard condition, the graph is practically straight. From this load stage, the middle span yield line begins, leading to quite great displacement variation even when small load increments are applied.

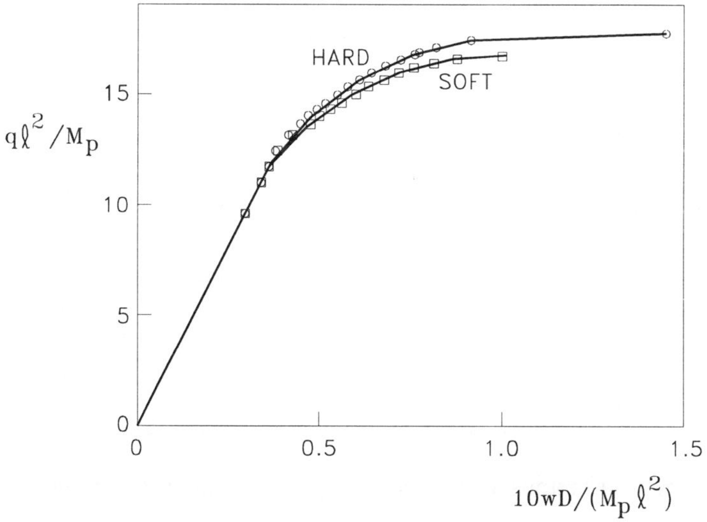

Figure 7. Non-layered model; load × displacement curve.

At this point, it is interesting to emphasize the result differences observed in both situations defined by hard or soft conditions. Assuming hard conditions makes the plate stiffer when compared with the soft case. In addition, the captured limit load is higher for the hard condition case. In both situations the limit loads have been achieved: $27.6\,\mathrm{kN/m^2}$ (hard condition) and $26.1\,\mathrm{kN/m^2}$ (soft condition). Assuming that the narrow plate is a 8. 0 m span beam with plastic ringes at the centre and the clamped ends, a limit load equal to 25.0 $\mathrm{kN/m^2}$ was obtained.

The same example is then analysed again, now to consider yielding through the thickness. For that analysis, yielding has been checked at 4, 8 and 12 Gauss points taken along the thickness. The plate was assumed to be constituted of an elastic-perfectly-plastic material that has yielding stress equal to 625 $\mathrm{kN/m^2}$. This value is equivalent to a yielding moment of 100 kNm/m assuming the von Mises hypothesis, already adopted for the previous analysis, and hard conditions along the clamped sides.

Table 3. Central point deflections. Example 3.

Increments	Load ($\mathrm{kN/m^2}$)	Non-layered model $w\,(10^3)$ m	Layered model $w\,(10^3)$ m *NG*=4	Layered model $w\,(10^3)$ m *NG*=12
1	12.4		0.1359	0.1360
2	12.5		0.1370	0.1371
3	13.0		0.1425	0.1427
4	13.5		0.1481	0.1483
5	14.0		0.1537	0.1542
6	14.5		0.1595	0.1603
7	15.0	0.1644	0.1665	0.1664
8	15.5		0.1738	0.1727
9	16.0		0.1811	0.1795
10	16.5		0.1885	0.1864
11	17.0	0.1865	0.1961	0.1934
12	17.5		0.2037	0.2004
13	18.0	0.1980	0.2114	0.2087
14	18.5		0.2192	0.2175
15	19.0	0.2115	0.2271	0.2266
16	19.5		0.2350	0.2361
17	19.6		0.2366	0.2380
18	20.1		0.2447	0.2479
19	20.6		0.2530	0.2582
20	21.1		0.2616	×
21	21.6			

Table 3 shows deflections at the plate central point for both analyses carried out.

For the four Gauss point case, the numerical limit load achieved was only 21.1 kN / m^2; lack of convergence was assumed after 5000 iterations. On the other hand, using 12 Gauss points gives a limit load of 20.6 kN / m^2. Note that a limit load of 27.6 kN / m^2 was achieved, for the analysis carried out without assuming yielding along the thickness.

The second elastoplastic example is a simply supported square plate, subjected to a uniformly distributed load q. This plate was also analyzed in two situations regarding boundary conditions along the edges: soft and hard conditions. The boundary discretization is given by 20 boundary elements with 44 nodes as illustrated in Figure 8. Two hundred triangular cells, with 81 internal points are taken to discretize the plate domain. The loading process was made in 6 increments as shown in Table 4.

Firstly, the analysis was conducted assuming the plastic model given in terms of bending moments. For that assumption the load × displacement curves, for the soft and hard condition cases, are given in Figure 9. The displacement computed at the third load increment was 0.08837 $D/(M_p\ell^2)$ when the hard condition was assumed, while Owen & Hinton[21] have obtained 0.08741 $D/(M_p\ell^2)$, and Corrêa[22] 0.08779 $D/(M_p\ell^2)$, both using FEM and $h/\ell = 0.01$. These values emphasize a very good accuracy of the BEM elastoplastic plate bending model. Moreover, the curves shown in Figure 9 stress this conclusion.

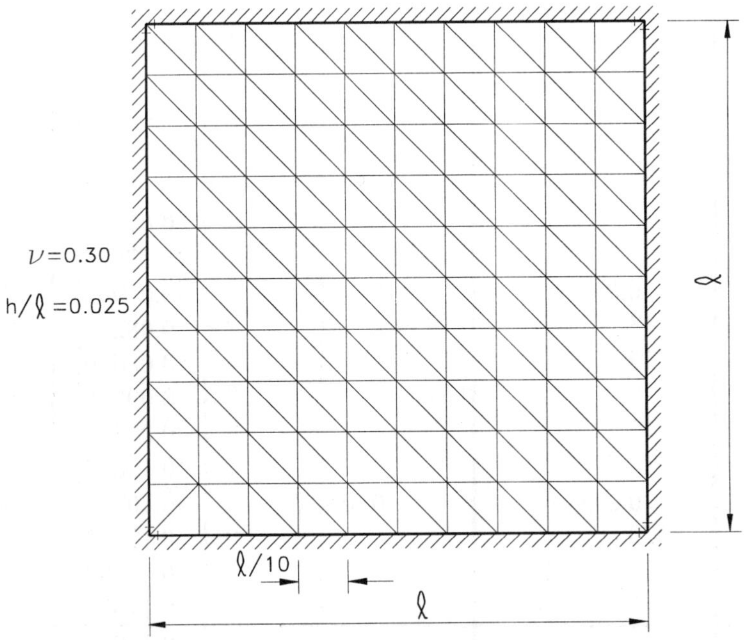

Figure 8. Boundary and domain discretization. Example 4.

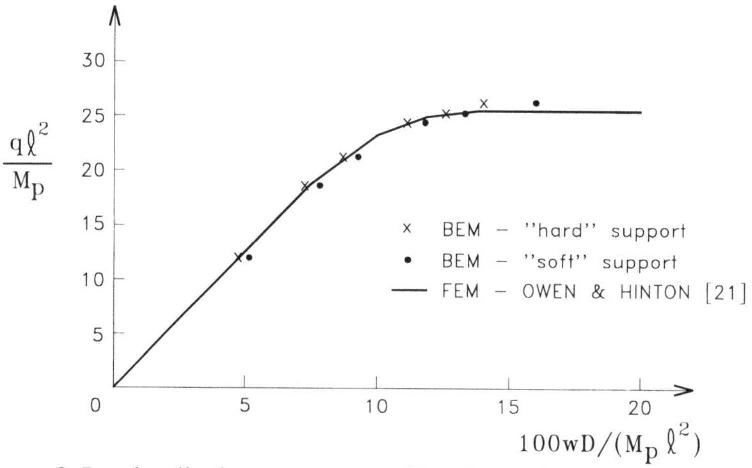

Figure 8. Load × displacement curve. Non-layered model. Example 4.

Table 4. Deflections at plate center point (values of $wD/(M_p\ell^2)$. Example 4

Load $q\,l^2/M_p$	Non-layered model "soft condition"	Non-layered model "hard condition"	Layered NG=6	Layered NG=12
12.00	0.05130	0.04888	0.04888	0.04888
13.00			0.05296	0.05296
14.00			0.05709	0.05708
14.35			0.05853	0.05853
15.00			0.06124	0.06127
15.50			0.06342	0.06346
16.00			0.06578	0.06574
16.50			0.06823	0.06813
17.00			0.07073	0.07063
17.50			0.07335	0.07321
18.00			0.07628	0.07614
18.60	0.07950	0.07587	0.07992	0.08016
19.00			0.08245	0.08292
19.50			0.08609	0.08645
20.00			0.09050	0.09073
20.50			0.09533	0.09569
21.00			0.10055	0.10171
21.40	0.09313	0.08837	0.10727	0.10691
22.00			0.12063	0.11681
22.50			0.13394	0.12759
23.00			0.14977	0.14273
23.50			0.16970	0.18310
24.20	0.12103	0.11282		
25.00	0.13422	0.12575		
25.80	0.14769	0.14028		

Again, the elastoplastic analysis assumes that yielding along the thickness has been carried out for this plate, for which only the hard condition case has been tested. Six and 12 Gauss point cases have been analysed applying the load in 22 increments with tolerance of 0.1%.

Deflections at the plate centre along the loading process are given in Table 4, together with the corresponding ones obtained for the non-layered case already described.

For both cases, in which six and 12 Gauss points were taken to integrate the stresses along the thickness, the convergence has been achieved at increment 22, corresponding to a limit load of $23.5\ell^2 / M_p$. Deflections computed for six and 12 Gauss point cases at the plate central point were $0.16970\, D/(M_p\ell^2)$ and $0.18310\, D/(M_p\ell^2)$, respectively. In both cases, after performing 1000 iterations, the convergence was not achieved for an applied load. After accumulating the load until the value $21.4\ell^2 / M_p$, the following deflections were achieved, $0.08837\, D/(M_p\ell^2)$, $0.10691\, D/(M_p\ell^2)$ and $0.10727\, D/(M_p\ell^2)$, for analysis carried out without assuming yielding along the thickness and assuming yielding at six and 12 Gauss points, respectively. From those values, one can see the strong influence of assuming yielding along the thickness. Differences in deflections of 21% between the layered and the simplified models have been observed. This tendency has been observed during the whole loading process and can be justified by the size of yielding zones which is significantly larger when yielding is assumed along the thickness. For the non-layered case, yielding occurs when the effective moment reaches the yielding surface, which happens only for rather high moment levels.

Assuming either six or 12 Gauss points along the thickness is important only near the limit load, where deflection differences of 7% have been observed, but without disturbing the limit load. In general, the differences between the values computed for those cases remain lower than 1% during almost the whole loading process.

8 Conclusion

An efficient approach to analyze elastoplastic Reissner's plates is presented in this paper, based on the well known initial stress or moment field procedure. The Reissner's plate model in conjunction with BEM enables the analysis of a rather wide range of plates, from thin to moderately thick plates. One can take into account the effects of the thickness, as well as the influences of soft and hard conditions assumed along the boundary, which may result in rather significant differences in comparison with the solutions obtained by using the classical Kirchhoff's theory. The boundary equations are achieved by adopting outside collocation points, which conveniently avoids carrying out some complex singular integrals. Using this procedure one can obtain very accurate results , but special attention must be paid when the analysed plate is very thin; this may lead matrix **G** to be ill-conditioned. In this case, the use of sub-element technique is the strategy to obtain good results.

The procedure adopted to integrate the initial moment field has been shown to be very efficient and accurate. The proposed cell subdivision allows the domain integrals to be performed accurately, even for that case where one has to deal with complex kernels. The good quality of elastoplastic results achieved by using the incremental and iterative strategy, where all matrices remain constant, pointed out the capability of the whole proposed procedure. In addition, the possibility of adapting the method to model more complex plastic behaviour has been shown. For instance, the classical case of layered models has been implemented, following a very simple integration scheme along the plate thickness, to give accurate numerical responses.

9 References

[1] Bezine, G. P. Boundary integral formulation for plate flexure with arbitrary boundary conditions, *Mech. Res. Comm.*, **5**, 197-206, 1978.

[2] Stern, M.A. A general boundary integral formulation for the numerical solution of plate bending problems, *Int. J. Solids Struct.*, **15**, 769-782, 1979.

[3] Altiero N.J.& Sikarskie, D.L. A boundary integral method applied to plates of arbitrary plan form, *Comp. & Struct.*, **9**, 13-168, 1978.

[4] Hartmann, F.& Zotemantel, R. The direct boundary element method in plate bending, *Int. J. Num. Meth. Engrg.*, **23**, 2049-2069, 1986.

[5] Ghosh, S. & Mukherjee, S. Boundary element method analysis of bending of elastic plates of arbitrary shape with general boundary conditions, *Engineering Analysis with Boundary Elements*, **3**, 2049-2069, 1986.

[6] Venturini, W.S. & Paiva, J.B. Boundary element for plate bending analysis, *Engineering Analysis with Boundary Elements*, **11**, 1-8, 1993.

[7] Katsikadelis , J.T. & Armenakas A.E. Plates on elastic foundation by BIE method, *J. Engrg. Mech.*, **110**, 1086-1104, 1984.

[8] Tanaka, M. Integral equation approach to small and large displacements of thin elastic plates, *Proc. of the 4th. Seminar on Boundary Element Method in Engineering*, University of Southampton, 1982.

[9] Morjaria, M. & Mukerjee, S. Numerical analysis of planar time dependent inelastic deformation of plates with cracks by the boundary element method, *Int. J. Solids Structures*, **17**, 127-135, 1981.

[10] Moshaiov, A. & Vorus, W.S. Elastoplastic bending analysis by a boundary method with initial plastic moments, *Int. J. Solids Structures*, **22**, 1213-1229, 1986.

[11] Providakis, C..P. & Beskos, D.E. Free and forced vibrations of plates by boundary and interior elements, *Int. J. Num. Meth. Engrg.*, **28**, 1977-1994, 1989.

[12] Vander Weeën, F. Application of boundary integral equation method to Reissner's plate model, *Int. J. Num. Meth. Engrg.*, **18**, 1-10, 1982.

[13] Vander Weeën, F. Application of the direct boundary element method Reissner's plate model, *Proc. of the 4th Seminar on Boundary Element Method in Engineering*, University of Southampton, 1982.

[14] Reissner, E. On the theory of bending of elastic plates, *J. Math. Physics*, **23**, 184-191, 1944.

[15] Ribeiro, G.O. & Venturini, W.S. Boundary element formulation for non-linear problems using the Reissner's theory, *Proc. of the Sixth Boundary Element Technology Conference*, ed. C.A. Brebbia, Computational Mechanics Publications, Southampton, U.K., 239-251, 1991.

[16] Karan, V.J & Telles, J.C.F The BEM applied to plate bending elastoplastic analysis using Reissner's theory. *Engineering Analysis with Boundary Elements*, **9**, 351-357, 1992.

[17] Miklin, S.G. Singular integral equations, *American Math. Soc. Trans.*, Series 1, **10**, 84-197, 1962.

[18] Telles, J.C.F.& Brebbia, C.A. On the application of the boundary element method to plasticity, *Appl. Math. Modeling*, **3**, 466-470, 1980.

[19] Bui, H.D. Some remarks about the formulation of three-dimensional thermoelastic problems by integral equations, *Int. J. Solids*, **14**, 935-939, 1978.

[20] Timoshenko, S.P. & Woynosky-Krieger, S. *Theory of Plates and Shells*, McGraw Hill Book Company, New York, 1976.

[21] Owen, D.R.J. & Hinton, E. *Finite Elements in Plasticity: Theory and Practice*, Pineridge Press, Swansea U.K., 1980.

[22] Corrêa, M.R.S. Usual up dated models for building structural design. Ph.D. Thesis, University of São Paulo, São Carlos, Brazil, 1991 (in Portuguese).

Chapter 4

Nonlinear material analysis of Reissner's plates

V.J.Karam[a] & J.C.F. Telles[b]
[a]*UENF, Universidade Estadual do Norte Fluminense, Laboratório de Ciências de Engenharia, Av. Alberto Lamego, 2000, CEP 28015-620, Campos, RJ, Brazil*
[b]*COPPE/UFRJ, Universidade Federal do Rio de Janeiro, Programa de Engenharia Civil, Cidade Universitária, Ilha do Fundão, CEP 21945-970, Caixa Postal 68506, Rio de Janeiro, RJ, Brazil*

Abstract

A formulation to nonlinear material analysis of Reissner's plates using the boundary element method is presented in this chapter. The governing equations with the consideration of bending inelastic strain rates are shown and integral equations are obtained, including those for moment and shear resultant rates at internal points. The expressions for the kernel functions are also shown, including the components of the tensors that multiply the inelastic terms and the free terms of the integral equations. An incremental-iterative procedure is employed to solve the elastoplastic problem and von Mises and Tresca yield criteria are considered. Numerical results are presented and compared to finite elements or to analytical solutions.

1 Introduction

Nonlinear material analysis by the boundary element method (BEM) has been the object of many researches over the last years and several kinds of problems, such as elastoplasticity, viscoplasticity and creep have been developed for two and three-dimensional analyses, including axisymmetric problems.[1–9]

In 1992, Karam & Telles[10] presented a formulation for Reissner's plate bending elastoplastic analysis, as an extension of the formulation previously presented for Reissner's plate bending elastic analysis.[11,12] Reissner's plate bending theory[13–15] holds for thin and thick plates, since the effects of transverse shear deformations are taken into account and admits the existence of three boundary conditions for each side of the plate.

This chapter presents a general formulation for nonlinear material Reissner's plate bending problems with emphasis on elastoplasticity. It introduces an initial stress formulation and the procedures adopted for the solution of the plate problem are the same as that adopted in Refs. 6-8,16 for other problem applications.

Basic equations for Reissner's theory with the consideration of bending inelastic strains are shown and integral equations are obtained for this case, including those for moments and shear forces at internal points. The expressions for the components of the tensors that multiply the inelastic moments are also presented, as well as the expressions for the free term of the integral equations.

The numerical implementation is carried out using continuous and discontinuous quadratic boundary elements, with linear geometry, and constant internal cells, also having linear geometry. Internal cells are used in the parts of the domain where the existence of inelastic strains is expected. Integrations over boundary elements and internal cells are evaluated numerically, using Gaussian quadrature when the corresponding integral is regular. The singular integrals are also evaluated numerically; however, special procedures are employed. After discretizing the integral equations, a system of equations is assembled.

To solve the elastoplastic problem, an incremental-iterative process is used together with the von Mises and Tresca yield criteria. The classical theory of plasticity is adopted, in which plastic strains are time rate independent.

Results obtained by the elastoplastic analysis with the BEM are presented and compared to analytical solution or to the finite element method (FEM).

The cartesian tensor notation is used throughout the text, with Greek indices varying from 1 to 2 and Latin ones from 1 to 3. It is important to mention that the repetition of an index in the same term indicates summation. In addition, the dot over some variables indicates time derivative.

2 Basic equations for nonlinear material Reissner's plates

In order to define a general formulation for nonlinear material Reissner's plates, let h be the constant thickness of a plate subjected to a transverse load q per unit area, and let the material of this plate be homogeneous and isotropic, but nonlinear.

It is considered that the cartesian axis x_α is defined in the undeformed midsurface of the plate and the axis x_3 in the transverse direction, as shown in Figure 1. The plate is considered to have a domain Ω and a boundary Γ.

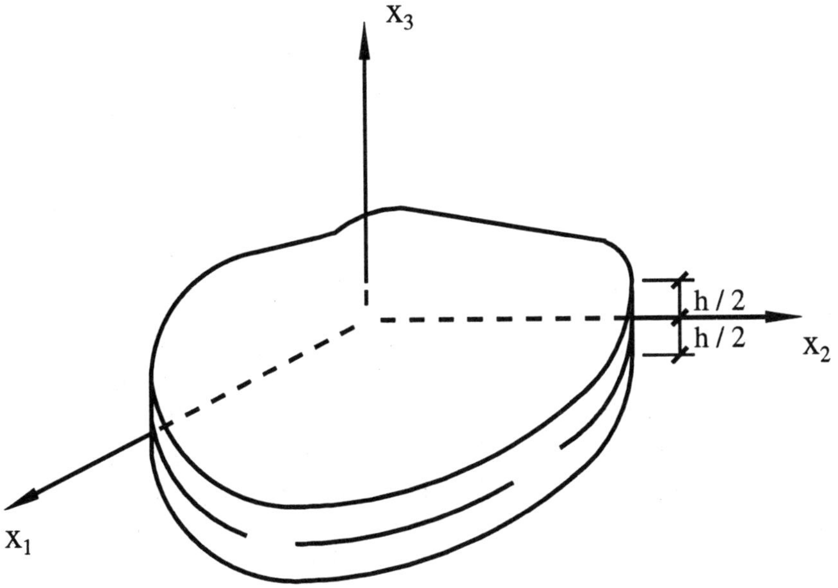

Figure 1. Coordinate system (x_1, x_2, x_3).

By considering inelastic strains only due to bending, one can define a total bending strain rate $\dot{\chi}_{\alpha\beta}$ and a total shear strain rate $\dot{\psi}_\alpha$ by:

$$\dot{\chi}_{\alpha\beta} = \dot{\chi}^e_{\alpha\beta} + \dot{\chi}^a_{\alpha\beta} \tag{1}$$

$$\dot{\psi}_\alpha = \dot{\psi}^e_\alpha \tag{2}$$

where $\dot{\chi}^e_{\alpha\beta}$ and $\dot{\psi}^e_\alpha$ represent the elastic parts and $\dot{\chi}^a_{\alpha\beta}$ is the inelastic contribution.

The total generalized strain rates $\dot{\chi}_{\alpha\beta}$ and $\dot{\psi}_\alpha$ can also be expressed in terms of the generalized displacement rates $\dot{u}_\alpha$ (rotations) and $\dot{u}_3$ (transverse deflection), as:

$$\dot{\chi}_{\alpha\beta} = \frac{1}{2}\left(\dot{u}_{\alpha,\beta} + \dot{u}_{\beta,\alpha}\right) \tag{3}$$

$$\dot{\psi}_\alpha = \dot{u}_\alpha + \dot{u}_{3,\alpha} \tag{4}$$

Positive directions for rotations and transverse deflection are shown in Figure 2.

The inelastic part of the total bending strain rate can be expressed as:

$$\dot{\chi}^a_{\alpha\beta} = \dot{\chi}^p_{\alpha\beta} + \dot{\chi}^c_{\alpha\beta} + \dot{\chi}^t_{\alpha\beta} \tag{5}$$

where $\dot{\chi}^p_{\alpha\beta}$ represents the plastic or viscoplastic bending strain rate, $\dot{\chi}^c_{\alpha\beta}$ is the creep bending strain rate and $\dot{\chi}^t_{\alpha\beta}$ represents the thermal bending strain rate.

By studying the equilibrium of an infinitesimal plate element, the following equations arise:

$$\dot{M}_{\alpha\beta,\beta} - \dot{Q}_{\alpha} = 0 \tag{6}$$

$$\dot{Q}_{\alpha,\alpha} + \dot{q} = 0 \tag{7}$$

Positive directions for moments and shear forces can be seen in Figure 3.

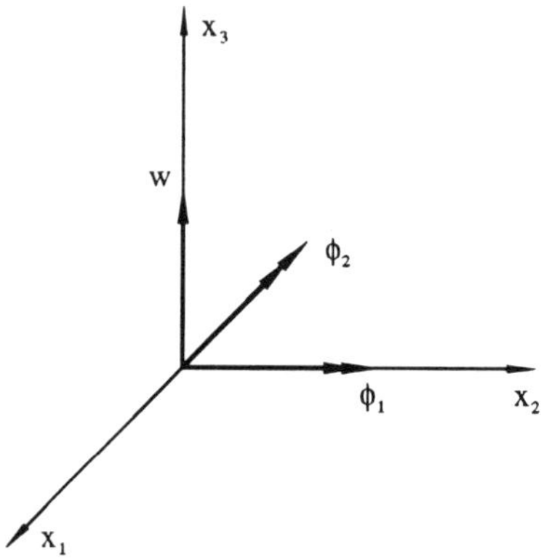

Figure 2. Rotations and transverse deflection.

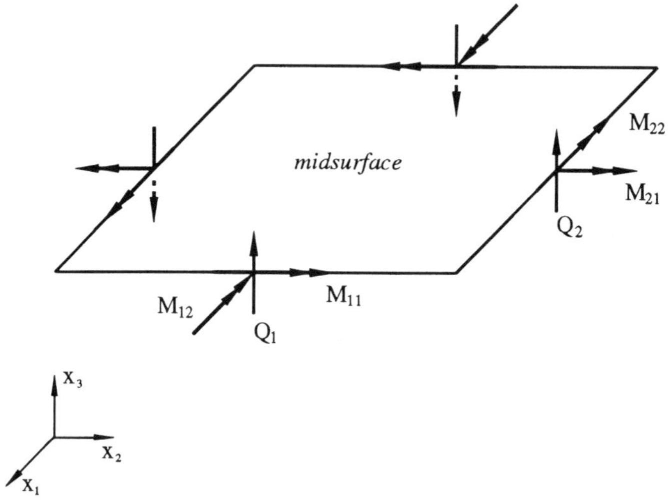

Figure 3. Moments and shear forces.

Moment rates and shear force rates are expressed by:

$$\dot{M}_{\alpha\beta} = \frac{D(1-\nu)}{2}\left[2\,\dot{\chi}_{\alpha\beta} + \frac{2\nu}{1-\nu}\,\dot{\chi}_{\gamma\gamma}\,\delta_{\alpha\beta}\right] + \frac{\nu\dot{q}}{(1-\nu)\lambda^2}\,\delta_{\alpha\beta} - \dot{M}^a_{\alpha\beta} \tag{8}$$

$$\dot{Q}_\alpha = \frac{D(1-\nu)\lambda^2}{2}\,\dot{\psi}_\alpha \tag{9}$$

where $\delta_{\alpha\beta}$ is the Kronecker delta, ν is Poisson's ratio, and

$$\lambda = \sqrt{10}/h \tag{10}$$

$$D = Eh^3/12(1-\nu^2) \tag{11}$$

with E being the Young's modulus. The term $\dot{M}^a_{\alpha\beta}$ represents the components of inelastic moment rates, defined in an initial stress formulation by:

$$\dot{M}^a_{\alpha\beta} = \frac{D(1-\nu)}{2}\left[2\,\dot{\chi}^a_{\alpha\beta} + \frac{2\nu}{1-\nu}\,\dot{\chi}^a_{\gamma\gamma}\,\delta_{\alpha\beta}\right] \tag{12}$$

Boundary conditions at Γ can be defined, considering that $\Gamma_1 \cup \Gamma_2 = \Gamma$, as:

$$\dot{u}_\alpha = \dot{\bar{u}}_\alpha \quad \text{and} \quad \dot{u}_3 = \dot{\bar{u}}_3 \quad \text{at} \quad \Gamma_1 \tag{13}$$

$$\dot{p}_\alpha = \dot{\bar{p}}_\alpha \quad \text{and} \quad \dot{p}_3 = \dot{\bar{p}}_3 \quad \text{at} \quad \Gamma_2 \tag{14}$$

where $\dot{p}_\alpha$ and $\dot{p}_3$ are generalized traction rates, defined as:

$$\dot{p}_\alpha = \dot{M}_{\alpha\beta}\, n_\beta \tag{15}$$

$$\dot{p}_3 = \dot{Q}_\beta\, n_\beta \tag{16}$$

with n_β representing the direction cosines of the outward normal to the boundary.

3 Integral equations for displacements

Integral equations for the generalized displacements at a point ξ (source point) of the domain Ω can be obtained from a weighted residual method.[16]

By considering the equilibrium eqns (6) and (7) and the boundary conditions (13) and (14), one can write the following equation, for an approximate solution u_k, using the fundamental solution u^*_k (where $u^*_k = u^*_{ik}P_i$, with $P_i = 1$), as a weighting function:

$$\int_\Omega \left[\left(\dot{M}_{\alpha\beta,\beta} - \dot{Q}_\alpha\right)u^*_\alpha + \left(\dot{Q}_{\alpha,\alpha} + \dot{q}\right)u^*_3\right]\mathrm{d}\Omega$$

$$= \int_{\Gamma_1}\left(\dot{\bar{u}}_k - \dot{u}_k\right)p^*_k\,\mathrm{d}\Gamma + \int_{\Gamma_2}\left(\dot{p}_k - \dot{\bar{p}}_k\right)u^*_k\,\mathrm{d}\Gamma \tag{17}$$

where $p^*_\alpha = M^*_{\alpha\beta} n_\beta$ and $p^*_3 = Q^*_\alpha n_\alpha$, with $M^*_{\alpha\beta}$ and Q^*_α being the moments and shear forces, respectively, related to the fundamental solution.

By integrating the first and the third terms of the first member by parts, and considering eqns (15) and (16), one has:

$$- \int_\Omega \dot{M}_{\alpha\beta}\, u^*_{\alpha,\beta}\, d\Omega - \int_\Omega \dot{Q}_\alpha \left(u^*_\alpha + u^*_{3,\alpha}\right) d\Omega + \int_\Omega \dot{q}\, u^*_3\, d\Omega$$
$$= - \int_{\Gamma_1} \dot{p}_k\, u^*_k\, d\Gamma - \int_{\Gamma_2} \dot{\bar{p}}_k\, u^*_k\, d\Gamma + \int_{\Gamma_1} \left(\dot{\bar{u}}_k - \dot{u}_k\right) p^*_k\, d\Gamma \tag{18}$$

Then, considering eqns (3) and (4), written for a region Ω^* that contains the region Ω, and also

$$\dot{M}_{\alpha\beta} = \dot{M}^e_{\alpha\beta} - \dot{M}^a_{\alpha\beta} \tag{19}$$

$$\dot{Q}_\alpha = \dot{Q}^e_\alpha \tag{20}$$

the following equation is obtained:

$$- \int_\Omega \dot{M}^e_{\alpha\beta}\, \chi^*_{\alpha\beta}\, d\Omega + \int_\Omega \dot{M}^a_{\alpha\beta}\, \chi^*_{\alpha\beta}\, d\Omega - \int_\Omega \dot{Q}^e_\alpha\, \psi^*_\alpha\, d\Omega + \int_\Omega \dot{q}\, u^*_3\, d\Omega$$
$$= - \int_{\Gamma_1} \dot{p}_k\, u^*_k\, d\Gamma - \int_{\Gamma_2} \dot{\bar{p}}_k\, u^*_k\, d\Gamma + \int_{\Gamma_1} \left(\dot{\bar{u}}_k - \dot{u}_k\right) p^*_k\, d\Gamma \tag{21}$$

Recalling now the reciprocity and integrating by parts once again:

$$\int_\Omega M^*_{\alpha\beta,\beta}\, \dot{u}_\alpha\, d\Omega + \int_\Omega \dot{M}^a_{\alpha\beta}\, \chi^*_{\alpha\beta}\, d\Omega - \int_\Omega Q^*_\alpha\, \dot{u}_\alpha\, d\Omega + \int_\Omega Q^*_{\alpha,\alpha}\, \dot{u}_3\, d\Omega$$
$$+ \int_\Omega \dot{q}\, u^*_3\, d\Omega - \frac{\nu}{(1-\nu)\lambda^2} \int_\Omega \dot{q}\, \delta_{\alpha\beta}\, u^*_{\alpha,\beta}\, d\Omega$$
$$= - \int_{\Gamma_1} \dot{p}_k\, u^*_k\, d\Gamma - \int_{\Gamma_2} \dot{\bar{p}}_k\, u^*_k\, d\Gamma + \int_{\Gamma_1} \dot{\bar{u}}_k\, p^*_k\, d\Gamma + \int_{\Gamma_2} \dot{u}_k\, p^*_k\, d\Gamma \tag{22}$$

By considering the equilibrium equations for region Ω^*:

$$M^*_{\alpha\beta,\beta} - Q^*_\alpha + F^*_\alpha = 0 \tag{23}$$

$$Q^*_{\alpha,\alpha} + F^*_3 = 0 \tag{24}$$

in which F^*_j is the generalized unit point load applied in each generalized direction of a point ξ situated within the domain Ω^*, one has:

$$\int_\Omega F^*_j\, \dot{u}_j\, d\Omega = \int_\Gamma u^*_j\, \dot{p}_j\, d\Gamma - \int_\Gamma p^*_j\, \dot{u}_j\, d\Gamma$$
$$+ \int_\Omega \dot{q} \left(u^*_3 - \frac{\nu}{(1-\nu)\lambda^2}\, u^*_{\alpha,\alpha} \right) d\Omega + \int_\Omega \chi^*_{\alpha\beta}\, \dot{M}^a_{\alpha\beta}\, d\Omega \tag{25}$$

Forces F_j^* can then be represented as:

$$F_j^* = \delta(x - \xi) P_j \tag{26}$$

where P_j is the unit point load intensity and $\delta(x - \xi)$ is the generalized Dirac delta function, which has the following property:

$$\int_{\Omega^*} g(x)\, \delta(x - \xi)\, \mathrm{d}\Omega(x) = \begin{cases} g(\xi) & \text{if} \quad \xi \in \Omega^* \\ 0 & \text{if} \quad \xi \notin \Omega^* \end{cases} \tag{27}$$

By considering eqn (26) together with the property given in eqn (27), the first integral of eqn (25) becomes:

$$\int_{\Omega} F_j^*\, \dot{u}_j\, \mathrm{d}\Omega = \dot{u}_j(\xi) P_j = \sum_{j=1}^{3} \dot{u}_j(\xi) \tag{28}$$

in which ξ is also a point of Ω.

By supposing that each unit point load acts independently, three equations of the following form can be written:

$$\begin{aligned} \dot{u}_i(\xi) = & \int_{\Gamma} u_{ij}^*(\xi, x)\, \dot{p}_j(x)\, \mathrm{d}\Gamma(x) - \int_{\Gamma} p_{ij}^*(\xi, x)\, \dot{u}_j(x)\, \mathrm{d}\Gamma(x) \\ & + \int_{\Omega} \left[u_{i3}^*(\xi, x) - \frac{\nu}{(1-\nu)\lambda^2}\, u_{i\alpha,\alpha}^*(\xi, x) \right] \dot{q}(x)\, \mathrm{d}\Omega(x) \\ & + \int_{\Omega} \chi_{\alpha\beta i}^*(\xi, x)\, \dot{M}_{\alpha\beta}^a(x)\, \mathrm{d}\Omega(x) \end{aligned} \tag{29}$$

Eqn (29) holds for a point ξ in the interior of Ω, with p_{ij}^* and $\chi_{\alpha\beta i}^*$ being the tensors corresponding to the fundamental solution which represent the response, at a point x (field point), due to a generalized unit point load that acts in direction i at point ξ.

In order to solve the problem by the BEM, it is necessary to write the integral eqn (29) considering that the point ξ is too near the boundary Γ and take the limit of the resulting integrals, as the point ξ tends to Γ. This process is considered in details in Ref. 17 for the first three integrals of eqn (29); the fourth integral do not present problems, since it has a weak singularity.

Hence, for a source point situated on the boundary, one can write:

$$\begin{aligned} C_{ij}(\xi)\, \dot{u}_j(\xi) = & \int_{\Gamma} u_{ij}^*(\xi, x)\, \dot{p}_j(x)\, \mathrm{d}\Gamma(x) - \int_{\Gamma} p_{ij}^*(\xi, x)\, \dot{u}_j(x)\, \mathrm{d}\Gamma(x) \\ & + \int_{\Omega} \left[u_{i3}^*(\xi, x) - \frac{\nu}{(1-\nu)\lambda^2}\, u_{i\alpha,\alpha}^*(\xi, x) \right] \dot{q}(x)\, \mathrm{d}\Omega(x) \\ & + \int_{\Omega} \chi_{\alpha\beta i}^*(\xi, x)\, \dot{M}_{\alpha\beta}^a(x)\, \mathrm{d}\Omega(x) \end{aligned} \tag{30}$$

It should be observed that eqn (30) can be made valid for internal points (with $C_{ij} = \delta_{ij}$) and for boundary points, with C_{ij} having the value $\delta_{ij}/2$ in the case of smooth boundaries.

Eqn (30) is the basic integral equation for the nonlinear material analysis of Reissner's plate bending with a formulation that considers initial inelastic moments.

Expressions for u^*_{ij} and p^*_{ij} were presented in Refs. 11,12 and 17 and for $\chi^*_{\alpha\beta i}$ in Refs. 10 and 18. These expressions are shown in the Appendix of this text.

The domain integral related to the transverse load rate $\dot{q}$, in eqn (30), can be transformed into a boundary integral for several kinds of loading. This transformation is performed here for uniformly distributed loads ($\dot{q}(x) = \dot{q} =$ constant), using the divergence theorem. In this case, eqn (30) becomes:

$$\begin{aligned} C_{ij}(\xi)\,\dot{u}_j(\xi) &= \int_\Gamma u^*_{ij}(\xi,x)\,\dot{p}_j(x)\,\mathrm{d}\Gamma(x) - \int_\Gamma p^*_{ij}(\xi,x)\,\dot{u}_j(x)\,\mathrm{d}\Gamma(x) \\ &+ \dot{q}\int_\Gamma \left[v^*_{i,\alpha}(\xi,x) - \frac{\nu}{(1-\nu)\lambda^2}\,u^*_{i\alpha}(\xi,x)\right] n_\alpha(x)\,\mathrm{d}\Gamma(x) \\ &+ \int_\Omega \chi^*_{\alpha\beta i}(\xi,x)\,\dot{M}^a_{\alpha\beta}(x)\,\mathrm{d}\Omega(x) \end{aligned} \tag{31}$$

where v^*_i satisfy the Poisson equation:

$$v^*_{i,\alpha\alpha}(\xi,x) = u^*_{i3}(\xi,x) \tag{32}$$

Expressions for v^*_i and its derivatives were presented in Refs. 11,12 and 17 and are given in the Appendix of this text.

4 Integral equations for moments and shear forces at internal points

Moment rates and shear force rates at internal points are calculated by substituting eqn (31) with $C_{ij} = \delta_{ij}$ into eqns (3) and (4), and then, substituting the resulting expressions in eqns (8) and (9). It should be noticed that the corresponding derivatives are taken with respect to the coordinates of point ξ.

For the first three integrals of eqn (31), the differentiation can be applied directly to the tensors related to the fundamental solution, whereas in the case of the last integral, special considerations are necessary. In order to represent this integral in a more formal manner, one should write:

$$V_i = \lim_{\varepsilon \to 0} \int_{\Omega_\varepsilon} \chi^*_{\alpha\beta i}\,\dot{M}^a_{\alpha\beta}\,\mathrm{d}\Omega \tag{33}$$

where Ω_ε is the domain that remains when one removes, from the domain Ω, a circle of radius ε centred at point ξ.

The derivative of V_i with respect to the coordinate x_θ of point ξ can be represented by:

$$\frac{\partial V_i}{\partial x_\theta} = \lim_{\varepsilon \to 0} \left\{ \frac{\partial}{\partial x_\theta} \int_{\Omega_\varepsilon} \chi^*_{\alpha\beta i} \, \dot{M}^a_{\alpha\beta} \, \mathrm{d}\Omega \right\} \tag{34}$$

After using a procedure analogous to that employed in Ref. 8, one concludes that:

$$\frac{\partial V_i}{\partial x_\theta} = \int_\Omega \frac{\partial \chi^*_{\alpha\beta i}}{\partial x_\theta} \, \dot{M}^a_{\alpha\beta} \, \mathrm{d}\Omega - \dot{M}^a_{\alpha\beta}(\xi) \int_{\Gamma'_1} \chi^*_{\alpha\beta i} \, r_{,\theta} \, \mathrm{d}\Gamma \tag{35}$$

where the first integral in the right hand side is in the sense of Cauchy principal value and the second one is calculated for a circle of unit radius centred at point ξ, with $r_{,\theta}$ being the derivative of r with respect to the coordinate x_θ of point x.

One can now obtain the following equation for the derivative of eqn (31) at an internal point, with respect to the coordinate x_θ of point ξ:

$$\begin{aligned} \frac{\partial \dot{u}_i}{\partial x_\theta} &= \int_\Gamma \frac{\partial u^*_{ij}}{\partial x_\theta} \, \dot{p}_j \, \mathrm{d}\Gamma - \int_\Gamma \frac{\partial p^*_{ij}}{\partial x_\theta} \, \dot{u}_j \, \mathrm{d}\Gamma \\ &+ \dot{q} \int_\Gamma \left[\frac{\partial v^*_{i,\alpha}}{\partial x_\theta} - \frac{\nu}{(1-\nu)\lambda^2} \frac{\partial u^*_{i\alpha}}{\partial x_\theta} \right] n_\alpha \, \mathrm{d}\Gamma \\ &+ \int_\Omega \frac{\partial \chi^*_{\alpha\beta i}}{\partial x_\theta} \, \dot{M}^a_{\alpha\beta} \, \mathrm{d}\Omega - \dot{M}^a_{\alpha\beta}(\xi) \int_{\Gamma'_1} \chi^*_{\alpha\beta i} \, r_{,\theta} \, \mathrm{d}\Gamma \end{aligned} \tag{36}$$

where the fourth integral is in the sense of Cauchy principal value and the fifth one is calculated at a unit circle centred at point ξ.

After solving the integral in Γ'_1, one has:

$$\begin{aligned} \frac{\partial \dot{u}_\gamma}{\partial x_\theta} &= \int_\Gamma \frac{\partial u^*_{\gamma j}}{\partial x_\theta} \, \dot{p}_j \, \mathrm{d}\Gamma - \int_\Gamma \frac{\partial p^*_{\gamma j}}{\partial x_\theta} \, \dot{u}_j \, \mathrm{d}\Gamma \\ &+ \dot{q} \int_\Gamma \left[\frac{\partial v^*_{\gamma,\alpha}}{\partial x_\theta} - \frac{\nu}{(1-\nu)\lambda^2} \frac{\partial u^*_{\gamma\alpha}}{\partial x_\theta} \right] n_\alpha \, \mathrm{d}\Gamma + \int_\Omega \frac{\partial \chi^*_{\alpha\beta\gamma}}{\partial x_\theta} \, \dot{M}^a_{\alpha\beta} \, \mathrm{d}\Omega \\ &+ \frac{1}{8D(1-\nu)} \left[2(3-\nu)\dot{M}^a_{\gamma\theta} - (1+\nu)\delta_{\gamma\theta}\dot{M}^a_{\Delta\Delta} \right] \end{aligned} \tag{37}$$

and

$$\begin{aligned} \frac{\partial \dot{u}_3}{\partial x_\theta} &= \int_\Gamma \frac{\partial u^*_{3j}}{\partial x_\theta} \, \dot{p}_j \, \mathrm{d}\Gamma - \int_\Gamma \frac{\partial p^*_{3j}}{\partial x_\theta} \, \dot{u}_j \, \mathrm{d}\Gamma \\ &+ \dot{q} \int_\Gamma \left[\frac{\partial v^*_{3,\alpha}}{\partial x_\theta} - \frac{\nu}{(1-\nu)\lambda^2} \frac{\partial u^*_{3\alpha}}{\partial x_\theta} \right] n_\alpha \, \mathrm{d}\Gamma + \int_\Omega \frac{\partial \chi^*_{\alpha\beta 3}}{\partial x_\theta} \, \dot{M}^a_{\alpha\beta} \, \mathrm{d}\Omega \end{aligned} \tag{38}$$

By taking into account eqns (3), (4), (8), (9), (37) and (38), moment rates and shear force rates at internal points can be calculated in the form:

$$\begin{aligned} \dot{M}_{\alpha\beta} &= \int_\Gamma u^*_{\alpha\beta k} \, \dot{p}_k \, \mathrm{d}\Gamma - \int_\Gamma p^*_{\alpha\beta k} \, \dot{u}_k \, \mathrm{d}\Gamma + \dot{q} \int_\Gamma w^*_{\alpha\beta} \, \mathrm{d}\Gamma + \int_\Omega \chi^*_{\alpha\beta\gamma\theta} \, \dot{M}^a_{\gamma\theta} \, \mathrm{d}\Omega \\ &+ \frac{\nu \dot{q}}{(1-\nu)\lambda^2} \, \delta_{\alpha\beta} - \frac{1}{8} \left[2(1+\nu)\dot{M}^a_{\alpha\beta} + (1-3\nu)\delta_{\alpha\beta}\dot{M}^a_{\theta\theta} \right] \end{aligned} \tag{39}$$

and

$$\dot{Q}_\beta = \int_\Gamma u^*_{3\beta k}\, \dot{p}_k \,\mathrm{d}\Gamma - \int_\Gamma p^*_{3\beta k}\, \dot{u}_k \,\mathrm{d}\Gamma + \dot{q} \int_\Gamma w^*_{3\beta} \,\mathrm{d}\Gamma + \int_\Omega \chi^*_{3\beta\gamma\theta}\, \dot{M}^a_{\gamma\theta} \,\mathrm{d}\Omega \tag{40}$$

It should be noticed that the last integral in eqns (39) and (40) should be calculated in the Cauchy principal value sense.

Expressions for $u^*_{i\beta k}$, $p^*_{i\beta k}$ and $w^*_{i\beta}$ were presented in Refs. 12 and 17 and for $\chi^*_{i\beta\gamma\theta}$ in Refs. 10 and 18. These expressions are given in the Appendix of this text.

5 Discretization and system of equations

In order to analyse the problem by the BEM, the integral eqns (31), (39) and (40) are written in discretized form, considering that the boundary is divided into the so-called boundary elements and the parts of the domain where the existence of plastic deformation is expected is divided into internal cells. This discretization is carried out, in this work, using quadratic boundary elements, continuous and discontinuous, having linear geometry; and constant triangular internal cells, also having linear geometry (see Figure 4).

The discretization of boundary integrals is performed by considering that the coordinates, the displacements and the surface tractions of an arbitrary point within the element Γ_j are calculated, respectively, by the equations:

$$\mathbf{x}^{(j)} = \mathbf{M}\,\mathbf{x}^{(m)} \tag{41}$$

$$\dot{\mathbf{U}}^{(j)} = \mathbf{N}\,\dot{\mathbf{U}}^{(n)} \tag{42}$$

$$\dot{\mathbf{P}}^{(j)} = \mathbf{N}\,\dot{\mathbf{P}}^{(n)} \tag{43}$$

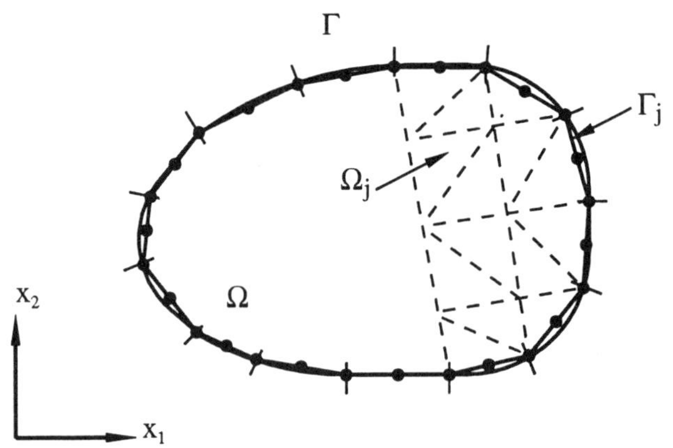

Figure 4. Discretization in boundary elements and internal cells.

where $\mathbf{x}^{(j)}$, $\dot{\mathbf{U}}^{(j)}$ and $\dot{\mathbf{P}}^{(j)}$ are vectors that contain the coordinates, the displacement rates and the surface traction rates, respectively, related to an arbitrary point within the considered element Γ_j; $\mathbf{N}$ and $\mathbf{M}$ are matrices that contain the respective interpolation functions; and $\mathbf{x}^{(m)}$, $\dot{\mathbf{U}}^{(n)}$ and $\dot{\mathbf{P}}^{(n)}$ are vectors containing the coordinates, the displacement rates and the surface traction rates, respectively, related to the nodal points of the element.

For the discretization of the domain integrals that contain the plastic moments, it is considered that the coordinates of an arbitrary interior point within the cell Ω_j are calculated by equation:

$$\mathbf{x}^{(j)} = \overline{\mathbf{M}}\,\mathbf{x}^{(m)} \tag{44}$$

where $\mathbf{x}^{(j)}$ is the vector that contains the cartesian coordinates of an arbitrary point within the cell, $\overline{\mathbf{M}}$ is the matrix containing the interpolation functions and $\mathbf{x}^{(m)}$ is the vector containing the cartesian coordinates of the points that define the cell geometry.

Plastic moments at an arbitrary point within the cell are calculated by equation:

$$\dot{\mathbf{M}}^{a(j)} = \overline{\mathbf{N}}\,\dot{\mathbf{M}}^{a(n)} \tag{45}$$

where $\dot{\mathbf{M}}^{a(j)}$ represents the vector containing the plastic moment rates of an interior point within the cell, $\overline{\mathbf{N}}$ is the matrix that contains the interpolation functions and $\dot{\mathbf{M}}^{a(n)}$ is the vector that contains the plastic moment rates at a certain number of "moment points".

Eqn (31) can then be written in discretized form, for each nodal point ξ_i at the boundary, using matrix notation, as:

$$\begin{aligned}\mathbf{C}_i\dot{\mathbf{U}}_i = {} & \sum_{j=1}^{L}\left(\int_{\Gamma_j}\mathbf{U}_i^*\mathbf{N}\,d\Gamma\right)\dot{\mathbf{P}}^{(n)} - \sum_{j=1}^{L}\left(\int_{\Gamma_j}\mathbf{P}_i^*\mathbf{N}\,d\Gamma\right)\dot{\mathbf{U}}^{(n)} \\ & + \sum_{j=1}^{L}\left(\dot{q}\int_{\Gamma_j}\mathbf{S}_i^*\,d\Gamma\right) + \sum_{j=1}^{Z}\left(\int_{\Omega_j}\boldsymbol{\chi}_i^*\overline{\mathbf{N}}\,d\Omega\right)\dot{\mathbf{M}}^{a(n)}\end{aligned} \tag{46}$$

with L being the number of boundary elements and Z the number of internal cells. It should be pointed out that the summation convention is not implicit now.

It can be observed that eqn (46) also holds for internal points and, in this case, $\mathbf{C}_i = \mathbf{I}$.

By applying eqn (46) to all nodal points, one obtains a system of equations with the following form:

$$\mathbf{H}\dot{\mathbf{U}} = \mathbf{G}\dot{\mathbf{P}} + \dot{\mathbf{B}} + \mathbf{T}\dot{\mathbf{M}}^a \tag{47}$$

where $\dot{\mathbf{U}}$ is the nodal displacement rates vector, $\dot{\mathbf{P}}$ is the nodal traction rates vector, $\dot{\mathbf{B}}$ is the vector that contains the influence of the distributed load rate,

$\dot{\mathbf{M}}^a$ is the vector containing the plastic moment rates at the cell points, $\mathbf{H}$ and $\mathbf{G}$ are square matrices generated by the boundary integrals and $\mathbf{T}$ is the matrix formed by the internal cell integrals.

Moments and shear forces at internal points are evaluated with eqns (39) and (40), in discretized form. This yields, for each moment point ξ_i, using matrix notation:

$$\begin{aligned}\dot{\mathbf{M}}_i = &\sum_{j=1}^{L}\left(\int_{\Gamma_j}\mathbf{U}_i^{*\prime}\mathbf{N}\,\mathrm{d}\Gamma\right)\dot{\mathbf{P}}^{(n)} - \sum_{j=1}^{L}\left(\int_{\Gamma_j}\mathbf{P}_i^{*\prime}\mathbf{N}\,\mathrm{d}\Gamma\right)\dot{\mathbf{U}}^{(n)}\\ &+\sum_{j=1}^{L}\left(\dot{q}\int_{\Gamma_j}\mathbf{W}_i^{*\prime}\,\mathrm{d}\Gamma\right)+\sum_{j=1}^{Z}\left(\int_{\Omega_j}\chi_i^{*\prime}\overline{\mathbf{N}}\,\mathrm{d}\Omega\right)\dot{\mathbf{M}}^{a(n)}\\ &+\frac{\nu\dot{q}}{(1-\nu)\lambda^2}\,\delta_{\alpha\beta}+\mathbf{E}_i^{\prime}\dot{\mathbf{M}}_i^{a}\end{aligned} \tag{48}$$

and

$$\begin{aligned}\dot{\mathbf{Q}}_i = &\sum_{j=1}^{L}\left(\int_{\Gamma_j}\mathbf{U}_i^{*\prime\prime}\mathbf{N}\,\mathrm{d}\Gamma\right)\dot{\mathbf{P}}^{(n)} - \sum_{j=1}^{L}\left(\int_{\Gamma_j}\mathbf{P}_i^{*\prime\prime}\mathbf{N}\,\mathrm{d}\Gamma\right)\dot{\mathbf{U}}^{(n)}\\ &+\sum_{j=1}^{L}\left(\dot{q}\int_{\Gamma_j}\mathbf{W}_i^{*\prime\prime}\,\mathrm{d}\Gamma\right)+\sum_{j=1}^{Z}\left(\int_{\Omega_j}\chi_i^{*\prime\prime}\overline{\mathbf{N}}\,\mathrm{d}\Omega\right)\dot{\mathbf{M}}^{a(n)}\end{aligned} \tag{49}$$

By applying eqns (48) and (49) to all moment internal points, one obtains:

$$\dot{\mathbf{M}} = \mathbf{G}'\dot{\mathbf{P}} - \mathbf{H}'\dot{\mathbf{U}} + (\dot{\mathbf{W}}' + \dot{\mathbf{V}}') + (\mathbf{T}' + \mathbf{E}')\dot{\mathbf{M}}^a \tag{50}$$

$$\dot{\mathbf{Q}} = \mathbf{G}''\dot{\mathbf{P}} - \mathbf{H}''\dot{\mathbf{U}} + \dot{\mathbf{W}}'' + \mathbf{T}''\dot{\mathbf{M}}^a \tag{51}$$

where matrices $\mathbf{G}'$, $\mathbf{G}''$, $\mathbf{H}'$ and $\mathbf{H}''$ and vectors $\dot{\mathbf{W}}'$ and $\dot{\mathbf{W}}''$ contain the boundary integrals related to the fundamental solution; $\dot{\mathbf{V}}'$ contains the free term related to the transverse load; $\mathbf{T}'$ and $\mathbf{T}''$ contain the domain integrals that multiply the plastic moments; and $\mathbf{E}'$ represents the free term related to the plastic moments.

After considering the boundary conditions, the system of eqn (47) can be reordered to the form:

$$\mathbf{A}\dot{\mathbf{y}} = \dot{\mathbf{f}} + \mathbf{T}\dot{\mathbf{M}}^a \tag{52}$$

and, analogously, eqns (50) and (51) become, respectively:

$$\dot{\mathbf{M}} = -\mathbf{A}'\dot{\mathbf{y}} + \dot{\mathbf{f}}' + \mathbf{T}^*\dot{\mathbf{M}}^a \tag{53}$$

and

$$\dot{\mathbf{Q}} = -\mathbf{A}''\dot{\mathbf{y}} + \dot{\mathbf{f}}'' + \mathbf{T}''\dot{\mathbf{M}}^a \tag{54}$$

where $\dot{\mathbf{f}}$, $\dot{\mathbf{f}}'$ and $\dot{\mathbf{f}}''$ are vectors that contain the prescribed values, including the influence of the transverse load, and

$$\mathbf{T}^* = \mathbf{T}' + \mathbf{E}' \tag{55}$$

The pre-multiplication of eqn (52) by $\mathbf{A}^{-1}$ leads to:

$$\dot{\mathbf{y}} = \mathbf{R}\dot{\mathbf{M}}^a + \dot{\mathbf{m}} \tag{56}$$

where

$$\mathbf{R} = \mathbf{A}^{-1}\mathbf{T} \tag{57}$$

$$\dot{\mathbf{m}} = \mathbf{A}^{-1}\dot{\mathbf{f}} \tag{58}$$

The substitution of eqn (56) into eqns (53) and (54) yields:

$$\dot{\mathbf{M}} = \mathbf{S}'\dot{\mathbf{M}}^a + \dot{\mathbf{n}}' \tag{59}$$

$$\dot{\mathbf{Q}} = \mathbf{S}''\dot{\mathbf{M}}^a + \dot{\mathbf{n}}'' \tag{60}$$

with

$$\mathbf{S}' = \mathbf{T}^* - \mathbf{A}'\mathbf{R} \tag{61}$$

$$\mathbf{S}'' = \mathbf{T}'' - \mathbf{A}''\mathbf{R} \tag{62}$$

$$\dot{\mathbf{n}}' = \dot{\mathbf{f}}' - \mathbf{A}'\dot{\mathbf{m}} \tag{63}$$

$$\dot{\mathbf{n}}'' = \dot{\mathbf{f}}'' - \mathbf{A}''\dot{\mathbf{m}} \tag{64}$$

It should be pointed out that the vectors $\dot{\mathbf{m}}$, $\dot{\mathbf{n}}'$ and $\dot{\mathbf{n}}''$ represent the elastic solution to the problem.

5.1 Boundary elements

The continuous and discontinuous quadratic boundary element used are shown in Figures 5 and 6, respectively.

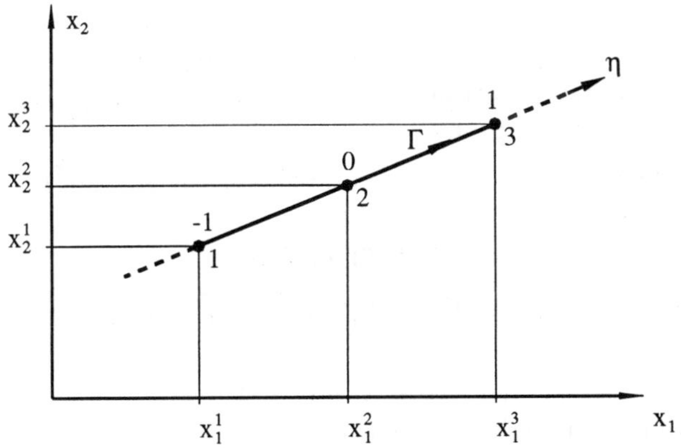

Figure 5. Continuous quadratic element.

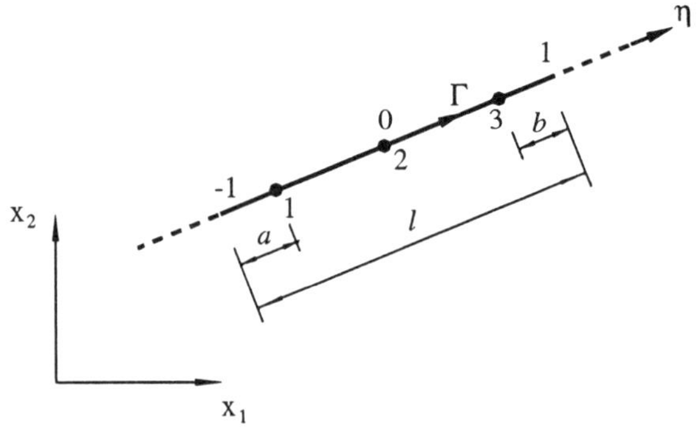

Figure 6. Discontinuous quadratic element.

The interpolation functions for coordinates, in the case of continuous elements with linear geometry, are:

$$M_1 = \frac{1}{2}(1-\eta) \tag{65a}$$

$$M_3 = \frac{1}{2}(1+\eta) \tag{65b}$$

where η is an intrinsic coordinate defined in the interval $[-1,1]$. In the case of discontinuous elements, also with linear geometry, the interpolation functions can be considered as:

$$M_1 = \frac{\ell(1-\eta)-2b}{2(\ell-a-b)} \tag{66a}$$

$$M_3 = \frac{\ell(1+\eta)-2a}{2(\ell-a-b)} \tag{66b}$$

where a and b are the distances from points 1 and 3 (see Figure 6), respectively, to the extremes of the element; and ℓ is the element length.

The coordinates of any point within the element are calculated by eqn (41), in which, in this case:

$$\mathbf{x}^{(j)} = [x_1 \; x_2]^{\mathrm{T}} \tag{67}$$

$$\mathbf{M} = [\mathbf{I}\, M_1 \quad \mathbf{I}\, M_3] \tag{68}$$

$$\mathbf{x}^{(m)} = [x_1^1 \; x_2^1 \; x_1^3 \; x_2^3]^{\mathrm{T}} \tag{69}$$

where $\mathbf{I}$ is the identity matrix of order two.

Since the interpolation functions are defined in terms of η, it is necessary to transform $\mathrm{d}\Gamma$ that appears in the boundary integrals of eqns (46), (48) and (49) to this intrinsic system of coordinates. This is performed using:

$$\mathrm{d}\Gamma = |\mathbf{J}|\mathrm{d}\eta \tag{70}$$

where the jacobian $|\mathbf{J}|$ is, in this case:

$$|\mathbf{J}| = \frac{\mathrm{d}\Gamma}{\mathrm{d}\eta} = \frac{\ell}{2} \tag{71}$$

The interpolation functions for displacements and tractions, in the case of continuous quadratic elments, are:

$$N_1 = \frac{1}{2}\eta(\eta - 1) \tag{72a}$$

$$N_2 = (1 - \eta)(1 + \eta) \tag{72b}$$

$$N_3 = \frac{1}{2}\eta(\eta + 1) \tag{72c}$$

and, in the case of discontinuous quadratic elements, can be considered as:

$$N_1 = \frac{\ell\eta(\ell\eta - \ell + 2b)}{2(\ell - a - b)(\ell - 2a)} \tag{73a}$$

$$N_2 = \frac{\ell\eta[2(a - b) - \ell\eta]}{(\ell - 2a)(\ell - 2b)} + 1 \tag{73b}$$

$$N_3 = \frac{\ell\eta(\ell\eta + \ell - 2a)}{2(\ell - a - b)(\ell - 2b)} \tag{73c}$$

The displacements and tractions are interpolated using eqns (42) and (43), respectively, where:

$$\mathbf{U}^{(j)} = [u_1\ u_2\ u_3]^{\mathrm{T}} \tag{74}$$

$$\mathbf{P}^{(j)} = [p_1\ p_2\ p_3]^{\mathrm{T}} \tag{75}$$

$$\mathbf{N} = [\mathbf{I}N_1\quad \mathbf{I}N_2\quad \mathbf{I}N_3] \tag{76}$$

$$\mathbf{U}^{(n)} = [u_1^1\ u_2^1\ u_3^1\ u_1^2\ u_2^2\ u_3^2\ u_1^3\ u_2^3\ u_3^3]^{\mathrm{T}} \tag{77}$$

$$\mathbf{P}^{(n)} = [p_1^1\ p_2^1\ p_3^1\ p_1^2\ p_2^2\ p_3^2\ p_1^3\ p_2^3\ p_3^3]^{\mathrm{T}} \tag{78}$$

with $\mathbf{I}$ being the identity matrix of order three.

Continuous elements can always be employed in cases where there are normal continuity between adjacent elements.

Discontinuous elements are employed in this work in cases of normal or boundary conditions discontinuities, when a same component of the boundary traction is discontinuous and unknown in two adjacent elements. When this fact does not occur, but there is normal discontinuity, continuous elements with double nodes between the adjacent elements can be used. It is worth mentioning that these elements can also be semicontinuous, i.e. continuous at one extreme and discontinuous at the other. In this case, one has $a = 0$ or $b = 0$ (see Figure 6) in eqns (73).

The integrals over the boundary elements are evaluated numerically, using the Gaussian quadrature, in selective form. For singular integrals, which occur when the field and the source points are situated over the same elements, special procedures are used.

In the case of matrix $\mathbf{G}$ and vector $\dot{\mathbf{B}}$ in eqn (47), the singularities are of order $\ln r$ and a coordinate transformation of the third degree,[22] involving the coordinates of the integration points, is used. This transformation produces a jacobian that eliminates the singularity at the considered point.

Matrix $\mathbf{H}$ in eqn (47) has singularities of order $\ln r$ and r^{-1}, but the corresponding elements can be calculated indirectly, by considering that, for rigid body motions, $\mathbf{HU} = \mathbf{0}$ and non trivial solutions for this equation can be expressed by $(1, 0, x_1(\xi) - x_1(x))$, $(0, 1, x_2(\xi) - x_2(x))$ and $(0, 0, 1)$, that correspond to three rigid body motions.

In addition, to solve the quasi-singular boundary integrals which occur in the evaluation of moments and shear forces at internal points, when these points are situated very near the boundary, the same third degree transformation mentioned above is used.

5.2 Internal cells

The internal cells employed in this work have triangular form, as can be seen in Figure 7, where the intrinsic coordinate system (ζ_1, ζ_2) is also represented.

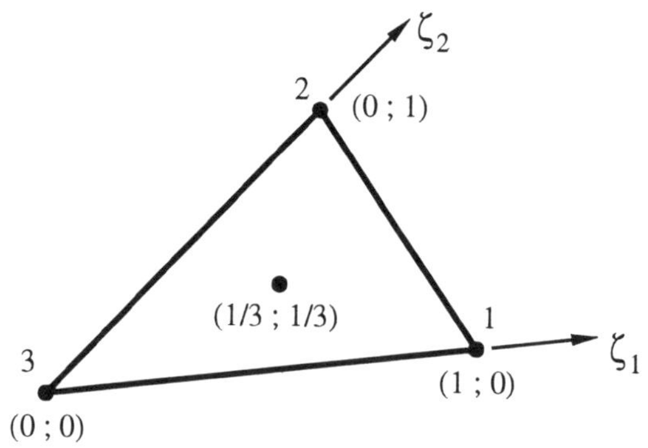

Figure 7. Triangular cell and intrinsic coordinate system (ζ_1, ζ_2).

The coordinates of any point within the cell are calculated by expression (44), where the interpolation functions matrix is represented by:

$$\overline{\mathbf{M}} = [\mathbf{I}\zeta_1 \; \mathbf{I}\zeta_2 \; \mathbf{I}\zeta_3] \tag{79}$$

in which $\mathbf{I}$ is the identity matrix of order two and $\zeta_3 = 1 - \zeta_1 - \zeta_2$. In addition, still in eqn (44), one has

$$\mathbf{x}^{(m)} = [x_1^1 \; x_2^1 \; x_1^2 \; x_2^2 \; x_1^3 \; x_2^3]^{\mathrm{T}} \tag{80}$$

which contains the coordinates x_1 and x_2 of each vertex of the triangular cell.

If A is the area of the triangular cell, the jacobian of the transformation is:

$$|\overline{\mathbf{J}}| = 2A \tag{81}$$

Plastic moment rates at any point within the cell are calculated from eqn (45). For constant cells $\overline{\mathbf{N}} = \mathbf{I}$, where $\mathbf{I}$ is the identity matrix of order three and

$$\dot{\mathbf{M}}^{a(n)} = [\dot{M}_{11}^a \; \dot{M}_{12}^a \; \dot{M}_{22}^a]^{\mathrm{T}} \tag{82}$$

in which the inelastic moments in a point located at the centre of gravity of the cell are represented.

The intrinsic coordinates ζ_α can be written in terms of cartesian coordinates x_1 and x_2, in the form:

$$\zeta_\alpha = \frac{1}{2A}\left(2A_\alpha^0 + b_\alpha x_1 + a_\alpha x_2\right) \tag{83}$$

in which α is the point to which the function is referred to, and

$$a_\alpha = x_1^\gamma - x_1^\beta \tag{84}$$

$$b_\alpha = x_2^\beta - x_2^\gamma \tag{85}$$

$$2A_\alpha^0 = x_1^\beta x_2^\gamma - x_1^\gamma x_2^\beta \tag{86}$$

$$A = \frac{1}{2}\left(b_1 a_2 - b_2 a_1\right) \tag{87}$$

with $\alpha = 1, 2, 3$ for $\beta = 2, 3, 1$ and $\gamma = 3, 1, 2$.

For the calculation of matrix $\mathbf{T}$, each cell contributes with a (3×3) matrix of the form:

$$\mathbf{t} = \int_{\Omega_j} \boldsymbol{\chi}_i^* \, \mathrm{d}\Omega \tag{88}$$

where

$$\boldsymbol{\chi}_i^* = \begin{bmatrix} {}_1\boldsymbol{\chi}_i^* \\ \\ {}_2\boldsymbol{\chi}_i^* \end{bmatrix} = \begin{bmatrix} \chi_{111}^* & 2\,\chi_{121}^* & \chi_{221}^* \\ \chi_{112}^* & 2\,\chi_{122}^* & \chi_{222}^* \\ \\ \chi_{113}^* & 2\,\chi_{123}^* & \chi_{223}^* \end{bmatrix} \tag{89}$$

For the calculation of matrices $\mathbf{T}'$ and $\mathbf{T}''$, each cell contributes with a (5×3) matrix of the form (considering the two matrices together):

$$\overline{\mathbf{t}} = \int_{\Omega_j} \overline{\chi_i^*} \, \mathrm{d}\Omega \tag{90}$$

where

$$\overline{\chi_i^*} = \begin{bmatrix} \chi_i^{*'} \\ \\ \chi_i^{*''} \end{bmatrix} = \begin{bmatrix} \chi_{1111}^* & 2\,\chi_{1112}^* & \chi_{1122}^* \\ \chi_{1211}^* & 2\,\chi_{1212}^* & \chi_{1222}^* \\ \chi_{2211}^* & 2\,\chi_{2212}^* & \chi_{2222}^* \\ \\ \chi_{3111}^* & 2\,\chi_{3112}^* & \chi_{3122}^* \\ \chi_{3211}^* & 2\,\chi_{3212}^* & \chi_{3222}^* \end{bmatrix} \tag{91}$$

Due to the singularity that occurs when the source point is coincident with a cell point, it becomes more convenient to define a polar coordinate system (r, ϕ) centred at the source point γ. In this case:

$$d\Omega = r \; dr \; d\phi \tag{92}$$

$$x_1(x) = x_1^\gamma + r \; \cos \, \phi \tag{93}$$

$$x_2(x) = x_2^\gamma + r \sin \phi \tag{94}$$

By considering eqns (93) and (94), the expressions corresponding to tensor χ_i^* can be written as:

$${}_1\chi_i^*(\xi, x) = \frac{1}{r} \; {}_1\Lambda(\xi, x) \tag{95}$$

$${}_2\chi_i^*(\xi, x) = \; {}_2\Lambda(\xi, x) \tag{96}$$

and the expressions corresponding to tensor $\overline{\chi_i^*}$, as:

$$\chi_i^{*'}(\xi, x) = \frac{1}{r^2} \; \Lambda'(\xi, x) \tag{97}$$

$$\chi_i^{*''}(\xi, x) = \frac{1}{r} \; \Lambda''(\xi, x). \tag{98}$$

In the case of matrix $\mathbf{t}$, when the singular point γ is located at one of the cell vertices, as depicted in Figure 8, the corresponding submatrices ${}_1\mathbf{t}$ and ${}_2\mathbf{t}$ can be represented in the form:

$$\begin{aligned} {}_1\mathbf{t} &= \lim_{\varepsilon\to0} \left(\int_{\phi_1}^{\phi_2} \int_{\varepsilon}^{R(\phi)} \frac{1}{r} \; {}_1\Lambda \; r \; \mathrm{d}r \; \mathrm{d}\phi \right) \\ &= \lim_{\varepsilon\to0} \left(\int_{\phi_1}^{\phi_2} \int_{\varepsilon}^{R(\phi)} {}_1\Lambda \; \mathrm{d}r \; \mathrm{d}\phi \right) \end{aligned} \tag{99}$$

$${}_2\mathbf{t} = \lim_{\varepsilon\to0} \left(\int_{\phi_1}^{\phi_2} \int_{\varepsilon}^{R(\phi)} {}_2\Lambda \; r \; \mathrm{d}r \; \mathrm{d}\phi \right) \tag{100}$$

where

$$R(\phi) = \frac{-2A}{b_\gamma \cos\ \phi + a_\gamma \sin\ \phi} \tag{101}$$

with

$$\cos\ \phi = \frac{\partial r}{\partial x_1} \tag{102}$$

$$\sin\ \phi = \frac{\partial r}{\partial x_2} \tag{103}$$

It can be observed that, in this case, the singularity is eliminated and integration can be evaluated using the Gaussian quadrature with respect to both r and ϕ. The variable ϕ is then expressed as:

$$\phi = \frac{\eta}{2}\,(\phi_2 - \phi_1) + \frac{1}{2}\,(\phi_1 + \phi_2) \tag{104}$$

and the variable r as:

$$r = \frac{R(\phi)}{2}\,\eta \tag{105}$$

where η is an intrinsic coordinate defined in the interval $[-1, 1]$.

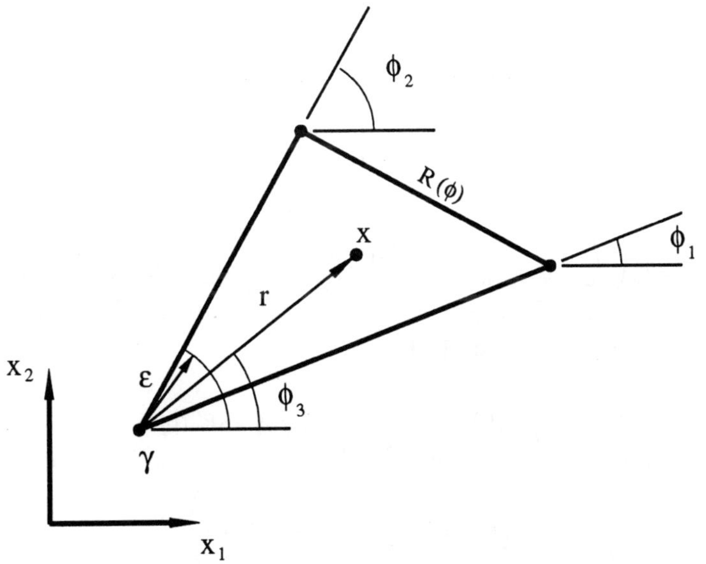

Figure 8. Triangular cell with singular point γ coincident with one of the triangular vertices.

The jacobians of these transformations are:

$$\frac{\mathrm{d}\phi}{\mathrm{d}\eta} = \frac{\phi_2 - \phi_1}{2} \tag{106}$$

$$\frac{\mathrm{d}r}{\mathrm{d}\eta} = \frac{R(\phi)}{2} \tag{107}$$

For matrix $\mathbf{t}$, when the singular point γ is situated at any point in one of the cell sides, as shown in Figure 9, one can consider the cell divided into two parts, integrate separately over each part and add the corresponding results. Then:

$${}_1\mathbf{t} = \lim_{\varepsilon \to 0} \left(\int_{\phi_1}^{\phi_2} \int_{\varepsilon}^{R'_\gamma(\phi)} {}_1\boldsymbol{\Lambda} \, \mathrm{d}r \, \mathrm{d}\phi + \int_{\phi_2}^{\phi_3} \int_{\varepsilon}^{R''_\gamma(\phi)} {}_1\boldsymbol{\Lambda} \, \mathrm{d}r \, \mathrm{d}\phi \right) \tag{108}$$

$${}_2\mathbf{t} = \lim_{\varepsilon \to 0} \left(\int_{\phi_1}^{\phi_2} \int_{\varepsilon}^{R'_\gamma(\phi)} {}_2\boldsymbol{\Lambda} \, r \, \mathrm{d}r \, \mathrm{d}\phi + \int_{\phi_2}^{\phi_3} \int_{\varepsilon}^{R''_\gamma(\phi)} {}_2\boldsymbol{\Lambda} \, r \, \mathrm{d}r \, \mathrm{d}\phi \right) \tag{109}$$

in which

$$R'_\gamma(\phi) = \frac{-2A'}{b'_\gamma \cos\,\phi + a'_\gamma \sin\,\phi} \tag{110}$$

$$R''_\gamma(\phi) = \frac{-2A''}{b''_\gamma \cos\,\phi + a''_\gamma \sin\,\phi} \tag{111}$$

In this case, the integration can also be fully evaluated numerically with Gaussian quadrature.

In case of matrix $\bar{\mathbf{t}}$, when the singular point γ coincide with the cell point where the integration is being evaluated, as depicted in Figure 10, one has to divide the cell into three parts. This leads to:

$$\mathbf{t}' = \lim_{\varepsilon \to 0} \left(\int_{\phi_1}^{\phi_2} \int_{\varepsilon}^{R'_\gamma(\phi)} \frac{1}{r} \boldsymbol{\Lambda}' \, \mathrm{d}r \, \mathrm{d}\phi + \int_{\phi_2}^{\phi_3} \int_{\varepsilon}^{R''_\gamma(\phi)} \frac{1}{r} \boldsymbol{\Lambda}' \, \mathrm{d}r \, \mathrm{d}\phi \right.$$
$$\left. + \int_{\phi_3}^{\phi_1} \int_{\varepsilon}^{R'''_\gamma(\phi)} \frac{1}{r} \boldsymbol{\Lambda}' \, \mathrm{d}r \, \mathrm{d}\phi \right) \tag{112}$$

$$\mathbf{t}'' = \lim_{\varepsilon \to 0} \left(\int_{\phi_1}^{\phi_2} \int_{\varepsilon}^{R'_\gamma(\phi)} \boldsymbol{\Lambda}'' \, \mathrm{d}r \, \mathrm{d}\phi + \int_{\phi_2}^{\phi_3} \int_{\varepsilon}^{R''_\gamma(\phi)} \boldsymbol{\Lambda}'' \, \mathrm{d}r \, \mathrm{d}\phi \right.$$
$$\left. + \int_{\phi_3}^{\phi_1} \int_{\varepsilon}^{R'''_\gamma(\phi)} \boldsymbol{\Lambda}'' \, \mathrm{d}r \, \mathrm{d}\phi \right) \tag{113}$$

with R'_γ, R''_γ and R'''_γ calculated analogously to the previous case.

All integrations required for matrix $\mathbf{t}''$, are regular and Gaussian quadrature is employed. As for matrix $\mathbf{t}'$, a mix of Gaussian and Kutt's[23] finite part numerical integration has been adopted for ϕ and r respectively.

For both matrices, $\mathbf{t}$ and $\bar{\mathbf{t}}$, in the general case when the singular point γ is not situated in the cell (see Figure 11), the coordinate transformation to the system (r, ϕ) can also be employed, yielding:

$$_1\mathbf{t} = \int_{\phi_1}^{\phi_3} \int_{R_2(\phi)}^{R_3(\phi)} {}_1\Lambda \; dr \; d\phi + \int_{\phi_3}^{\phi_2} \int_{R_1(\phi)}^{R_3(\phi)} {}_1\Lambda \; dr \; d\phi \tag{114}$$

$$_2\mathbf{t} = \int_{\phi_1}^{\phi_3} \int_{R_2(\phi)}^{R_3(\phi)} {}_2\Lambda \; r \; dr \; d\phi + \int_{\phi_3}^{\phi_2} \int_{R_1(\phi)}^{R_3(\phi)} {}_2\Lambda \; r \; dr \; d\phi \tag{115}$$

$$\mathbf{t}' = \int_{\phi_1}^{\phi_3} \int_{R_2(\phi)}^{R_3(\phi)} \frac{1}{r} \Lambda' \; dr \; d\phi + \int_{\phi_3}^{\phi_2} \int_{R_1(\phi)}^{R_3(\phi)} \frac{1}{r} \Lambda' \; dr \; d\phi \tag{116}$$

$$\mathbf{t}'' = \int_{\phi_1}^{\phi_3} \int_{R_2(\phi)}^{R_3(\phi)} \Lambda'' \; dr \; d\phi + \int_{\phi_3}^{\phi_2} \int_{R_1(\phi)}^{R_3(\phi)} \Lambda'' \; dr \; d\phi \tag{117}$$

where

$$R_\alpha(\phi) = \frac{-2A \; {}^\gamma\zeta_\alpha}{b_\alpha \cos \phi + a_\alpha \sin \phi} \tag{118}$$

with ${}^\gamma\zeta_\alpha$ being the value of the interpolation function at the source point.

The integration can then be performed numerically with respect to both r and ϕ, using Gaussian quadrature.

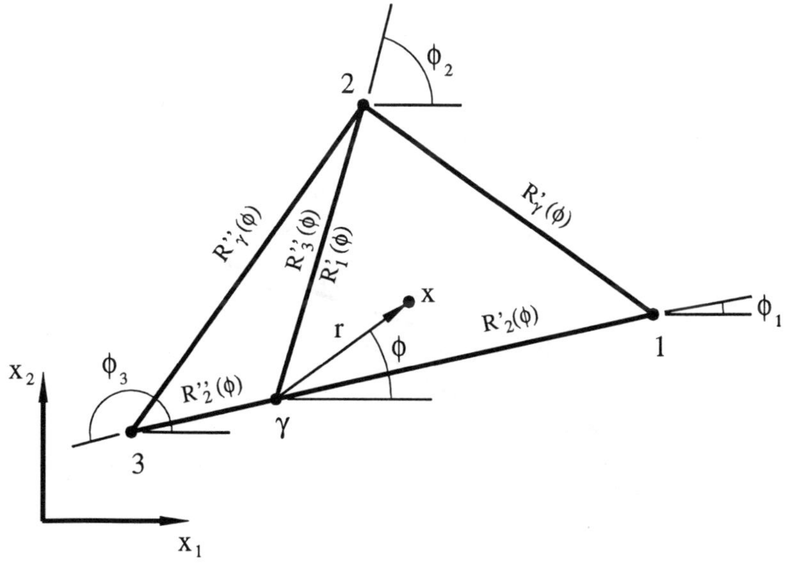

Figure 9. Triangular cell with singular point γ coincident with one of the triangular vertices.

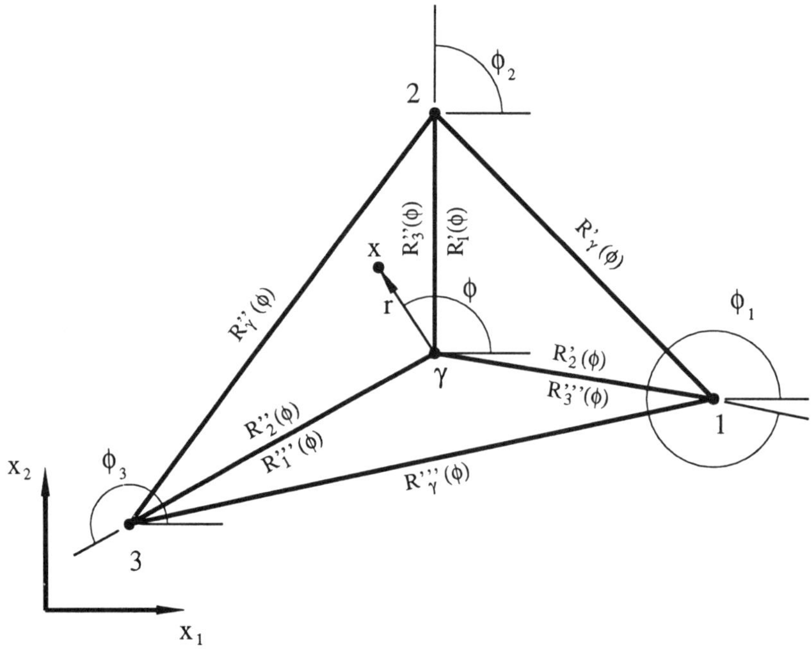

Figure 10. Triangular cell with singular point γ coincident with the cell internal point.

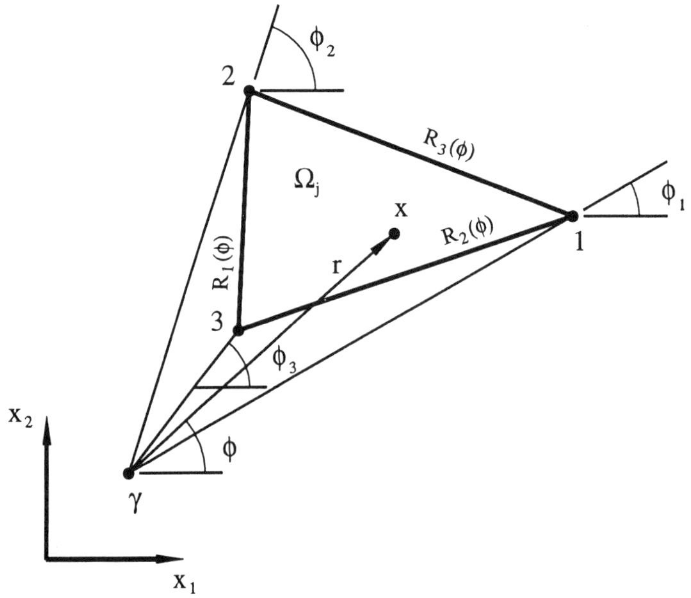

Figure 11. Triangular cell with source point γ out of the cell.

6 Elastoplastic constitutive equations

In elastoplastic analysis, it is admitted the existence of a yield function Φ, which is expressed in terms of the moments $M_{\alpha\beta}$ and of a hardening parameter k. During the loading that produces yielding, the moments $M_{\alpha\beta}$ must remain at the yield surface, so that the following equation be satisfied:

$$\Phi(M_{\alpha\beta}, k) = f(M_{\alpha\beta}) - \Psi(k) = M_e - M_0 = 0 \tag{119}$$

where M_e is the equivalent moment calculated, in this work, using either von Mises or Tresca yield criteria, and M_0 is the uniaxial yield moment.

In von Mises yield criterion, M_e is calculated as:

$$M_e = \sqrt{3J_2'} \tag{120}$$

and, in Tresca yield criterion, as:

$$M_e = 2\cos\theta\sqrt{J_2'} \tag{121}$$

where θ is similar to the Lode parameter, defined in Ref. 19, and expressed by:

$$\theta = \frac{1}{3}\text{arc}\sin\left(-\frac{3\sqrt{3}}{2}\,\frac{J_3'}{J_2'\sqrt{J_2'}}\right) \tag{122}$$

In addition, J_2' and J_3' are analogous to the second and third invariants of the deviatoric stresses,[19–21] written here in terms of the "deviatoric moments" M_{ij}', as:

$$J_2' = \frac{1}{2}\,M_{ij}'M_{ij}' \tag{123}$$

$$J_3' = \frac{1}{3}\,M_{ij}'M_{jk}'M_{ki}' \tag{124}$$

with

$$M_{ij}' = M_{ij} - \frac{1}{3}\,\delta_{ij}M_{kk} \tag{125}$$

The uniaxial yield moment is calculated by:

$$M_0 = \frac{\sigma_0\,h^2}{4} \tag{126}$$

in which σ_0 is the uniaxial yield stress of the material.

It is considered here that, whenever the equivalent moment, at any point, reaches the yield moment M_0, the whole cross section plastifies simultaneously.

In addition, an equivalent plastic strain χ_e^p can be defined, so that:

$$M_e\, \mathrm{d}\chi_e^p = M_{\alpha\beta}\, \mathrm{d}\chi_{\alpha\beta}^p = \mathrm{d}k \tag{127}$$

In order to determine moment-strain relations for the behaviour after yielding, let eqn (8) be written in incremental form as:

$$\mathrm{d}M_{\alpha\beta} = C_{\alpha\beta\gamma\theta}\left(\mathrm{d}\chi_{\gamma\theta} - \mathrm{d}\chi_{\gamma\theta}^p\right) + \frac{\nu\, \mathrm{d}q}{(1-\nu)\lambda^2}\, \delta_{\alpha\beta} \tag{128}$$

where $C_{\alpha\beta\gamma\theta}$ represents the components of the fourth order isotropic tensor of elastic constants. From the normality principle,[19] one has:

$$\mathrm{d}\chi_{\alpha\beta}^p = \mathrm{d}\lambda\, \frac{\partial\Phi}{\partial M_{\alpha\beta}} \tag{129}$$

where $\mathrm{d}\lambda$ is a proportionality factor called plastic multiplier.

Substituting eqn (129) into expression (128) results in:

$$\mathrm{d}M_{\alpha\beta} = C_{\alpha\beta\gamma\theta}\left(\mathrm{d}\chi_{\gamma\theta} - a_{\gamma\theta}\, \mathrm{d}\lambda\right) + \frac{\nu\, \mathrm{d}q}{(1-\nu)\lambda^2}\, \delta_{\alpha\beta} \tag{130}$$

with

$$a_{\gamma\theta} = \frac{\partial\Phi}{\partial M_{\gamma\theta}} = \frac{\partial f}{\partial M_{\gamma\theta}} \tag{131}$$

When plastic strain happens, the moments satisfy eqn (123). By differentiating this equation, one obtains:

$$\mathrm{d}\Phi = a_{\alpha\beta}\, \mathrm{d}M_{\alpha\beta} - \frac{\mathrm{d}\Psi}{\mathrm{d}k}\, \mathrm{d}k = 0 \tag{132}$$

or, considering eqn (127):

$$a_{\alpha\beta}\, \mathrm{d}M_{\alpha\beta} - \frac{\mathrm{d}\Psi}{\mathrm{d}k}\, M_{\alpha\beta}\, \mathrm{d}\chi_{\alpha\beta}^p = 0 \tag{133}$$

By applying the normality principle to eqn (133),

$$a_{\alpha\beta}\, \mathrm{d}M_{\alpha\beta} - \frac{\mathrm{d}\Psi}{\mathrm{d}k}\, M_{\alpha\beta}\, a_{\alpha\beta}\, \mathrm{d}\lambda = 0 \tag{134}$$

Now, substituting $\mathrm{d}M_{\alpha\beta}$ expressed by eqn (130) into eqn (134) and solving for $\mathrm{d}\lambda$, results:

$$\mathrm{d}\lambda = \frac{1}{\gamma'}\left(a_{\alpha\beta}\, C_{\alpha\beta\gamma\theta}\, \mathrm{d}\chi_{\gamma\theta} + a_{\alpha\beta}\, \frac{\nu\, \mathrm{d}q}{(1-\nu)\lambda^2}\, \delta_{\alpha\beta}\right) \tag{135}$$

in which

$$\gamma' = a_{\alpha\beta}\, C_{\alpha\beta\gamma\theta}\, a_{\gamma\theta} + \frac{\mathrm{d}\Psi}{\mathrm{d}k}\, M_{\alpha\beta}\, a_{\alpha\beta} \tag{136}$$

By using Euler's theorem, valid for any homogeneous function of order one, it can be written:

$$M_{\alpha\beta}\,\frac{\partial f}{\partial M_{\alpha\beta}} = f(M_{\alpha\beta}) = M_e \tag{137}$$

The substitution of eqns (127) and (137) into eqn (136) leads to:

$$\gamma' = a_{\alpha\beta}\, C_{\alpha\beta\gamma\theta}\, a_{\gamma\theta} + \frac{\mathrm{d}\Psi}{\mathrm{d}\chi_e^p} \tag{138}$$

where, for the case of Ψ be defined as the uniaxial yield moment, one has:

$$\frac{\mathrm{d}\Psi}{\mathrm{d}\chi_e^p} = H' \tag{139}$$

with H' being the slope of the moment-plastic strain curve.

Now, substituting eqn (135) into eqn (130), the following incremental moment-strain relation arises:

$$\mathrm{d}M_{\alpha\beta} = C^{ep}_{\alpha\beta\gamma\theta}\,\mathrm{d}\chi_{\gamma\theta} + \frac{\nu\,\mathrm{d}q}{(1-\nu)\lambda^2}\left[\delta_{\alpha\beta} - \frac{1}{\gamma'}\, C_{\alpha\beta\mu\rho}\, a_{\mu\rho}\, a_{\eta\zeta}\,\delta_{\eta\zeta}\right] \tag{140}$$

where

$$C^{ep}_{\alpha\beta\gamma\theta} = C_{\alpha\beta\gamma\theta} - \frac{1}{\gamma'}\, C_{\alpha\beta\mu\rho}\, a_{\mu\rho}\, a_{\eta\zeta}\, C_{\eta\zeta\gamma\theta} \tag{141}$$

In order to apply these equations to a formulation that considers an initial moment (initial stress procedure), a modification becomes convenient. Let an incremental fictitious "elastic moment" be written as:

$$\mathrm{d}M^e_{\alpha\beta} = C_{\alpha\beta\gamma\theta}\,\mathrm{d}\chi_{\gamma\theta} + \frac{\nu\,\mathrm{d}q}{(1-\nu)\lambda^2}\,\delta_{\alpha\beta} \tag{142}$$

which represents the incremental moment as if a pure elastic problem were being solved.

By taking into account eqn (142), one can rewrite eqn (140) as:

$$\mathrm{d}M_{\alpha\beta} = \mathrm{d}M^e_{\alpha\beta} - \frac{1}{\gamma'}\, C_{\alpha\beta\mu\rho}\, a_{\mu\rho}\, a_{\eta\zeta}\,\mathrm{d}M^e_{\eta\zeta} \tag{143}$$

which signifies that the true incremental moments can be calculated from the corresponding incremental "elastic moments". In addition, the initial moment increments can also be calculated by:

$$\mathrm{d}M^p_{\alpha\beta} = \mathrm{d}M^e_{\alpha\beta} - \mathrm{d}M_{\alpha\beta} \tag{144}$$

or, substituting eqn (143) into eqn (144):

$$\mathrm{d}M^p_{\alpha\beta} = \frac{1}{\gamma'} C_{\alpha\beta\mu\rho}\, a_{\mu\rho}\, a_{\eta\zeta}\, \mathrm{d}M^e_{\eta\zeta} \tag{145}$$

7 Solution technique for the elastoplastic analysis

By considering that

$$\dot{M}^e_{\alpha\beta} = \dot{M}_{\alpha\beta} + \dot{M}^p_{\alpha\beta} \tag{146}$$

one can apply eqn (50) to calculate $\dot{\mathbf{M}}^e$ if $\mathbf{E}'$ is substituted by

$$\overline{\mathbf{E}} = \mathbf{E}' + \mathbf{I} \tag{147}$$

where $\mathbf{I}$ is the identity matrix. This leads to:

$$\dot{\mathbf{M}}^e = \mathbf{G}'\dot{\mathbf{P}} - \mathbf{H}'\dot{\mathbf{U}} + \left(\dot{\mathbf{W}}' + \dot{\mathbf{V}}'\right) + \mathbf{T}^*\mathbf{M}^p \tag{148}$$

where

$$\mathbf{T}^* = \mathbf{T}' + \overline{\mathbf{E}} \tag{149}$$

In elastoplastic analysis, eqns (47), (148) and (51) can be written, respectively, as:

$$\mathbf{H}\,\mathrm{d}\mathbf{U} = \mathbf{G}\,\mathrm{d}\mathbf{P} + \mathrm{d}\mathbf{B} + \mathbf{T}\,\mathrm{d}\mathbf{M}^p \tag{150}$$

$$\mathrm{d}\mathbf{M}^e = \mathbf{G}'\,\mathrm{d}\mathbf{P} - \mathbf{H}'\,\mathrm{d}\mathbf{U} + (\mathrm{d}\mathbf{W}' + \mathrm{d}\mathbf{V}') + \mathbf{T}^*\,\mathrm{d}\mathbf{M}^p \tag{151}$$

$$\mathrm{d}\mathbf{Q} = \mathbf{G}''\,\mathrm{d}\mathbf{P} - \mathbf{H}''\,\mathrm{d}\mathbf{U} + \mathrm{d}\mathbf{W}'' + \mathbf{T}''\,\mathrm{d}\mathbf{M}^p \tag{152}$$

and these three last equations can be manipulated as in Section 5, resulting in:

$$\mathrm{d}\mathbf{y} = \mathbf{R}\,\mathrm{d}\mathbf{M}^p + \mathrm{d}\mathbf{m} \tag{153}$$

$$\mathrm{d}\mathbf{M}^e = \mathbf{S}'\,\mathrm{d}\mathbf{M}^p + \mathrm{d}\mathbf{n}' \tag{154}$$

$$\mathrm{d}\mathbf{Q} = \mathbf{S}''\,\mathrm{d}\mathbf{M}^p + \mathrm{d}\mathbf{n}'' \tag{155}$$

where $\mathbf{R}$, $\mathbf{S}'$ and $\mathbf{S}''$ are given by the same expressions of Section 5 and $\mathrm{d}\mathbf{m}$, $\mathrm{d}\mathbf{n}'$ and $\mathrm{d}\mathbf{n}''$ have expressions analogous to $\dot{\mathbf{m}}$, $\dot{\mathbf{n}}'$ and $\dot{\mathbf{n}}''$, respectively.

The incremental process begins with the reduction of the maximum equivalent moment calculated at the cell points, represented by $M_e^{\max}$, to the initial yield moment M_0. An initial load factor is, then, evaluated as:

$$\lambda_0 = \frac{M_0}{M_e^{\max}} \tag{156}$$

Subsequent values of the load factor for the incremental process are calculated by the following recursive equation:

$$\lambda_i = \lambda_{i-1} + \Delta\lambda_i \tag{157}$$

where $\Delta\lambda_i$ is the increment defined as a given percentage β_i in terms of the load at first yielding:

$$\Delta\lambda_i = \beta_i \, \lambda_0 \tag{158}$$

In addition, eqns (153), (154) and (155) can be written, for a pure incremental process, as:

$$\Delta\mathbf{y} = \mathbf{R} \, \Delta\mathbf{M}^p + \Delta\lambda_i \, \mathbf{m} \tag{159}$$

$$\Delta\mathbf{M}^e = \mathbf{S}' \, \Delta\mathbf{M}^p + \Delta\lambda_i \, \mathbf{n}' \tag{160}$$

$$\Delta\mathbf{Q} = \mathbf{S}'' \, \Delta\mathbf{M}^p + \Delta\lambda_i \, \mathbf{n}'' \tag{161}$$

where $\mathbf{m}$, $\mathbf{n}'$ and $\mathbf{n}''$ correspond to the application of the total load and $\Delta\mathbf{M}^p$ is the current initial moment increment.

For each value of λ_i, the initial moment increment is calculated iteratively, following a solution technique for an initial stress formulation analogous to that used in Ref. 8. The steps are described in what follows:

(1) Calculate elastic moment increment, by:

$$\Delta\mathbf{M}^e = \mathbf{S}' \, \Delta\mathbf{M}^p + \Delta\lambda_i \, \mathbf{n}' \quad \text{in the first iteration} \tag{162}$$

or

$$\Delta\mathbf{M}^e = \mathbf{S}' \, \Delta\mathbf{M}^p \quad \text{in the following iterations} \tag{163}$$

(2) Calculate true moment increment, by:

$$\Delta M_{\alpha\beta} = \Delta M^e_{\alpha\beta} - \frac{1}{\gamma'} \, C_{\alpha\beta\mu\rho} \, a_{\mu\rho} \, a_{\eta\zeta} \, \Delta M^e_{\eta\zeta} \tag{164}$$

(3) Verify convergence, by comparing $\Delta\chi^p_e$ that was calculated with its acumulated value obtained during the current load increment to see if it can be neglected;

(4) Calculate initial moment increment, by:

$$\Delta M^p_{\alpha\beta} = \Delta M^e_{\alpha\beta} - \Delta M_{\alpha\beta} \tag{165}$$

(5) Accumulate values of initial moments and actual moments:

$$M^p_{\alpha\beta} = M^p_{\alpha\beta} + \Delta M^p_{\alpha\beta} \tag{166}$$

$$M_{\alpha\beta} = M_{\alpha\beta} + \Delta M_{\alpha\beta} \tag{167}$$

(6) Continue with next point and start the process from item (2), until all points have been considered;
(7) Start a new iteration from item (1).

The iterations are carried out until convergence is obtained for the prescribed tolerance.

In order to avoid cumulative errors, the values of $\Delta \mathbf{M}^p$ obtained at the end of iterations are applied together with $\lambda_i \mathbf{n}'$ in eqn (162) for the first iteration of the next load increment.

It can be observed that only the equation related to the moments is used in the incremental-iterative process and also, that matrices $\mathbf{R}$, $\mathbf{S}'$ and $\mathbf{S}''$, as well as vectors $\mathbf{m}$, $\mathbf{n}'$ and $\mathbf{n}''$, are assembled only once, at the beginning of the process.

Finally, it is worth mentioning that the explicit process of solution employed here can be easily substituted by any of the implicit solution techniques presented in Ref. 24.

8 Applications

In order to test the formulation, some examples involving elastoplastic analysis are presented and results are compared with finite elements or analytical solutions.

8.1 Example 1: simply supported square plate

Consider a simply supported square plate, having side $\ell = 1$ and thickness $h = 0.01$, subjected to a uniformly distributed load q. Due to symmetry conditions, only a quarter of the plate is discretized, as shown in Figure 12a.

The same plate was analysed by Owen and Hinton,[25] using the finite element method (FEM), with von Mises yield criterion and the discretization of Figure 12b. It was assumed an ideally plastic material ($H' = 0$) with: $\nu = 0.3$, $E = 10.92$ and $\sigma_0 = 1,600$. The results that were obtained by both methods are presented in Figure 13, where the load-deflection curve is shown for the point situated at the centre of the plate.

8.2 Example 2: simply supported Timoshenko beam

This example consists of a simply supported rectangular beam, subjected to a uniformly distributed load, with: length $l = 3,000$ mm; width $b = 150$ mm and thickness $h = 900$ mm. The problem was analysed with the discretization shown in Figure 14. It should be noticed that, since $h/l = 0.3$, the effects of transverse shear strains are expected.

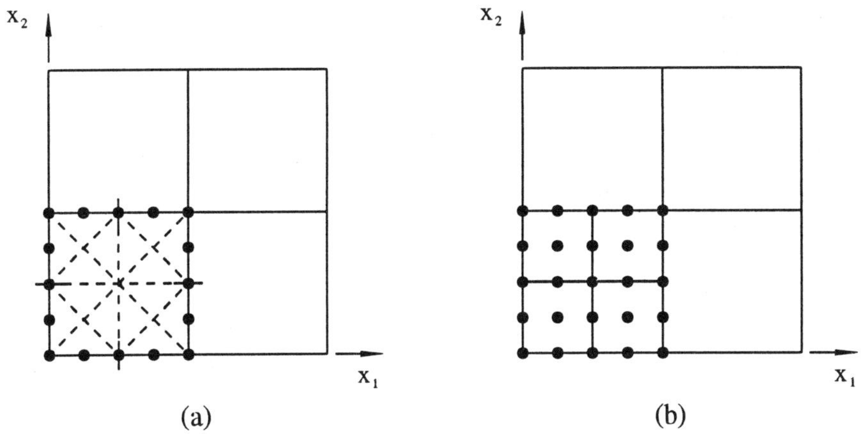

Figure 12. Square plate. (a) Discretization in boundary elements and internal cells; (b) discretization in finite elements.

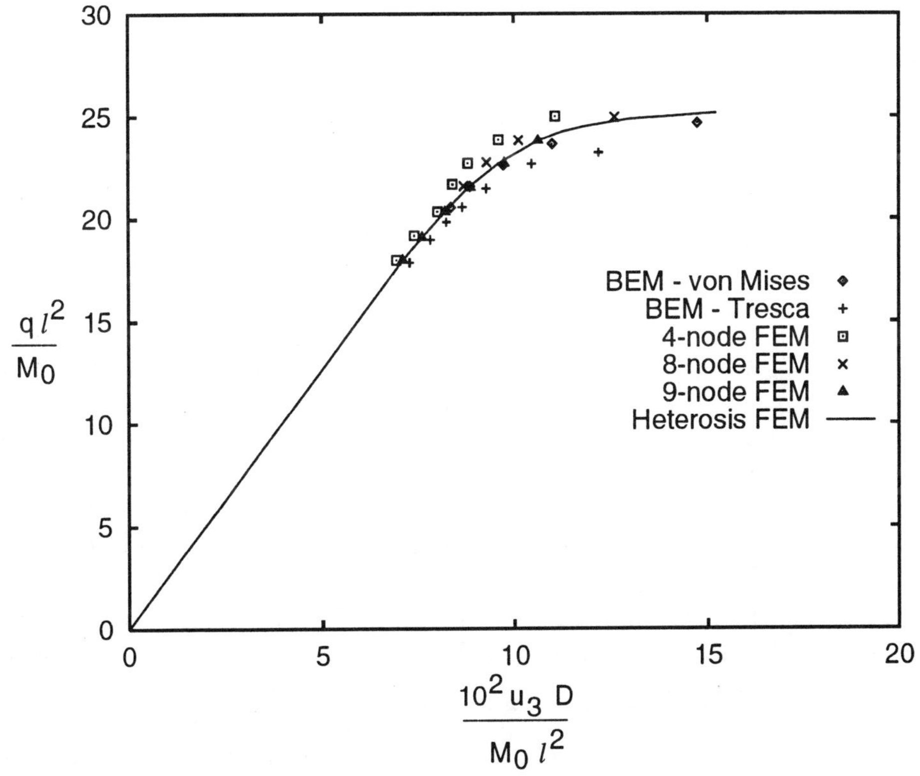

Figure 13. Square plate. Load-deflection curve.

The material parameters are: $H' = 0$, $\nu = 0.3$, $E = 210\,\mathrm{kN/mm^2}$ and $\sigma_0 = 0.25\,\mathrm{kN/mm^2}$.

The results obtained with the BEM are presented in Figure 15. The central deflection of 4.46 mm, which occurs in the middle surface when the corresponding whole cross section becomes plastic, is in accordance with the value of 4.48 mm calculated from the Timoshenko beam theory,[26] which takes into account transverse shear strains.

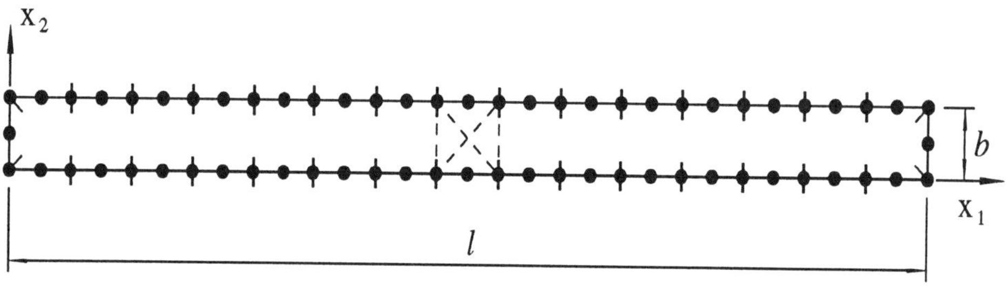

Figure 14. Rectangular beam. Discretization in boundary elements and internal cells.

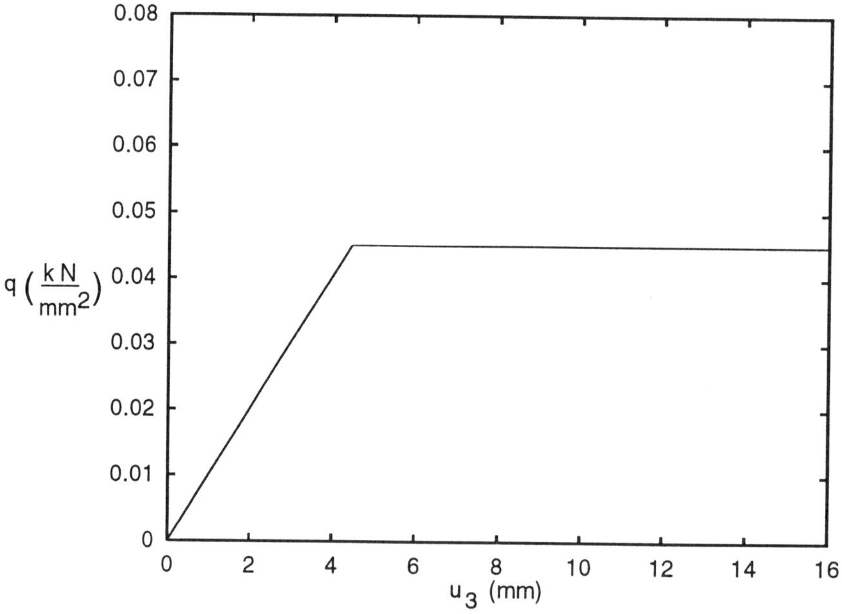

Figure 15. Rectangular beam. Load-deflection curve.

8.3 Example 3: simply supported circular plate

A simply supported circular plate of radius $a = 10.0$ in and thickness $h = 1.0$ in, subjected to a uniformly distributed load, is analysed in this example. The von Mises yield criterion is used, ideal plasticity ($H' = 0$) is considered and the material properties are: $\nu = 0.24$, $\sigma_0 = 16.0$ ksi and $E = 10^4$ ksi.

The load-deflection curve obtained with the BEM for the point at the centre of the circle, using the discretization shown in Figure 16, is represented by curve I in Figure 17. The last value of ρ (see Figure 17) for which the solution converges is 6.62 and it diverges for $\rho = 6.67$. By employing a much more refined discretization, in which smaller elements and cells were utilized near the centre of the plate and larger ones near its boundary, curve II in Figure 17 was obtained. In this case, the program converges until the value of $\rho = 6.49$ and diverges for $\rho = 6.54$. This result is in a good agreement with the limit value of $\rho = 6.51$ presented in Ref. 27 employing limit analysis. For the discretization depicted in Figure 16, the error is about 2%, indicating that the corresponding result is already very good. This problem was also analysed by Armen Jr. *et al.*[28] using FEM and a value of $\rho = 6.50$ was obtained for the colapse load.

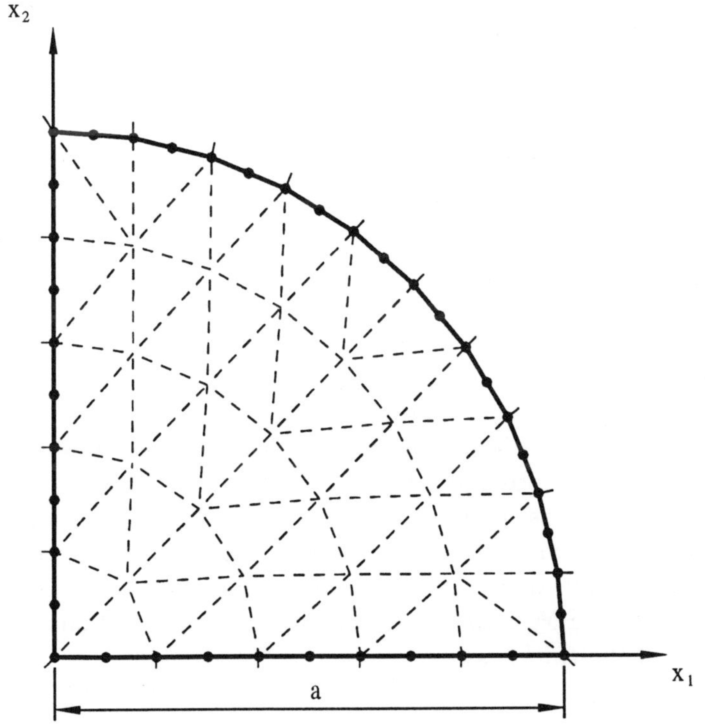

Figure 16. Circular plate. Discretization in boundary elements and internal cells.

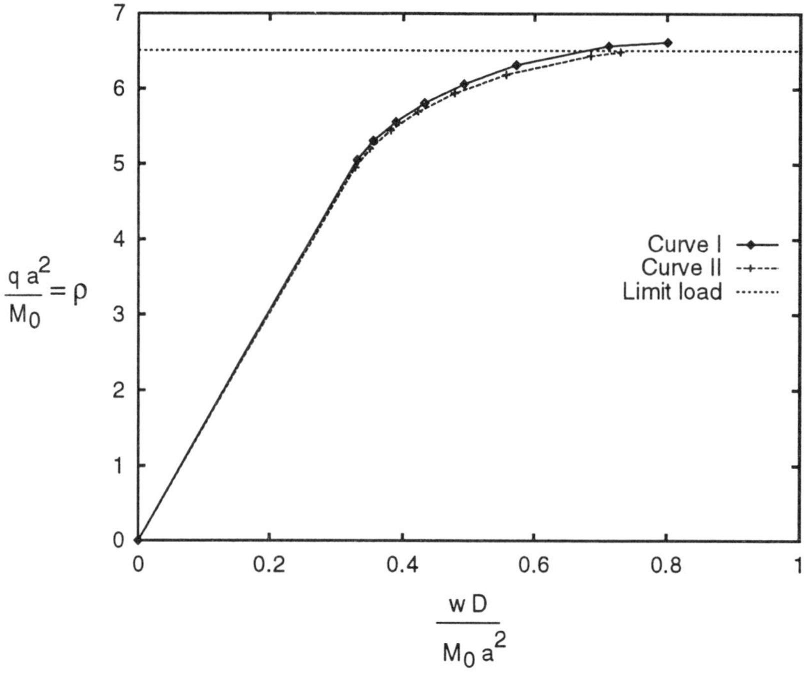

Figure 17. Circular plate. Load-deflection curve.

9 Conclusions

The present chapter describes an application of the BEM to nonlinear material plate bending analysis using Reissner's theory with emphasis on the elastoplastic application. The complete integral equations were obtained, including those for moments and shear forces at internal points. The implementation was performed using quadratic boundary elements and constant internal cells. The main feature of the inelastic formulation is that now domain discretization is required, but only over localized regions where inelastic strains may exist.

The elastoplastic problem was solved by using an incremental-iterative method which has been verified to be very efficient for both von Mises and Tresca yield criteria. It was also observed that the hypothesis of simultaneous complete plastification over the cross section led to results with a good approximation in comparison to that obtained by other analytical or numerical methods. This approximation, however, does not constitute an impediment to a more sophisticated procedure via BEM, such as the consideration of gradual distribution of plasticity through the thickness.

References

[1] Swedlow, J.L. & Cruse, T.A., Formulation of boundary integral equations for three dimensional elastoplastic flow, *International Journal of Solids and Structures*, **7**, pp. 1673–1683, 1971.

[2] Mendelson, A., Boundary integral methods in elasticity and plasticity, Report no. NASA TN D-7418, NASA, 1973.

[3] Riccardella, P.C., An implementation of the boundary integral technique for planar problems in elasticity and elastoplasticity, Report no. SM-73-10, Department of Mechanical Engineering, Carnegie Mellon University, Pittsburg, 1973.

[4] Mukherjee, S., Corrected boundary integral equations in planar thermoelastoplasticity, *International Journal of Solids and Structures*, **13**, pp. 331–335, 1977.

[5] Bui, H.D., Some remarks about the formulation of three dimensional thermoelastoplastic problems by integral equations, *International Journal of Solids and Structures*, **14**, pp. 935–939, 1978.

[6] Telles, J.C.F. & Brebbia, C.A., On the application of the boundary element method to plasticity, *Applied Mathematical Modelling*, **3**, pp. 466–470, 1979.

[7] Telles, J.C.F. & Brebbia, C.A., Elastic/viscoplastic problems using boundary elements, *International Journal of Mechanical Science*, **24** (10), pp. 605–618, 1982.

[8] Telles, J.C.F., *The Boundary Element Method Applied to Inelastic Problems.* Springer-Verlag, Berlin, 1983.

[9] Mukherjee, S., *Boundary Element Methods in Creep and Fracture.* Applied Science Publishers, London, 1982.

[10] Karam, V.J. & Telles, J.C.F., The BEM applied to plate bending elastoplastic analysis using Reissner's theory, *Engineering Analysis with Boundary Elements*, **9** (4), pp. 351–357, 1992.

[11] Vander Weeën, F., Application of the boundary integral equation method to Reissner's plate model, *International Journal for Numerical Methods in Engineering*, **18**, pp. 1–10, 1982.

[12] Karam, V.J. & Telles, J.C.F., On boundary elements for Reissner's plate theory, *Engineering Analysis*, **5** (1), pp. 21-27, 1988.

[13] Reissner, E., On the theory of bending of elastic plates, *Journal of Mathematics and Physics*, **23**, pp. 184–191, 1944.

[14] Reissner, E., The effect of transverse shear deformation on the bending of elastic plates, *Journal of Applied Mechanics*, **12**, pp. A69–A77, 1945.

[15] Reissner, E., On bending of elastic plates, *Quarterly of Applied Mechanics*, **5**, pp. 55–68, 1947.

[16] Brebbia, C.A., Telles, J.C.F. & Wrobel, L.C., *Boundary Element Techniques: Theory and Applications in Engineering*. Springer-Verlag, Berlin, 1984.

[17] Karam, V.J., Application of the boundary element method to Reissner's theory for plate bending, M.Sc. Dissertation (in Portuguese), COPPE/UFRJ, 1986.

[18] Karam, V.J., Plate bending analysis by the BEM including physical nonlinearity, Ph.D. Thesis (in Portuguese), COPPE/UFRJ, 1992.

[19] Hill, R., *The Mathematical Theory of Plasticity*. Oxford University Press, Oxford, 1950.

[20] Prager, W., *An Introduction to Plasticity*. Addison-Wesley, Amsterdam, 1959.

[21] Mendelson, A., *Plasticity: Theory and Application*. Macmillan, New York, 1968.

[22] Telles, J.C.F., A self-adaptive coordinate transformation for efficient numerical evaluation of general boundary element integrals, *International Journal for Numerical Methods in Engineering*, **24**, pp. 959–973, 1987.

[23] Kutt, H.R., Quadrature formulae for finite part integrals, Report Wisk 178, The National Research Institute for Mathematical Sciences, Pretoria, 1975.

[24] Telles, J.C.F. & Carrer, J.A.M., Implicit procedures for the solution of elastoplastic problems by the boundary element method, *Mathematical and Computer Modelling*, **15**, pp. 303–311, 1991.

[25] Owen, D.R.J. & Hinton, E., *Finite Elements in Plasticity: Theory and Practice*. Pineridge Press, Swansea, 1980.

[26] Timoshenko, S.P. & Goodier, J.N., *Theory of Elasticity*. McGraw-Hill Kogakusha, 1970.

[27] Hopkins, H.G. & Wang, A.J., Load-carrying capacities for circular plates of perfectly-plastic material with arbitrary yield condition, *Journal of the Mechanics and Physics of Solids*, **3**, pp. 117–129, 1954.

[28] Armen, Jr., H., Pifko, A.B., Levine, H.S. & Isakson, G., Plasticity, *Finite Element Techniques in Structural Mechanics*, eds. H. Tottenham & C.A. Brebbia, Chap. 8. Stress Analysis Publisher, pp. 209–257, 1970.

[29] Abramowitz, M. & Stegun, I.A., *Handbook of Mathematical Functions*. Dover Publications, New York, 1965.

Appendix

Kernel functions

The displacements of the fundamental solution are expressed by:

$$u^*_{\alpha\beta} = \frac{1}{8\pi D}\left[\left(\frac{8B(z)}{(1-\nu)} - 2\ln z + 1\right)\delta_{\alpha\beta} - \left(\frac{8A(z)}{(1-\nu)} + 2\right)r_{,\alpha}\, r_{,\beta}\right] \quad \text{(A.1)}$$

$$u^*_{\alpha 3} = -u^*_{3\alpha} = \frac{1}{8\pi D}\left(2\ln z - 1\right) r\, r_{,\alpha} \quad \text{(A.2)}$$

$$u^*_{33} = \frac{1}{8\pi D}\left[r^2(\ln\, z - 1) - \frac{8\ln z}{(1-\nu)\lambda^2}\right] \quad \text{(A.3)}$$

where r is the distance between the points ξ and x, defined by $r = \sqrt{r_\alpha r_\alpha}$, with $r_\alpha = x_\alpha(x) - x_\alpha(\xi)$, and one also has:

$$r_{,\alpha} = \frac{\partial r}{\partial x_\alpha(x)} = \frac{r_\alpha}{r} \quad \text{(A.4)}$$

$$z = \lambda r \quad \text{(A.5)}$$

$$A(z) = K_0(z) + 2z^{-1}\left[K_1(z) - z^{-1}\right] \quad \text{(A.6)}$$

$$B(z) = K_0(z) + z^{-1}\left[K_1(z) - z^{-1}\right] \quad \text{(A.7)}$$

with K_0 and K_1 being modified Bessel functions of order 0 and 1, respectively, and are calculated by polynomial expansions given in Ref. 29.

The surface tractions corresponding to the fundamental solution are:

$$p^*_{\alpha\beta} = -\frac{1}{4\pi r}\left[\left(4A + 2zK_1 + 1 - \nu\right)\left(\delta_{\alpha\beta}\, r_{,n} + r_{,\beta}\, n_\alpha\right)\right.$$
$$\left. + (4A + 1 + \nu)\, r_{,\alpha}\, n_\beta - 2\left(8A + 2zK_1 + 1 - \nu\right) r_{,\alpha}\, r_{,\beta}\, r_{,n}\right] \quad \text{(A.8)}$$

$$p^*_{\alpha 3} = \frac{\lambda^2}{2\pi}\left[B\, n_\alpha - A\, r_{,\alpha}\, r_{,n}\right] \quad \text{(A.9)}$$

$$p^*_{3\alpha} = -\frac{1}{8\pi}\left[\left(2(1+\nu)\ln z - 1 + \nu\right) n_\alpha + 2(1-\nu)\, r_{,\alpha}\, r_{,n}\right] \quad \text{(A.10)}$$

$$p^*_{33} = -\frac{1}{2\pi r}\, r_{,n} \quad \text{(A.11)}$$

where $r_{,n}$ is the derivative of r with respect to the normal at point x, defined by:

$$r_{,n} = r_{,\alpha}\, n_\alpha \quad \text{(A.12)}$$

Other expressions used in this chapter are:

$$v_{\alpha}^{*} = \frac{1}{128\pi D\lambda^{2}} r_{,\alpha}\, r z^{2}(4\ln z - 5) \tag{A.13}$$

$$v_{3}^{*} = -\frac{1}{256\pi D\lambda^{4}(1-\nu)} z^{2}\left[64(\ln z - 1) - z^{2}(1-\nu)(2\ln z - 3)\right] \tag{A.14}$$

$$v_{\alpha,\beta}^{*} = \frac{r^{2}}{128\pi D}\left[\delta_{\alpha\beta}(4\ln z - 5) + 2(4\ln z - 3) r_{,\alpha}\, r_{,\beta}\right] \tag{A.15}$$

$$v_{3,\beta}^{*} = -\frac{r\, r_{,\beta}}{128\pi D\lambda^{2}(1-\nu)}\left[32(2\ln z - 1) - z^{2}(1-\nu)(4\ln z - 5)\right] \tag{A.16}$$

$$u_{\alpha\beta\gamma}^{*} = \frac{1}{4\pi r}\Big[(4A + 2zK_{1} + 1 - \nu)\,(\delta_{\beta\gamma}\, r_{,\alpha} + r_{,\beta}\,\delta_{\alpha\gamma})$$
$$- 2\,(8A + 2zK_{1} + 1 - \nu)\, r_{,\alpha}\, r_{,\beta}\, r_{,\gamma} + (4A + 1 + \nu)\delta_{\alpha\beta} r_{,\gamma}\Big] \tag{A.17}$$

$$u_{\alpha\beta 3}^{*} = -\frac{1}{8\pi}\left[(2(1+\nu)\ln z - 1 + \nu)\,\delta_{\alpha\beta} + 2(1-\nu) r_{,\alpha}\, r_{,\beta}\right] \tag{A.18}$$

$$u_{3\beta\gamma}^{*} = \frac{\lambda^{2}}{2\pi}\left[B\,\delta_{\gamma\beta} - A\, r_{,\beta}\, r_{,\gamma}\right] \tag{A.19}$$

$$u_{3\beta 3}^{*} = \frac{1}{2\pi r} r_{,\beta} \tag{A.20}$$

$$p_{\alpha\beta\gamma}^{*} = \frac{D(1-\nu)}{4\pi r^{2}}\Big\{(4A + 2zK_{1} + 1 - \nu)\,(\delta_{\gamma\alpha}\, n_{\beta} + \delta_{\gamma\beta}\, n_{\alpha})$$
$$+ (4A + 1 + 3\nu)\,\delta_{\alpha\beta}\, n_{\gamma} - (16A + 6zK_{1} + z^{2}K_{0} + 2 - 2\nu)$$
$$\cdot\left[(n_{\alpha}\, r_{,\beta} + n_{\beta}\, r_{,\alpha}) r_{,\gamma} + (\delta_{\gamma\alpha}\, r_{,\beta} + \delta_{\gamma\beta}\, r_{,\alpha}) r_{,n}\right]$$
$$- 2(8A + 2zK_{1} + 1 + \nu)(\delta_{\alpha\beta}\, r_{,\gamma}\, r_{,n} + n_{\gamma}\, r_{,\alpha}\, r_{,\beta})$$
$$+ 4(24A + 8zK_{1} + z^{2}K_{0} + 2 - 2\nu)\, r_{,\alpha}\, r_{,\beta}\, r_{,\gamma}\, r_{,n}\Big\} \tag{A.21}$$

$$p_{\alpha\beta 3}^{*} = \frac{D(1-\nu)\lambda^{2}}{4\pi r}\Big[(2A + zK_{1})(r_{,\beta}\, n_{\alpha} + r_{,\alpha}\, n_{\beta})$$
$$- 2(4A + zK_{1})(r_{,\alpha}\, r_{,\beta}\, r_{,n} + 2A\,\delta_{\alpha\beta}\, r_{,n})\Big] \tag{A.22}$$

$$p^*_{3\beta\gamma} = -\frac{D(1-\nu)\lambda^2}{4\pi r}\Big[(2A + zK_1)(\delta_{\gamma\beta}\, r_{,n} + r_{,\gamma}\; n_\beta)$$

$$- 2(4A + zK_1)\; r_{,\gamma}\; r_{,\beta}\; r_{,n} + 2A\; n_\gamma\; r_{,\beta}\Big] \qquad \text{(A.23)}$$

$$p^*_{3\beta 3} = \frac{D(1-\nu)\lambda^2}{4\pi r^2}\Big[(z^2B + 1)n_\beta - (z^2A + 2)\; r_{,\beta}\; r_{,n}\Big] \qquad \text{(A.24)}$$

$$w^*_{\alpha\beta} = -\frac{r}{64\pi}\Big\{(4\ln z - 3)\Big[(1-\nu)(r_{,\beta}\; n_\alpha + r_{,\alpha}\; n_\beta) + (1+3\nu)\;\delta_{\alpha\beta}\; r_{,n}\Big]$$

$$+ 4\Big[(1-\nu)\; r_{,\alpha}\; r_{,\beta} + \nu\;\delta_{\alpha\beta}\Big] r_{,n}\Big\} - \frac{\nu}{(1-\nu)\lambda^2}\, u^*_{\alpha\beta\gamma}\; n_\gamma \qquad \text{(A.25)}$$

$$w^*_{3\beta} = \frac{1}{8\pi}\Big[(2\ln z - 1)n_\beta + 2\; r_{,\beta}\; r_{,n}\Big] - \frac{\nu}{(1-\nu)\lambda^2}\, u^*_{3\beta\gamma}\; n_\gamma \qquad \text{(A.26)}$$

$$\chi^*_{\alpha\beta\gamma} = -\frac{1}{8\pi D(1-\nu) r}\Big[(8A + 4zK_1 + 2 - 2\nu)(r_{,\beta}\;\delta_{\alpha\gamma} + r_{,\alpha}\;\delta_{\beta\gamma})$$

$$- 4(8A + 2zK_1 + 1 - \nu)\; r_{,\alpha}\; r_{,\beta}\; r_{,\gamma} + 2(4A + 1 - \nu)\;\delta_{\alpha\beta}\; r_{,\gamma}\Big] \qquad \text{(A.27)}$$

$$\chi^*_{\alpha\beta 3} = -\frac{1}{8\pi D}\Big[\delta_{\alpha\beta}(2\ln z - 1) + 2\; r_{,\alpha}\; r_{,\beta}\Big] \qquad \text{(A.28)}$$

$$\chi^*_{\alpha\beta\gamma\theta} = \frac{1}{16\pi r^2}\Big[4(4A + 2zK_1 + 1 - \nu)(\delta_{\gamma\alpha}\;\delta_{\theta\beta} + \delta_{\theta\alpha}\;\delta_{\gamma\beta})$$

$$+ 4(4A + 1 + \nu)\;\delta_{\gamma\theta}\;\delta_{\alpha\beta} - 8(8A + 2zK_1 + 1 - \nu)\;\delta_{\gamma\theta}\; r_{,\alpha}\; r_{,\beta}$$

$$- 8(8A + 2zK_1 + 1 + \nu)\;\delta_{\alpha\beta}\; r_{,\gamma}\; r_{,\theta}$$

$$- 4(16A + 6zK_1 + z^2K_0 + 2 - 2\nu)(\delta_{\gamma\alpha}\; r_{,\theta}\; r_{,\beta} + \delta_{\theta\alpha}\; r_{,\gamma}\; r_{,\beta}$$

$$+ \delta_{\gamma\beta}\; r_{,\alpha}\; r_{,\theta} + \delta_{\theta\beta}\; r_{,\gamma}\; r_{,\alpha})$$

$$+ 16(24A + 8zK_1 + z^2K_0 + 2 - 2\nu)\; r_{,\alpha}\; r_{,\beta}\; r_{,\gamma}\; r_{,\theta}\Big] \qquad \text{(A.29)}$$

$$\chi^*_{3\beta\gamma\theta} = -\frac{\lambda^2}{16\pi r}\Big[(8A + 4zK_1)(r_{,\theta}\;\delta_{\gamma\beta} + r_{,\gamma}\;\delta_{\theta\beta})$$

$$- 4(8A + 2zK_1)\; r_{,\theta}\; r_{,\gamma}\; r_{,\beta} + 8A\;\delta_{\gamma\theta}\; r_{,\beta}\Big] \qquad \text{(A.30)}$$

Chapter 5

Stress resultant based integral equation formulation for plate bending analysis

Y.F. Rashed[a], M.H. Aliabadi[b], C.A. Brebbia[c]
[a]*Department of Structural Engineering, Cairo University, Giza, Egypt*
[b]*Department of Engineering, Queen Mary College, University of London, Mile End, London, EL 4NS, UK*
[c]*Wessex Institute of Technology, Ashurst, Southampton, SO40 7AA, UK*

Abstract

This chapter presents the derivation of the stress resultant based integral equation for plate bending analysis. The plate is modelled using the Reissner plate derived. The numerical implementation is also discussed.

The boundary stress resultants are evaluated using the proposed integral equations and compared to the evaluation using the derivatives of the shape functions. Several examples are solved to demonstrate the accuracy of the proposed formulations.

1 Introduction

Recently, the stress based or the hypersingular integral equations have been very popular among the boundary element community. These integral equations are based on the stresses (unlike the standard BEM, it is based on displacements). The main applications of these integral equations are in the dual boundary element method for modeling cracks in single domain (see for example Portela *et al.* [1] and Mi and Aliabadi [2]) and the direct evaluation of the boundary stress tensor (see for example Huber *et al.* [3]). Although other applications such as error analysis (see for example Paulino *et al.* [4]) and generating additional equations for multi connected bodies at the common node (see for example Bialecki *et al.* [5]) have also been reported.

In the classical plate theory, the slope equation is a hypersingular eqn (see Bézine [6] and Stern [7]). Recently, Knöpke [8] derived a hypersingular integral equation for the bending moments in the classical plate theory. In his work, the shear force equation was ignored and no numerical results were presented.

In 1947, Reissner [9] introduced a six order theory for bending of plates. The theory is more refined than the classical plate theory and more accurate results can be obtained for the problems of stress concentrations. The boundary integral equations for Reissner plate model was reported by Vander Weeën [10].

This chapter mainly demonstrates the derivation of the stress resultant based integral equations (the hypersingular integral equations) for Reissner plates. A singularity subtraction method based on the Taylor series expansion is used to deal with singular integrals. As an application of the new type of integral equations, the evaluation of the boundary stress resultants is described and compared to the conventional way of the evaluation using the derivatives of the shape functions. Several numerical examples are solved to demonstrate the accuracy and the use of the new set of integral equations.

The source of information of this chapter is the work by Rashed *et al.* [11-13].

2 Basic theory

In this section, the basic equations of the Reissner plate bending theory are reviewed. Throughout this chapter, the indicial notation is used. Greek indices will vary from 1 to 2 whereas, Roman indices from 1 to 3. The comma subscript denotes differentiation, such as $(\cdot)_{,\alpha}$, stands for the derivative of $(\cdot)$ with respect to the coordinate x_α.

Consider an arbitrary plate of thickness h, with a domain Ω and boundary Γ in the x_i space. The $x_1 - x_2$ plane is assumed to be located at the middle surface $x_3 = 0$. The generalized displacement are denoted as u_i, where, u_α denotes rotations (ϕ_{x_1} and ϕ_{x_2}) and u_3 denotes the transverse deflection w in x_3 direction. According to Reissner [9], the generalized stress resultant-displacement relationships can be written as follows:

$$\begin{aligned} M_{\alpha\beta} &= D\frac{1-\nu}{2}\left(u_{\alpha,\beta} + u_{\beta,\alpha} + \frac{2\nu}{1-\nu}u_{\gamma,\gamma}\delta_{\alpha\beta}\right) + \frac{\nu q}{(1-\nu)\lambda^2}\delta_{\alpha\beta} \\ Q_\alpha &= D\frac{1-\nu}{2}\lambda^2(u_\alpha + u_{3,\alpha}) \end{aligned} \quad (1)$$

where $M_{\alpha\beta}$ and Q_α are the bending and shear stress resultants respectively, and $D = Eh^3/12(1-\nu^2)$ is the plate flexural rigidity, E is Young's modulus, ν is Poisson's ratio, $\lambda = \sqrt{10}/h$ is the shear factor and q is the distributed load per unit area.

The equilibrium equations can be written as follows:

$$M_{\alpha\beta,\beta} - Q_\alpha = 0$$
$$Q_{\alpha,\alpha} + q = 0 \tag{2}$$

The generalized tractions at a boundary point can be defined as:

$$\begin{aligned} p_\alpha &= M_{\alpha\beta} n_\beta \\ p_3 &= Q_\alpha n_\alpha \end{aligned} \tag{3}$$

where n_β are the components of the outward normal vector to the plate boundary Γ.

The generalized Navier equations can be formed by substituting (1) into (2) to give:

$$L_{ij}^* u_j + b_i = 0 \tag{4}$$

where L_{ij}^* is the generalized Navier differential operator and b_i is the generalized body force components.

The fundamental solution of eqns (4) is obtained by taking $b_{ki} = \delta(\mathbf{X}', \mathbf{X})\delta_{ki}$ where, $\delta(\mathbf{X}', \mathbf{X})$ is the Dirac delta distribution, k is an arbitrary direction of the applied unit load, and $\mathbf{X}, \mathbf{X}' \in \Omega$ are the field and source points respectively in an infinite plate. By using the Hörmender method [14], the fundamental solution U_{ij}^* can be obtained. The traction fundamental solution P_{ij}^* can be obtained by utilizing eqn (1) together with eqn (3), noting that the differentiation here is with respect to the coordinates of the field point $\mathbf{X}$. The expressions for U_{ij}^* and P_{ij}^* are given by Vander Weeën [10] as follows:

$$\begin{aligned} U_{\alpha\beta}^* &= \frac{1}{8\pi D(1-\nu)}\{[8B(z) - (1-\nu)(2\ln z - 1)]\delta_{\alpha\beta} \\ &- [8A(z) + 2(1-\nu)]r_{,\alpha}r_{,\beta}\} \\ U_{\alpha 3}^* &= -U_{3\alpha}^* = \frac{1}{8\pi D}(2\ln z - 1)rr_{,\alpha} \\ U_{33}^* &= \frac{1}{8\pi D(1-\nu)\lambda^2}[(1-\nu)z^2(\ln z - 1) - 8\ln z] \end{aligned} \tag{5}$$

$$
\begin{aligned}
P^*_{\gamma\alpha} &= \frac{-1}{4\pi r}[(4A(z) + 2zK_1(z) + 1 - \nu)(\delta_{\alpha\gamma} r_{,n} + r_{,\alpha} n_\gamma) \\
&+ (4A(z) + 1 + \nu) r_\gamma n_\alpha - 2(8A(z) + 2zK_1(z) + 1 - \nu) r_{,\alpha} r_{,\gamma} r_{,n}] \\
P^*_{\gamma 3} &= \frac{\lambda^2}{2\pi}[B(z) n_\gamma - A(z) r_{,\gamma} r_{,n}] \\
P^*_{3\alpha} &= \frac{-(1-\nu)}{8\pi}\left[\left(2\frac{(1+\nu)}{(1-\nu)}\ln z - 1\right) n_\alpha + 2 r_{,\alpha} r_{,n}\right] \\
P^*_{33} &= \frac{-1}{2\pi r} r_{,n}
\end{aligned}
\qquad (6)
$$

where

$$
\begin{aligned}
A(z) &= K_0(z) + \frac{2}{z}\left[K_1(z) - \frac{1}{z}\right] \\
B(z) &= K_0(z) + \frac{1}{z}\left[K_1(z) - \frac{1}{z}\right]
\end{aligned}
\qquad (7)
$$

in which $K_0(z)$ and $K_1(z)$ are modified Bessel functions [15], $z = \lambda r$, r is the absolute distance between the source and the field points and $r_{,n} = r_{,\alpha} n_\alpha$. By expanding the modified Bessel functions for small arguments, it can be seen that $A(z)$ is a smooth function, whereas $B(z)$ is a weakly singular $O(\ln r)$. Therefore U^*_{ij} is weakly singular and P^*_{ij} has a strong singularity $O(\frac{1}{r})$.

3 Displacement boundary integral equation

The displacement boundary integral equation can be derived by considering the following integral identity:

$$
\int_\Omega [(M_{\alpha\beta,\beta} - Q_\alpha)\, U^*_\alpha + (Q_{\alpha,\alpha} + q)\, U^*_3]\, \mathrm{d}\Omega = 0 \qquad (8)
$$

where $U^*_i (i = \alpha, 3)$ are weighting functions.

Integrating by parts (i.e. applying Green second identity) suitable number of times and introducing the fundamental fields as the weighting functions, the displacement boundary integral equation can be obtained as follows:

$$\begin{aligned} c_{ij}(\mathbf{x}')u_j(\mathbf{x}') &+ \int\!\!\!\!\!\!-_{\Gamma} P^*_{ij}(\mathbf{x}',\mathbf{x})u_j(\mathbf{x})d\Gamma(\mathbf{x}) = \int_{\Gamma} U^*_{ij}(\mathbf{x}',\mathbf{x})p_j(\mathbf{x})d\Gamma(\mathbf{x}) \\ &+ \int_{\Omega}\left(U^*_{i3}(\mathbf{x}',\mathbf{X}) - \frac{\nu}{(1-\nu)\lambda^2}U^*_{i\alpha,\alpha}(\mathbf{x}',\mathbf{X})\right)q(\mathbf{X})d\Omega(\mathbf{X}) \quad (9) \end{aligned}$$

where $\int\!\!\!\!-$ denotes a Cauchy principal value integral, $\mathbf{x}', \mathbf{x} \in \Gamma$ are source point and field point on the boundary respectively, and $c_{ij}(\mathbf{x}')$ are the jump terms from the principal value integral of the strongly singular integral in the kernel P^*_{ij}. The value of $c_{ij}(\mathbf{x}')$ is equal to $\frac{1}{2}\delta_{ij}$ for $\mathbf{x}'$ located on a smooth boundary, however it can be evaluated for a general case from the consideration of the generalized rigid body movements. The domain integral in eqn (9) can be transferred to the boundary integral, in the case of a uniform load (q =constant) to give [10]:

$$\begin{aligned} &\int_{\Omega}\left(U^*_{i3}(\mathbf{x}',\mathbf{X}) - \frac{\nu}{(1-\nu)\lambda^2}U^*_{i\alpha,\alpha}(\mathbf{x}',\mathbf{X})\right)q(\mathbf{X})d\Omega(\mathbf{X}) = \\ &\qquad q\int_{\Gamma}\left(V^*_{i,\alpha}(\mathbf{x}',\mathbf{x}) - \frac{\nu}{(1-\nu)\lambda^2}U^*_{i\alpha}(\mathbf{x}',\mathbf{x})\right)n_\alpha(\mathbf{x})d\Gamma(\mathbf{x}) \quad (10) \end{aligned}$$

where V^*_i are the particular solutions of the equation $V^*_{i,\theta\theta} = U^*_{i3}$. The expression for $V^*_{i,\beta}$ is given as follows:

$$\begin{aligned} V^*_{\alpha,\beta} &= \frac{r^2}{128\pi D}[(4\ln z - 5)\delta_{\alpha\beta} + 2(4\ln z - 3)r_{,\alpha}r_{,\beta}] \\ V^*_{3,\beta} &= \frac{-rr_{,\beta}}{128\pi D(1-\nu)\lambda^2}[32(2\ln z - 1) - z^2(1-v)(4\ln z - 5)] \quad (11) \end{aligned}$$

Eqn (9) represents three integral equations (two ($i = \alpha = 1,2$) for rotations and one ($i = 3$) for deflection).

Bending and shear stress resultants at any internal point $\mathbf{X}'$ can be computed by differentiating eqn (9) with respect to the coordinate of the source point $\mathbf{X}'$ and then substituting in eqn (1) to give:

$$\begin{aligned} M_{\alpha\beta}(\mathbf{X}') &= \int_{\Gamma} U^*_{\alpha\beta k}(\mathbf{X}',\mathbf{x})p_k(\mathbf{x})d\Gamma(\mathbf{x}) - \int_{\Gamma} P^*_{\alpha\beta k}(\mathbf{X}',\mathbf{x})u_k(\mathbf{x})d\Gamma(\mathbf{x}) \\ &+ q\int_{\Gamma} W^*_{\alpha\beta}(\mathbf{X}',\mathbf{x})d\Gamma(\mathbf{x}) + \frac{\nu}{(1-\nu)\lambda^2}q\delta_{\alpha\beta} \\ Q_\beta(\mathbf{X}') &= \int_{\Gamma} U^*_{3\beta k}(\mathbf{X}',\mathbf{x})p_k(\mathbf{x})d\Gamma(\mathbf{x}) - \int_{\Gamma} P^*_{3\beta k}(\mathbf{X}',\mathbf{x})u_k(\mathbf{x})d\Gamma(\mathbf{x}) + \\ &+ q\int_{\Gamma} W^*_{3\beta}(\mathbf{X}',\mathbf{x})d\Gamma(\mathbf{x}) \quad (12) \end{aligned}$$

where the kernels U^*_{ijk}, P^*_{ijk} and $W^*_{i\beta}$ are given in the Appendix.

4 Stress resultant boundary integral equation

Eqn (9) is also valid for an external point $\mathbf{X}'' \notin \Omega, \Gamma$ with $c_{ij} = 0$. The corresponding stress resultant identity may be obtained in similar way as eqn (12) to give:

$$
\begin{aligned}
\int_\Gamma P^*_{\alpha\beta k}(\mathbf{X}'', \mathbf{x}) u_k(\mathbf{x}) d\Gamma(\mathbf{x}) &= \int_\Gamma U^*_{\alpha\beta k}(\mathbf{X}'', \mathbf{x}) p_k(\mathbf{x}) d\Gamma(\mathbf{x}) \\
&+ q \int_\Gamma W^*_{\alpha\beta}(\mathbf{X}'', \mathbf{x}) d\Gamma(\mathbf{x}) \\
\int_\Gamma P^*_{3\beta k}(\mathbf{X}'', \mathbf{x}) u_k(\mathbf{x}) d\Gamma(\mathbf{x}) &= \int_\Gamma U^*_{3\beta k}(\mathbf{X}'', \mathbf{x}) p_k(\mathbf{x}) d\Gamma(\mathbf{x}) \\
&+ q \int_\Gamma W^*_{3\beta}(\mathbf{X}'', \mathbf{x}) d\Gamma(\mathbf{x}) \qquad (13)
\end{aligned}
$$

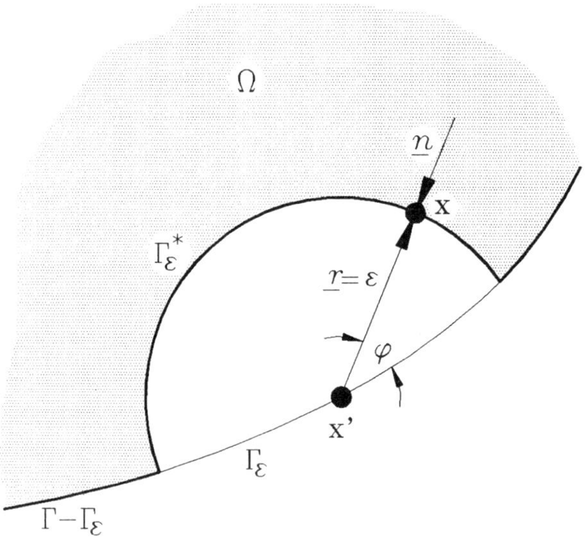

Figure 1. The semicircular region around the source point.

The stress resultant boundary integral equation is formed by considering the behavior of eqn (13) when the point $\mathbf{X}''$ tends to the boundary Γ at $\mathbf{x}'$. To satisfy the continuity requirements, the point $\mathbf{x}'$ is assumed to be on a smooth boundary.

A semi-circular domain with boundary Γ_ε^* is constructed around the point $\mathbf{x}'$ as shown in Figure 1.

Taking the limit as $\mathbf{X}'' \to \mathbf{x}'$, eqn (13) can be rewritten as follows:

$$\begin{aligned}
&\lim_{\varepsilon\to 0}\int_{\Gamma-\Gamma_\varepsilon+\Gamma_\varepsilon^*} P^*_{\alpha\beta\gamma}(\mathbf{x}',\mathbf{x})u_\gamma(\mathbf{x})d\Gamma(\mathbf{x}) + \lim_{\varepsilon\to 0}\int_{\Gamma-\Gamma_\varepsilon+\Gamma_\varepsilon^*} P^*_{\alpha\beta 3}(\mathbf{x}',\mathbf{x})u_3(\mathbf{x})d\Gamma(\mathbf{x}) \\
= &\lim_{\varepsilon\to 0}\int_{\Gamma-\Gamma_\varepsilon+\Gamma_\varepsilon^*} U^*_{\alpha\beta\gamma}(\mathbf{x}',\mathbf{x})p_\gamma(\mathbf{x})d\Gamma(\mathbf{x}) + \lim_{\varepsilon\to 0}\int_{\Gamma-\Gamma_\varepsilon+\Gamma_\varepsilon^*} U^*_{\alpha\beta 3}(\mathbf{x}',\mathbf{x})p_3(\mathbf{x})d\Gamma(\mathbf{x}) \\
+ &\ q\lim_{\varepsilon\to 0}\int_{\Gamma-\Gamma_\varepsilon+\Gamma_\varepsilon^*} W^*_{\alpha\beta}(\mathbf{x}',\mathbf{x})d\Gamma(\mathbf{x}) \qquad (14)
\end{aligned}$$

and

$$\begin{aligned}
&\lim_{\varepsilon\to 0}\int_{\Gamma-\Gamma_\varepsilon+\Gamma_\varepsilon^*} P^*_{3\beta\gamma}(\mathbf{x}',\mathbf{x})u_\gamma(\mathbf{x})d\Gamma(\mathbf{x}) + \lim_{\varepsilon\to 0}\int_{\Gamma-\Gamma_\varepsilon+\Gamma_\varepsilon^*} P^*_{3\beta 3}(\mathbf{x}',\mathbf{x})u_3(\mathbf{x})d\Gamma(\mathbf{x}) \\
= &\lim_{\varepsilon\to 0}\int_{\Gamma-\Gamma_\varepsilon+\Gamma_\varepsilon^*} U^*_{3\beta\gamma}(\mathbf{x}',\mathbf{x})p_\gamma(\mathbf{x})d\Gamma(\mathbf{x}) + \lim_{\varepsilon\to 0}\int_{\Gamma-\Gamma_\varepsilon+\Gamma_\varepsilon^*} U^*_{3\beta 3}(\mathbf{x}',\mathbf{x})p_3(\mathbf{x})d\Gamma(\mathbf{x}) \\
+ &\ q\lim_{\varepsilon\to 0}\int_{\Gamma-\Gamma_\varepsilon+\Gamma_\varepsilon^*} W^*_{3\beta}(\mathbf{x}',\mathbf{x})d\Gamma(\mathbf{x}) \qquad (15)
\end{aligned}$$

Eqns (14) and (15) represent the bending and shear stress resultant boundary integral equations, respectively, at the boundary point $\mathbf{x}'$.

4.1 The bending stress resultant integral equation

Eqn (14) can be written in the following form:

$$I_1^* + I_2^* = I_3^* + I_4^* + I_5^* \qquad (16)$$

Assume the boundary values of u_i are $C^{1,\alpha}$, $(0 < \alpha < 1)$ to allow for the Taylor expansion for the integrands up to two terms. Now, each of the integrals in eqn (16) will be considered individually.

The integral I_1^*

The integral I_1^* can be written as:

$$\begin{aligned}
I_1^* = &\lim_{\varepsilon\to 0}\int_{\Gamma-\Gamma_\varepsilon+\Gamma_\varepsilon^*} P^*_{\alpha\beta\gamma}(\mathbf{x}',\mathbf{x})u_\gamma(\mathbf{x})d\Gamma(\mathbf{x}) = \lim_{\varepsilon\to 0}\int_{\Gamma-\Gamma_\varepsilon} P^*_{\alpha\beta\gamma}(\mathbf{x}',\mathbf{x})u_\gamma(\mathbf{x})d\Gamma(\mathbf{x}) \\
+ &\lim_{\varepsilon\to 0}\int_{\Gamma_\varepsilon^*} P^*_{\alpha\beta\gamma}(\mathbf{x}',\mathbf{x})[u_\gamma(\mathbf{x}) - u_\gamma(\mathbf{x}') - u_{\gamma,\theta}(\mathbf{x}')(x_\theta(\mathbf{x}) - x_\theta(\mathbf{x}'))]d\Gamma(\mathbf{x})
\end{aligned}$$

$$+ \quad u_\gamma(\mathbf{x}') \lim_{\varepsilon \to 0} \int_{\Gamma_\varepsilon^*} P_{\alpha\beta\gamma}^*(\mathbf{x}', \mathbf{x}) d\Gamma(\mathbf{x})$$

$$+ \quad u_{\gamma,\theta}(\mathbf{x}') \lim_{\varepsilon \to 0} \int_{\Gamma_\varepsilon^*} P_{\alpha\beta\gamma}^*(\mathbf{x}', \mathbf{x})(x_\theta(\mathbf{x}) - x_\theta(\mathbf{x}')) d\Gamma(\mathbf{x}) \tag{17}$$

It has to be noted that the integral I_1^* contains the kernel $P_{\alpha\beta\gamma}^*$ which is hypersingular of $O(\frac{1}{r^2})$ so that two terms of the Taylor expansion for the integrand are appropriate.

In the above integrals, the second term of the right hand side (RHS) is zero in the limit as $\varepsilon \to 0$. The first and third RHS terms together form a Hadamard finite part integral. From Figure 1, the following relationships can be observed:

$$r = \varepsilon, \qquad r_{,n} = -1, \qquad d\Gamma = \varepsilon d\varphi,$$

$$r_{,1} = -n_1 = \cos\varphi, \qquad r_{,2} = -n_2 = \sin\varphi,$$

and

$$\int_{\Gamma_\varepsilon^*} \cdots d\Gamma_\varepsilon^* = \int_0^\pi \cdots \varepsilon d\varphi \tag{18}$$

By considering the above relationships, together with the following limits [15]:

$$\begin{aligned} & \lim_{\varepsilon \to 0} A(\lambda\varepsilon) = \frac{-1}{2}, \\ & \lim_{\varepsilon \to 0} \lambda^2\varepsilon^2 K_0(\lambda\varepsilon) = 0, \\ & \lim_{\varepsilon \to 0} \lambda\varepsilon K_1(\lambda\varepsilon) = 1, \\ \text{and} \quad & \lim_{\varepsilon \to 0} \lambda^2\varepsilon^2 B(\lambda\varepsilon) = 0, \end{aligned} \tag{19}$$

The last term on the RHS leads to the following jump terms:

$$\begin{aligned} u_{\gamma,\theta}(\mathbf{x}') \lim_{\varepsilon \to 0} \int_{\Gamma_\varepsilon^*} & P_{\alpha\beta\gamma}^*(\mathbf{x}', \mathbf{x})(x_\theta(\mathbf{x}) - x_\theta(\mathbf{x}')) d\Gamma(\mathbf{x}) \\ &= \frac{D(1+\nu)(1-\nu)}{16}(u_{\beta,\alpha}(\mathbf{x}') + u_{\alpha,\beta}(\mathbf{x}') + u_{\gamma,\gamma}(\mathbf{x}')\delta_{\alpha\beta}) \end{aligned} \tag{20}$$

Now, the integral I_1^* can be written as follow:

$$\begin{aligned} I_1^* \;&=\; \neq\!\!\!\!\!\!\int_\Gamma P_{\alpha\beta\gamma}^*(\mathbf{x}', \mathbf{x}) u_\gamma(\mathbf{x}) d\Gamma(\mathbf{x}) \\ &+ \; \frac{D(1+\nu)(1-\nu)}{16}(u_{\beta,\alpha}(\mathbf{x}') + u_{\alpha,\beta}(\mathbf{x}') + u_{\gamma,\gamma}(\mathbf{x}')\delta_{\alpha\beta}) \end{aligned} \tag{21}$$

where $\neq\!\!\!\!\!\!\int$ denotes the Hadamard finite part integral.

The integral I_2^*

The integral I_2^* can be treated in similar way as that of the integral I_1^*. Only one term of the Taylor expansion is needed for I_2^* as the kernel $P^*_{\alpha\beta 3}$ contains a strong singularity of $O(\frac{1}{r})$. So that the integral I_2^* can be written as follows:

$$\begin{aligned} I_2^* &= \lim_{\varepsilon\to 0}\int_{\Gamma-\Gamma_\varepsilon+\Gamma_\varepsilon^*} P^*_{\alpha\beta 3}(\mathbf{x}',\mathbf{x})u_3(\mathbf{x})d\Gamma(\mathbf{x}) = \lim_{\varepsilon\to 0}\int_{\Gamma-\Gamma_\varepsilon} P^*_{\alpha\beta 3}(\mathbf{x}',\mathbf{x})u_3(\mathbf{x})d\Gamma(\mathbf{x}) \\ &+ \lim_{\varepsilon\to 0}\int_{\Gamma_\varepsilon^*} P^*_{\alpha\beta 3}(\mathbf{x}',\mathbf{x})[u_3(\mathbf{x}) - u_3(\mathbf{x}')]d\Gamma(\mathbf{x}) \\ &+ u_3(\mathbf{x}')\lim_{\varepsilon\to 0}\int_{\Gamma_\varepsilon^*} P^*_{\alpha\beta 3}(\mathbf{x}',\mathbf{x})d\Gamma(\mathbf{x}) \end{aligned} \tag{22}$$

In the above integrals, the second term on the RHS is zero in the limit as $\varepsilon \to 0$. The first term on the RHS forms a Cauchy principal value integral. By considering the relationships in eqn (18) and the limits in (19), the jump terms that appear from the last term on the RHS vanish. So that the integral I_2^* can be written as follows:

$$I_2^* = \fint_\Gamma P^*_{\alpha\beta 3}(\mathbf{x}',\mathbf{x})u_3(\mathbf{x})d\Gamma(\mathbf{x}) \tag{23}$$

The integral I_3^*

The integral I_3^* contains $U^*_{\alpha\beta\gamma}$ which is strongly singular. Using the first term of the Taylor expansion of $M_{\gamma\theta}$ the following form can be written:

$$\begin{aligned} I_3^* &= \lim_{\varepsilon\to 0}\int_{\Gamma-\Gamma_\varepsilon+\Gamma_\varepsilon^*} U^*_{\alpha\beta\gamma}(\mathbf{x}',\mathbf{x})p_\gamma(\mathbf{x})d\Gamma(\mathbf{x}) \\ &= \lim_{\varepsilon\to 0}\int_{\Gamma-\Gamma_\varepsilon} U^*_{\alpha\beta\gamma}(\mathbf{x}',\mathbf{x})p_\gamma(\mathbf{x})d\Gamma(\mathbf{x}) + \\ &+ \lim_{\varepsilon\to 0}\int_{\Gamma_\varepsilon^*} U^*_{\alpha\beta\gamma}(\mathbf{x}',\mathbf{x})n_\theta(\mathbf{x})[M_{\gamma\theta}(\mathbf{x}) - M_{\gamma\theta}(\mathbf{x}')]d\Gamma(\mathbf{x}) + \\ &+ M_{\gamma\theta}(\mathbf{x}')\lim_{\varepsilon\to 0}\int_{\Gamma_\varepsilon^*} U^*_{\alpha\beta\gamma}(\mathbf{x}',\mathbf{x})n_\theta(\mathbf{x})d\Gamma(\mathbf{x}) \end{aligned} \tag{24}$$

In the above integrals, the second term on the RHS is zero in the limit as $\varepsilon \to 0$. The first term on the RHS forms a Cauchy principal value integral. Considering the relationships in eqn (18) and the limits in eqn (19), the last term on the RHS leads to the following jump terms:

$$M_{\gamma\theta}(\mathbf{x}') \lim_{\varepsilon \to 0} \int_{\Gamma_\varepsilon^*} U^*_{\alpha\beta\gamma}(\mathbf{x}', \mathbf{x}) n_\theta(\mathbf{x}) \mathrm{d}\Gamma(\mathbf{x})$$
$$= -\left(\frac{(3\nu - 1)}{16} M_{\gamma\gamma}(\mathbf{x}')\delta_{\alpha\beta} - \frac{2(\nu - 3)}{16} M_{\alpha\beta}(\mathbf{x}') \right) \tag{25}$$

Now, the integral I_3^* can be written as follows:

$$\begin{aligned} I_3^* &= \oint_\Gamma U^*_{\alpha\beta\gamma}(\mathbf{x}', \mathbf{x}) p_\gamma(\mathbf{x}) \mathrm{d}\Gamma(\mathbf{x}) \\ &- \left(\frac{(3\nu - 1)}{16} M_{\gamma\gamma}(\mathbf{x}')\delta_{\alpha\beta} - \frac{2(\nu - 3)}{16} M_{\alpha\beta}(\mathbf{x}') \right) \end{aligned} \tag{26}$$

The integral I_4^*

This integral contains $U^*_{\alpha\beta3}$ which is a weakly singular kernel. This integral will lead to no jump term and it exists in the limit, so that the integral I_4^* can be directly written as follows:

$$I_4^* = \int_\Gamma U^*_{\alpha\beta3}(\mathbf{x}', \mathbf{x}) p_3(\mathbf{x}) \mathrm{d}\Gamma(\mathbf{x}) \tag{27}$$

The integral I_5^*

This integral contains $W^*_{\alpha\beta}$ which is a strongly singular kernel. Directly one can write:

$$\begin{aligned} I_5^* &= q \lim_{\varepsilon \to 0} \int_{\Gamma - \Gamma_\varepsilon + \Gamma_\varepsilon^*} W^*_{\alpha\beta}(\mathbf{x}', \mathbf{x}) \mathrm{d}\Gamma(\mathbf{x}) \\ &= q \lim_{\varepsilon \to 0} \int_{\Gamma - \Gamma_\varepsilon} W^*_{\alpha\beta}(\mathbf{x}', \mathbf{x}) \mathrm{d}\Gamma(\mathbf{x}) \\ &+ q \lim_{\varepsilon \to 0} \int_{\Gamma_\varepsilon^*} W^*_{\alpha\beta}(\mathbf{x}', \mathbf{x}) \mathrm{d}\Gamma(\mathbf{x}) \end{aligned} \tag{28}$$

In the above integrals, the first term of the RHS forms a Cauchy principal value integral. Considering the relationships in eqn (18) and the limits in eqn (19), the last term on the RHS leads to the following jump terms:

$$q \lim_{\varepsilon \to 0} \int_{\Gamma_\varepsilon^*} W^*_{\alpha\beta}(\mathbf{x}', \mathbf{x}) \mathrm{d}\Gamma(\mathbf{x}) = \frac{\nu q}{(1 - \nu)\lambda^2} \frac{(1 + \nu)}{4} \delta_{\alpha\beta} \tag{29}$$

Now, the integral I_5^* can be written as follows:

$$I_5^* = q \fint_\Gamma W^*_{\alpha\beta}(\mathbf{x}', \mathbf{x}) \mathrm{d}\Gamma(\mathbf{x}) + \frac{\nu q}{(1-\nu)\lambda^2} \frac{(1+\nu)}{4} \delta_{\alpha\beta} \tag{30}$$

Substituting from eqns (21), (23), (26), (27), and (30) into eqn (14), and using eqn (1), gives:

$$\begin{aligned} \frac{1}{2} M_{\alpha\beta}(\mathbf{x}') &+ \fint_\Gamma P^*_{\alpha\beta\gamma}(\mathbf{x}', \mathbf{x}) u_\gamma(\mathbf{x}) \mathrm{d}\Gamma(\mathbf{x}) + \fint_\Gamma P^*_{\alpha\beta 3}(\mathbf{x}', \mathbf{x}) u_3(\mathbf{x}) \mathrm{d}\Gamma(\mathbf{x}) \\ &= \fint_\Gamma U^*_{\alpha\beta\gamma}(\mathbf{x}', \mathbf{x}) p_\gamma(\mathbf{x}) \mathrm{d}\Gamma(\mathbf{x}) + \int_\Gamma U^*_{\alpha\beta 3}(\mathbf{x}', \mathbf{x}) p_3(\mathbf{x}) \mathrm{d}\Gamma(\mathbf{x}) \\ &+ q \fint_\Gamma W^*_{\alpha\beta}(\mathbf{x}', \mathbf{x}) \mathrm{d}\Gamma(\mathbf{x}) + \frac{1}{2} \frac{q\nu}{(1-\nu)\lambda^2} \delta_{\alpha\beta} \end{aligned} \tag{31}$$

which is the bending stress resultant boundary integral equation.

4.2 The shear stress resultant integral equation

Eqn (15) can be written in the following form:

$$I_6^* + I_7^* = I_8^* + I_9^* + I_{10}^* \tag{32}$$

Now, each of the integrals I_6^* to I_{10}^* will be considered individually as follows:

The integral I_6^*

This integral contains $P^*_{3\beta\gamma}$ which is strongly singular. Using the first term of the Taylor expansion of u_γ the following form can be written:

$$\begin{aligned} I_6^* &= \lim_{\varepsilon \to 0} \int_{\Gamma - \Gamma_\varepsilon + \Gamma^*_\varepsilon} P^*_{3\beta\gamma}(\mathbf{x}', \mathbf{x}) u_\gamma(\mathbf{x}) \mathrm{d}\Gamma(\mathbf{x}) \\ &= \lim_{\varepsilon \to 0} \int_{\Gamma - \Gamma_\varepsilon} P^*_{3\beta\gamma}(\mathbf{x}', \mathbf{x}) u_\gamma(\mathbf{x}) \mathrm{d}\Gamma(\mathbf{x}) \\ &+ \lim_{\varepsilon \to 0} \int_{\Gamma^*_\varepsilon} P^*_{3\beta\gamma}(\mathbf{x}', \mathbf{x}) [u_\gamma(\mathbf{x}) - u_\gamma(\mathbf{x}')] \mathrm{d}\Gamma(\mathbf{x}) \\ &+ u_\gamma(\mathbf{x}') \lim_{\varepsilon \to 0} \int_{\Gamma^*_\varepsilon} P^*_{3\beta\gamma}(\mathbf{x}', \mathbf{x}) \mathrm{d}\Gamma(\mathbf{x}) \end{aligned} \tag{33}$$

In the above integrals, the second term on the RHS is zero in the limit $\varepsilon \to 0$. The first term forms a Cauchy principal value integral. Considering the relationships in eqn (18) and the limits in eqn (19), the jump terms that appear from the last term on the RHS can be written in the following form:

$$u_\gamma(\mathbf{x}') \lim_{\varepsilon \to 0} \int_{\Gamma_\varepsilon^*} P_{3\beta\gamma}^*(\mathbf{x}', \mathbf{x}) d\Gamma(\mathbf{x}) = \frac{D(1-\nu)\lambda^2}{8} u_\beta(\mathbf{x}') \tag{34}$$

Then the integral I_6^* can be written in the following form:

$$I_6^* = \fint_\Gamma P_{3\beta\gamma}^*(\mathbf{x}', \mathbf{x}) u_\gamma(\mathbf{x}) d\Gamma(\mathbf{x}) + \frac{D(1-\nu)\lambda^2}{8} u_\beta(\mathbf{x}') \tag{35}$$

The integral I_7^*

This integral contains $P_{3\beta 3}^*$ which is a hypersingular kernel. Using the first two terms of the Taylor expansion of u_3 the following form can be written as:

$$\begin{aligned} I_7^* &= \lim_{\varepsilon \to 0} \int_{\Gamma - \Gamma_\varepsilon + \Gamma_\varepsilon^*} P_{3\beta 3}^*(\mathbf{x}', \mathbf{x}) u_3(\mathbf{x}) d\Gamma(\mathbf{x}) = \lim_{\varepsilon \to 0} \int_{\Gamma - \Gamma_\varepsilon} P_{3\beta 3}^*(\mathbf{x}', \mathbf{x}) u_3(\mathbf{x}) d\Gamma(\mathbf{x}) \\ &+ \lim_{\varepsilon \to 0} \int_{\Gamma_\varepsilon^*} P_{3\beta 3}^*(\mathbf{x}', \mathbf{x}) [u_3(\mathbf{x}) - u_3(\mathbf{x}') - u_{3,\alpha}(\mathbf{x}')(x_\alpha(\mathbf{x}) - x_\alpha(\mathbf{x}'))] d\Gamma(\mathbf{x}) \\ &+ u_3(\mathbf{x}') \lim_{\varepsilon \to 0} \int_{\Gamma_\varepsilon^*} P_{3\beta 3}^*(\mathbf{x}', \mathbf{x}) d\Gamma(\mathbf{x}) \\ &+ u_{3,\alpha}(\mathbf{x}') \lim_{\varepsilon \to 0} \int_{\Gamma_\varepsilon^*} P_{3\beta 3}^*(\mathbf{x}', \mathbf{x}) (x_\alpha(\mathbf{x}) - x_\alpha(\mathbf{x}')) d\Gamma(\mathbf{x}) \end{aligned} \tag{36}$$

In the above integrals, the second term on the RHS is zero in the limit $\varepsilon \to 0$. The first and third terms together form a Hadamard finite part integral. Considering the relationships in eqn (18) and the limits in eqn (19), the last term on the RHS leads to the following jump terms:

$$\begin{aligned} & u_{3,\alpha}(\mathbf{x}') \lim_{\varepsilon \to 0} \int_{\Gamma_\varepsilon^*} P_{3\beta 3}^*(\mathbf{x}', \mathbf{x}) (x_\alpha(\mathbf{x}) - x_\alpha(\mathbf{x}')) d\Gamma(\mathbf{x}) \\ &= \frac{D(1-\nu)\lambda^2}{8} u_{3,\beta}(\mathbf{x}') \end{aligned} \tag{37}$$

Now, the integral I_7^* can be written as follows:

$$I_7^* = \oint\!\!\!\!\!\!=_\Gamma P_{3\beta 3}^*(\mathbf{x}', \mathbf{x}) u_3(\mathbf{x}) d\Gamma(\mathbf{x}) + \frac{D(1-\nu)\lambda^2}{8} u_{3,\beta}(\mathbf{x}') \tag{38}$$

The integral I_8^*

Similar to the integral I_4^*, The integral I_8 contains $U_{3\beta\gamma}^*$ which is weakly singular. This integral will lead to no jump term and it exists in the limit. The integral I_8^* can be written directly as follows:

$$I_8^* = \int_\Gamma U_{3\beta\gamma}^*(\mathbf{x}', \mathbf{x}) p_\gamma(\mathbf{x}) \mathrm{d}\Gamma(\mathbf{x}) \tag{39}$$

The integral I_9^*

This integral contains $U_{3\beta 3}^*$ which is a strongly singular kernel. Using the first term of the Taylor expansion of Q_θ the following form can be written as:

$$\begin{aligned} I_9^* &= \lim_{\varepsilon\to 0} \int_{\Gamma-\Gamma_\varepsilon+\Gamma_\varepsilon^*} U_{3\beta 3}^*(\mathbf{x}', \mathbf{x}) p_3(\mathbf{x}) \mathrm{d}\Gamma(\mathbf{x}) = \lim_{\varepsilon\to 0} \int_{\Gamma-\Gamma_\varepsilon} U_{3\beta 3}^*(\mathbf{x}', \mathbf{x}) p_3(\mathbf{x}) \mathrm{d}\Gamma(\mathbf{x}) \\ &+ \lim_{\varepsilon\to 0} \int_{\Gamma_\varepsilon^*} U_{3\beta 3}^*(\mathbf{x}', \mathbf{x}) n_\theta(\mathbf{x}) [Q_\theta(\mathbf{x}) - Q_\theta(\mathbf{x}')] \mathrm{d}\Gamma(\mathbf{x}) \\ &+ Q_\theta(\mathbf{x}') \lim_{\varepsilon\to 0} \int_{\Gamma_\varepsilon^*} U_{3\beta 3}^*(\mathbf{x}', \mathbf{x}) n_\theta(\mathbf{x}) \mathrm{d}\Gamma(\mathbf{x}) \end{aligned} \tag{40}$$

In the above integrals, the second term on the RHS is zero in the limit $\varepsilon \to 0$. The first term on the RHS forms a Cauchy principal value integral. By considering the relationships in eqn (18) and the limits in eqn (19), the jump terms that appear from the last term on the RHS can be written in the following form:

$$Q_\theta(\mathbf{x}') \lim_{\varepsilon\to 0} \int_{\Gamma_\varepsilon^*} U_{3\beta 3}^*(\mathbf{x}', \mathbf{x}) n_\theta(\mathbf{x}) \mathrm{d}\Gamma(\mathbf{x}) = -\frac{Q_\beta(\mathbf{x}')}{4} \tag{41}$$

Then the integral I_9^* can be written in the following form:

$$I_9^* = ⨍_\Gamma U_{3\beta 3}^*(\mathbf{x}', \mathbf{x}) p_3(\mathbf{x}) \mathrm{d}\Gamma(\mathbf{x}) - \frac{Q_\beta(\mathbf{x}')}{4} \tag{42}$$

The integral I_{10}^*

This integral contains $W_{3\beta}^*$ which is weakly singular of $O(\ln r)$. This integral will not lead to a jump term and it exists in the limit. So that, the integral I_{10}^* can be written as:

$$I_{10}^* = q \int_\Gamma W_{3\beta}^*(\mathbf{x}', \mathbf{x}) \mathrm{d}\Gamma(\mathbf{x}) \tag{43}$$

Substituting from eqns (35), (38), (39), (42), and (43) into eqn (15), and using eqn (1) gives

$$
\begin{aligned}
\frac{1}{2}Q_\beta(\mathbf{x}') &+ \fint_\Gamma P^*_{3\beta\gamma}(\mathbf{x}',\mathbf{x})u_\gamma(\mathbf{x})\mathrm{d}\Gamma(\mathbf{x}) + \ffint_\Gamma P^*_{3\beta 3}(\mathbf{x}',\mathbf{x})u_3(\mathbf{x})\mathrm{d}\Gamma(\mathbf{x}) \\
&= \int_\Gamma U^*_{3\beta\gamma}(\mathbf{x}',\mathbf{x})p_\gamma(\mathbf{x})\mathrm{d}\Gamma(\mathbf{x}) + \fint_\Gamma U^*_{3\beta 3}(\mathbf{x}',\mathbf{x})p_3(\mathbf{x})\mathrm{d}\Gamma(\mathbf{x}) \\
&+ q\int_\Gamma W^*_{3\beta}(\mathbf{x}',\mathbf{x})\mathrm{d}\Gamma(\mathbf{x}) \qquad (44)
\end{aligned}
$$

which is the shear stress resultant boundary integral equation.

Eqns (31) and (44) represent three stress resultant integral equations at a boundary point $\mathbf{x}'$.

5 The traction integral equations

If eqns (31) and (44) are dotted by n_β at the collocation point $\mathbf{x}'$, the following two expressions can be written:

$$
\begin{aligned}
\frac{1}{2}p_\alpha(\mathbf{x}') &+ n_\beta(\mathbf{x}')\ffint_\Gamma P^*_{\alpha\beta\gamma}(\mathbf{x}',\mathbf{x})u_\gamma(\mathbf{x})\mathrm{d}\Gamma(\mathbf{x}) \\
&+ n_\beta(\mathbf{x}')\fint_\Gamma P^*_{\alpha\beta 3}(\mathbf{x}',\mathbf{x})u_3(\mathbf{x})\mathrm{d}\Gamma(\mathbf{x}) \\
&= n_\beta(\mathbf{x}')\fint_\Gamma U^*_{\alpha\beta\gamma}(\mathbf{x}',\mathbf{x})p_\gamma(\mathbf{x})\mathrm{d}\Gamma(\mathbf{x}) \\
&+ n_\beta(\mathbf{x}')\int_\Gamma U^*_{\alpha\beta 3}(\mathbf{x}',\mathbf{x})p_3(\mathbf{x})\mathrm{d}\Gamma(\mathbf{x}) \\
&+ qn_\beta(\mathbf{x}')\fint_\Gamma W^*_{\alpha\beta}(\mathbf{x}',\mathbf{x})\mathrm{d}\Gamma(\mathbf{x}) + \frac{1}{2}\frac{q\nu}{(1-\nu)\lambda^2}n_\alpha(\mathbf{x}') \qquad (45)
\end{aligned}
$$

and

$$
\begin{aligned}
\frac{1}{2}p_3(\mathbf{x}') &+ n_\beta(\mathbf{x}')\fint_\Gamma P^*_{3\beta\gamma}(\mathbf{x}',\mathbf{x})u_\gamma(\mathbf{x})\mathrm{d}\Gamma(\mathbf{x}) \\
&+ n_\beta(\mathbf{x}')\ffint_\Gamma P^*_{3\beta 3}(\mathbf{x}',\mathbf{x})u_3(\mathbf{x})\mathrm{d}\Gamma(\mathbf{x}) \\
&= n_\beta(\mathbf{x}')\int_\Gamma U^*_{3\beta\gamma}(\mathbf{x}',\mathbf{x})p_\gamma(\mathbf{x})\mathrm{d}\Gamma(\mathbf{x}) \\
&+ n_\beta(\mathbf{x}')\fint_\Gamma U^*_{3\beta 3}(\mathbf{x}',\mathbf{x})p_3(\mathbf{x})\mathrm{d}\Gamma(\mathbf{x})
\end{aligned}
$$

$$+ \quad qn_\beta(\mathbf{x}') \int_\Gamma W^*_{3\beta}(\mathbf{x}', \mathbf{x}) \mathrm{d}\Gamma(\mathbf{x}) \tag{46}$$

Eqns (45) and (46) represent three integral equations (the hypersingular equations) in terms of boundary tractions, and can either replace or work together with the three displacement integral equations in eqn (9) to form the dual boundary integral formulation.

6 Numerical implementation

In the present work, quadratic isoparametric elements have been used to discretize the boundary of the plate. The local position of the element nodes are general, to allow for the use of continuous or discontinuous elements. The displacement or the traction boundary integral equations can be used. In case of using the traction boundary integral equations, only discontinuous elements are employed, to satisfy the assumed continuity requirements for the boundary variables.

After this discretization, eqn (9) or eqns (45) and (46) can be written in the following form:

$$[H]\{u\} = [G]\{p\} + \{Q\} \tag{47}$$

where $[H]$ and $[G]$ are the well-known boundary element influence matrices [16], $\{u\}$ and $\{p\}$ are the boundary displacement and traction vectors respectively, and $\{Q\}$ is the domain load vector. By imposing the boundary conditions, eqn (47) can be written as:

$$[A]\{x\} = \{b\} \tag{48}$$

where $[A]$ is the system matrix, $\{x\}$ is the vector of unknowns and $\{b\}$ is the vector of prescribed boundary values. This system of algebraic equations can be solved for the boundary unknowns.

It has to be noted that when the displacement integral equation is used, the determinant of the matrix $[A]$ is very small and pivoting is desirable in performing the solution of the equations. Whereas, when the traction integral equation is used, the determinant of the matrix $[A]$ is big, due to the higher singularity order of the kernels. To avoid numerical inaccuracies, the $[H]$ matrix is scaled by the modulus of elasticity. This scaling process is useful when both of the displacement and the traction boundary integral equations are used.

7 Singularities

In this section, the treatment of the singularities that occur in both of the displacement and traction boundary integral equations will be discussed. Singularities in the displacement BIE will be reviewed in a quick manner, as it can be found in other references, whereas singularities in the traction BIE will be discussed in detail.

7.1 Singularities in the displacement BIE

In the displacement boundary integral equation, the influence $[G]$ matrix and the load vector $\{Q\}$ contain weakly singular kernels, which can be canceled using a nonlinear coordinate transformation as in Telles [17]. The influence matrix $[H]$, on the other hand, contains a strongly singular kernel, which can be evaluated indirectly using the so called generalized rigid body movements. This can be achieved as follows. If a traction-free problem is considered, three independent cases may be considered:

- $u_1 = C$ then $u_2 = 0$ and $u_3 = -r_1 C$
- $u_2 = C$ then $u_1 = 0$ and $u_3 = -r_2 C$
- $u_3 = C$ then $u_1 = 0$ and $u_2 = 0$

where C is an arbitrary constant and $r_\alpha = x_\alpha(\mathbf{x}) - x_\alpha(\mathbf{x}')$.

By applying these cases to the system of eqns (47), the following expressions can be written:

$$\begin{aligned} H^{i\alpha}(\mathbf{x}') &= -\int_\Gamma [P^*_{i\alpha}(\mathbf{x}',\mathbf{x}) + (-r_\alpha)P^*_{i3}(\mathbf{x}',\mathbf{x})]\mathrm{d}\Gamma(\mathbf{x}) \\ H^{i3}(\mathbf{x}') &= -\int_\Gamma P^*_{i3}(\mathbf{x}',\mathbf{x})\mathrm{d}\Gamma(\mathbf{x}) \end{aligned} \tag{49}$$

where $H^{ij}(\mathbf{x}')$ (i.e. $H^{i\alpha}(\mathbf{x}')$ and $H^{i3}(\mathbf{x}')$) includes the diagonal sub-matrix and the jump term c_{ij} in the influence matrix $[H]$. The first term in the first integral and the second integral were already computed, however, the second term in the first integral remains to be computed. Fortunately, in this term, the distance r_α cancels the weak singularity in $P^*_{\alpha 3}$ and the strong singularity in P^*_{33} in the singular element under consideration.

7.2 Singularities in the traction BIE

In the traction BIE, the singularity order is higher than the displacement BIE. In this section, each element of eqn (47) will be individually discussed. Three orders of singularity will appear in this formulation. The weak singularity is treated using a nonlinear coordinate transformation as in Telles [17]. The strong singularity is treated using a first order Taylor expansion of the fundamental solution terms around the singular point, as in Aliabadi *et al.* [18]. In this case, the singular term is isolated and integrated analytically. The hypersingular kernels will be computed indirectly using generalized rigid body movements together with first order Taylor expansion.

The influence matrix $[H]$

In the $[H]$ matrix, the kernels $P^*_{\alpha\beta 3}$ and $P^*_{3\beta\gamma}$ are strongly singular, whereas, the kernels $P^*_{\alpha\beta\gamma}$ and $P^*_{3\beta 3}$ are hypersingular.

In the off-diagonal sub-matrices (see Figure 2), the element shape function, will reduce the order of singularities by one. This means that, the elements corresponding to the kernels $P^*_{\alpha\beta 3}$ and $P^*_{3\beta\gamma}$ become smooth, whereas, the elements of the kernels $P^*_{\alpha\beta\gamma}$ and $P^*_{3\beta 3}$ still remain strongly singular.

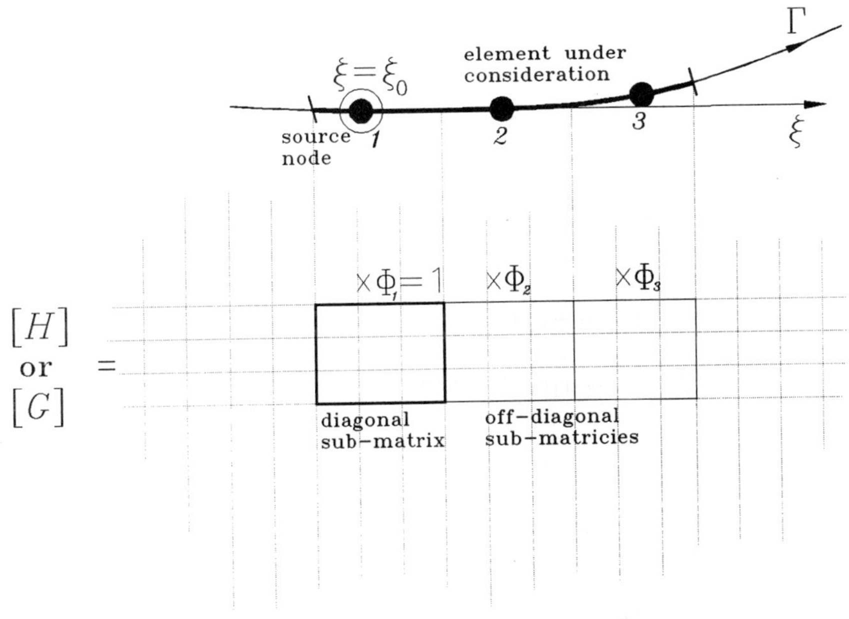

Figure 2. Sub-matrices for the influence matrices.

Now, consider the kernels $P^*_{\alpha\beta\gamma}$, for any off-diagonal sub-matrix, the corresponding integral can be written in the local coordinate system as:

$$\int_{\Gamma_e} P^*_{\alpha\beta\gamma}\Phi^i \mathrm{d}\Gamma = \int_{-1}^{+1} P^*_{\alpha\beta\gamma}\Phi^i(\xi)J(\xi)\mathrm{d}\xi \tag{50}$$

where Γ_e denotes the boundary of the singular element, Φ^i is the element shape function corresponding to the node i in the element under consideration and J is the Jacobian of the transformation form x_α coordinate system to the local coordinate system ξ (i.e. $\mathrm{d}\Gamma = J(\xi)\mathrm{d}\xi$). To deal with this strong singularity, consider the Taylor expansion about the singular point ξ_0 in the local coordinate system, as follows:

$$\begin{aligned}
\Phi^i(\xi) &= \Phi^i(\xi_0) + \Phi^{i\prime}(\xi_0)\delta\xi + \frac{1}{2}\Phi^{i\prime\prime}(\xi_0)\delta\xi^2 + \cdots \\
J(\xi) &= J(\xi_0) + J'(\xi_0)\delta\xi + \frac{1}{2}J''(\xi_0)\delta\xi^2 + \cdots \\
x_\alpha(\xi) &= x_\alpha(\xi_0) + x_\alpha^{'}(\xi_0)\delta\xi + \frac{1}{2}x_\alpha^{''}(\xi_0)\delta\xi^2 + \cdots
\end{aligned} \tag{51}$$

in which

$$x_\alpha(\xi) = \sum_{i=1}^{3} M^i(\xi)x_\alpha^i \tag{52}$$

where, x_α^i is the coordinate x_α at the nodal point i, M^i are the geometric shape functions, $(\cdot)'$ denotes the derivatives of $(\cdot)$ with respect to ξ and $\delta\xi = \xi - \xi_0$. Using eqn (51), the following expressions for the quadratic element can be written [18]:

$$\begin{aligned}
r_\alpha &= x_\alpha^{'}(\xi_0)\delta\xi + \frac{1}{2}x_\alpha^{''}(\xi_0)\delta\xi^2 \\
r &= |\delta\xi|\sqrt{d_0 + d_1\delta\xi + d_2\delta\xi^2}
\end{aligned} \tag{53}$$

where

$$\begin{aligned}
d_0 &= x_\alpha^{'}(\xi_0)x_\alpha^{'}(\xi_0) \\
d_1 &= x_\alpha^{'}(\xi_0)x_\alpha^{''}(\xi_0) \\
d_2 &= \frac{1}{4}x_\alpha^{''}(\xi_0)x_\alpha^{''}(\xi_0)
\end{aligned} \tag{54}$$

Noting that in the case of the off-diagonal sub-matrices $\Phi^i(\xi^{node}) = 0$ for $\xi^{node} \neq \xi_0$. Using the expressions in eqns (51) and (53), together with the limits in (19), eqn (50) can be written in the following form:

$$\begin{aligned}\int_{-1}^{+1} P^*_{\alpha\beta\gamma}\Phi^i(\xi)J(\xi)\mathrm{d}\xi &= \int_{-1}^{+1} [P^*_{\alpha\beta\gamma}\Phi^i(\xi)J(\xi) - S^H_{\alpha\beta\gamma}(\xi_0)]\mathrm{d}\xi \\ &+ \int_{-1}^{+1} S^H_{\alpha\beta\gamma}(\xi_0)\mathrm{d}\xi \end{aligned} \tag{55}$$

and

$$\begin{aligned} S^H_{\alpha\beta\gamma}(\xi_0) &= \frac{D(1-\nu)}{4\pi}\Phi^{i'}J\{[(1-\nu)(\delta_{\gamma\alpha}n_\beta + \delta_{\gamma\beta}n_\alpha) + (-1+3\nu)\delta_{\alpha\beta}n_\gamma]\mathcal{T}_1 \\ &+ [-2(1-\nu)(n_\alpha x'_\beta x'_\gamma + n_\beta x'_\alpha x'_\gamma + \delta_{\gamma\alpha}x'_\beta x'_\theta n_\theta \\ &+ \delta_{\gamma\beta}x'_\alpha x'_\theta n_\theta) - 2(\nu-1)(\delta_{\alpha\beta}x'_\gamma x'_\theta n_\theta + n_\gamma x'_\alpha x'_\beta)]\mathcal{T}_2 \\ &- [8(1+\nu)(x'_\alpha x'_\beta x'_\gamma x'_\theta n_\theta)]\mathcal{T}_3\} \end{aligned} \tag{56}$$

where $S^H_{\alpha\beta\gamma}$ is the isolated singular term, in which all of the involved functions are computed at the singular point ξ_0. The terms $\mathcal{T}_i$ can be defined as follows:

$$\mathcal{T}_i = \frac{1}{\delta\xi(d_0 + d_1\delta\xi + d_2\delta\xi^2)^i} \qquad i = 1,2,3 \tag{57}$$

As can be seen, the first integral on the RHS of eqn (55) is not singular and can be evaluated using Gauss-Legendre scheme. The second integral is computed in the Cauchy principal value sense and is computed analytically.

Similar procedures can be followed for the kernel $P^*_{3\beta3}$, then the following expression can be written:

$$\int_{\Gamma_e} P^*_{3\beta3}\Phi^i\mathrm{d}\Gamma = \int_{-1}^{+1} [P^*_{3\beta3}\Phi^i(\xi)J(\xi) - S^H_{3\beta3}(\xi_0)]\mathrm{d}\xi + \int_{-1}^{+1} S^H_{3\beta3}(\xi_0)\mathrm{d}\xi \tag{58}$$

and

$$S^H_{3\beta3}(\xi_0) = \frac{D(1-\nu)\lambda^2}{4\pi}\Phi^{i'}J[n_\beta\mathcal{T}_1 - 2x'_\beta x'_\theta n_\theta \mathcal{T}_2] \tag{59}$$

On the other hand, the diagonal sub-matrices contain strongly singular and hyper singular terms. Herein, the generalized rigid body movements will be considered in a similar way to that of the displacement BIE, to compute these terms indirectly. By considering the generalized body movements, a similar equation to that in (49) is obtained, i.e.

$$
\begin{aligned}
H^{i\alpha}(\mathbf{x}') &= -n_\beta(\mathbf{x}')\int_\Gamma [P^*_{i\beta\alpha}(\mathbf{x}',\mathbf{x}) + (-r_\alpha)P^*_{i\beta 3}(\mathbf{x}',\mathbf{x})]\mathrm{d}\Gamma(\mathbf{x}) \\
H^{i3}(\mathbf{x}') &= -n_\beta(\mathbf{x}')\int_\Gamma P^*_{i\beta 3}(\mathbf{x}',\mathbf{x})\mathrm{d}\Gamma(\mathbf{x}) \qquad (60)
\end{aligned}
$$

Unlike the displacement BIE, the second term in the first integral of eqn (60) is still singular. The distance r_α cancels the strong singularity in the kernel $P^*_{\alpha\beta 3}$; but it only reduces the singularity order of the kernel $P^*_{3\beta 3}$ to that of a strong singular kernel. This term is considered in Cauchy principal value sense. By employing a first order Taylor expansion as in eqns (51) together with the limits in eqn (19) for the variables in the integral, and noting that on the diagonal sub-matrices $\Phi^i(\xi^{node}) = 1$ for $\xi^{node} = \xi_0$, the following expression can be written:

$$
\begin{aligned}
\fint_{\Gamma_e}(-r_\alpha)P^*_{3\beta 3}\Phi^i\mathrm{d}\Gamma &= \int_{-1}^{+1}[(-r_\alpha)P^*_{3\beta 3}\Phi^i(\xi)J(\xi) - S^{RB}_{\alpha\beta}(\xi_0)]\mathrm{d}\xi \\
&+ \fint_{-1}^{+1} S^{RB}_{\alpha\beta}(\xi_0)\mathrm{d}\xi \qquad (61)
\end{aligned}
$$

and

$$
S^{RB}_{\alpha\beta}(\xi_0) = \frac{-D(1-\nu)\lambda^2}{4\pi}J\{n_\beta x'_\alpha T_1 - 2x'_\beta x'_\alpha x'_\theta n_\theta T_2\} \qquad (62)
$$

where the first integral on the RHS of eqn (61) is not singular, and the second integral contains the singular term and is computed analytically.

The influence matrix $[G]$

In this matrix, the off-diagonal sub-matrices are smooth due to the element shape functions. The diagonal matrices, on the other hand, contain the kernels $U^*_{\alpha\beta 3}$ and $U^*_{3\beta\gamma}$ which are weakly singular. A nonlinear coordinate transformation [17] is used to cancel these singularities. On the other hand, the two kernels, $U^*_{\alpha\beta\gamma}$ and $U^*_{3\beta 3}$ are strongly singular. By employing a first order Taylor expansion (eqn (51)) together with the limits in eqn (19) for the variables in these integrals, and noting that on the diagonal sub-matrices $\Phi^i(\xi^{node}) = 1$ for $\xi^{node} = \xi_0$, the following expressions can be written:

$$
\fint_{\Gamma_e} U^*_{\alpha\beta\gamma}\Phi^i\mathrm{d}\Gamma = \int_{-1}^{+1}[U^*_{\alpha\beta\gamma}\Phi^i(\xi)J(\xi) - S^G_{\alpha\beta\gamma}(\xi_0)]\mathrm{d}\xi + \fint_{-1}^{+1} S^G_{\alpha\beta\gamma}(\xi_0)\mathrm{d}\xi \qquad (63)
$$

$$\int_{\Gamma_e} U^*_{3\beta 3}\Phi^i \mathrm{d}\Gamma = \int_{-1}^{+1} [U^*_{3\beta 3}\Phi^i(\xi)J(\xi) - S^G_{3\beta 3}(\xi_0)]\mathrm{d}\xi + \int_{-1}^{+1} S^G_{3\beta 3}(\xi_0)\mathrm{d}\xi \quad (64)$$

and

$$\begin{aligned} S^G_{\alpha\beta\gamma}(\xi_0) &= \frac{1}{4\pi}J\{[(1-\nu)(\delta_{\beta\gamma}x'_{\alpha} + \delta_{\alpha\gamma}x'_{\beta}) + (\nu-1)\delta_{\alpha\beta}x'_{\gamma}]\mathcal{T}_1 \\ &+ [2(1+\nu)x'_{\alpha}x'_{\beta}x'_{\gamma}]\mathcal{T}_2\} \quad (65) \\ S^G_{3\beta 3}(\xi_0) &= \frac{1}{2\pi}Jx'_{\beta}\mathcal{T}_1 \quad (66) \end{aligned}$$

where the first integral on the RHS of eqn (63) and eqn (64) is not singular, and the second integral contains the singular term and is computed analytically. The terms $\mathcal{T}_i$ are defined in eqn (57).

The domain load vector $\{Q\}$

There are two kernels in the consideration of the domain load vector term. The first kernel is $W^*_{3\beta}$ which is weakly singular, hence, a nonlinear coordinate transformation may be used to treat this kernel. The second one is $W^*_{\alpha\beta}$, which is strongly singular. By employing a first order Taylor expansion (eqns (51) and (53)) together with the limits in eqn (19) for the variables in this integral, the singular term is isolated as follows:

$$\int_{\Gamma_e} W^*_{\alpha\beta}\Phi^i \mathrm{d}\Gamma = \int_{-1}^{+1} [W^*_{\alpha\beta}\Phi^i(\xi)J(\xi) - S^Q_{\alpha\beta}(\xi_0)]\mathrm{d}\xi + \int_{-1}^{+1} S^Q_{\alpha\beta}(\xi_0)\mathrm{d}\xi \quad (67)$$

and

$$\begin{aligned} S^Q_{\alpha\beta}(\xi_0) &= \frac{-\nu}{(1-\nu)\lambda^2}\frac{1}{4\pi}J\{[(1-\nu)(x'_{\alpha}n_{\beta} + x'_{\beta}n_{\alpha}) + (\nu-1)\delta_{\alpha\beta}x'_{\gamma}n_{\gamma}]\mathcal{T}_1 \\ &+ [2(1+\nu)x'_{\alpha}x'_{\beta}x'_{\gamma}n_{\gamma}]\mathcal{T}_2\} \quad (68) \end{aligned}$$

where the first integral on the RHS of eqn (67) is not singular, and the second integral contains the singular term and is computed analytically. The terms $\mathcal{T}_i$ are defined in eqn (57).

8 Boundary stress resultants

The boundary stress resultant computation can be achieved either by computing the generalized local strains on the boundary using displacement derivatives, and then make use of the generalized Hooke's law or by using the stress resultant integral identity (eqn(12)) at a boundary point.

8.1 Numerical differentiation of local variables (SFD)

The values of the stress resultant tensor at a boundary point can be evaluated in terms of the local tractions and tangential generalized strains as in two- or three-dimensional elasticity [19]. Consider the local coordinate system shown in Figure 3, the boundary generalized displacements and tractions at the point $\mathbf{x}'$ in the system are given as:

$$\begin{aligned} u_i^o &= e_{ij}^o u_j \\ p_i^o &= e_{ij}^o p_j \end{aligned} \tag{69}$$

where the superscript $(^o)$ denotes the local boundary values, and the tensor e_{ij}^o is the rotation matrix and given by:

$$e_{ij}^o = \begin{bmatrix} n_1 & n_2 & 0 \\ -n_2 & n_1 & 0 \\ 0 & 0 & 1 \end{bmatrix} \tag{70}$$

By considering the equilibrium of stress resultants in the local coordinate system, it can be seen that:

$$\begin{aligned} M_{1\alpha}^o &= p_\alpha^o \\ Q_1^o &= p_3^o \end{aligned} \tag{71}$$

Other components of the local boundary stress resultant tensor can be evaluated using eqn (1) as follows:

$$\begin{aligned} M_{22}^o &= D[u_{2,2}^o + \nu u_{1,1}^o] + \frac{\nu q}{(1-\nu)\lambda^2} \\ Q_2^o &= \frac{D(1-\nu)\lambda^2}{2}[u_2^o + u_{3,2}^o] \end{aligned} \tag{72}$$

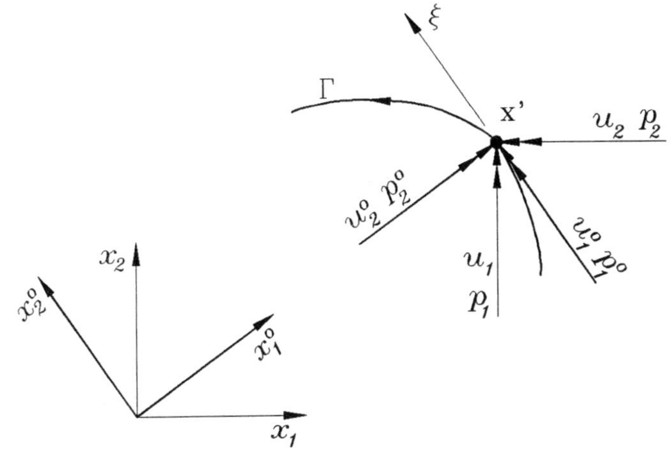

Figure 3. Local and global coordinate system at $\mathbf{x}'$.

By considering Figure 3, one can find [19] the following relationships:

$$\begin{aligned} \frac{\partial \xi}{\partial x_2^o(\mathbf{x}')} &= \frac{1}{J(\mathbf{x}')} \\ \frac{\partial \xi}{\partial x_1^o(\mathbf{x}')} &= 0 \end{aligned} \tag{73}$$

The displacement approximation in terms of the element shape functions Φ^k may be written as:

$$u_j^o = \Phi^k u_j^{ok} \tag{74}$$

where u_j^{ok} are the local boundary displacements at the boundary node j.
The displacement derivatives in eqn (72) can be rewritten in the following form:

$$u_{j,\theta}^o = \Phi_{,\theta}^k u_j^{ok} = \frac{\partial \Phi^k}{\partial \xi} e_{ji}^o u_i^k \frac{\partial \xi}{\partial x_\theta^o(\mathbf{x}')} \tag{75}$$

Using eqn (75), the displacement derivatives can be computed and hence, the local boundary stress tensor can be evaluated.
The global stress tensor can be evaluated via the following transformation:

$$\begin{aligned} M_{\alpha\beta} &= e_{\theta\alpha}^o e_{\gamma\beta}^o M_{\theta\gamma}^o \\ Q_\alpha &= e_{\beta\alpha}^o Q_\beta^o \end{aligned} \tag{76}$$

As can be seen this method is based on the computation of the tangential strains by approximating the strains in terms of the element shape function derivatives. In this chapter, this method will be referred to as (SFD).

It is worth noting that eqn (72) cannot be easily used to evaluate the tangential bending stresses for the case of the clamped edge as $u_\alpha = 0$.

8.2 Stress resultant integral equations (SIE)

The boundary stress resultants can be computed directly from eqns (31) and (44). As can be seen this method contains no approximations and can be used for any type of boundary conditions. However it is time consuming as it is based on computing hypersingular kernels. In this chapter the symbol (SIE) is used to refer to this method.

9 Numerical examples

In this section, several numerical examples are analyzed using the present boundary element formulations. In all examples quadratic discontinuous boundary elements are used for both the traction boundary integral equation (TBIE) and the displacement boundary integral equation (DBIE) to allow the comparison with the same discretization scheme; otherwise it will be stated. The collocation points are placed at $\xi = 0.7, 0, -0.7$. It was found that the numerical results are not significantly affected by the use continuous or discontinuous elements for the DBIE, or even by the location of the collocation points in the discontinuous elements for both the DBIE or the TBIE. The Gauss-Legendre scheme is used to compute the regular integrals with 10 points. The results obtained by the TBIE are compared with analytical solutions and the results from the standard DBIE.

9.1 Circular clamped thick plate

In this example, a clamped circular plate is considered. A uniform load of intensity q is applied over the plate domain. Due to symmetry, only one quarter of the plate is modeled. The plate is of radius a and with a thickness of $0.2a$. The exact solution for this problem is given by Vander Weeën [10]. Table 1 shows the results for the deflection and the radial rotation at different internal points. Table 2 shows the results for the tangential and the radial bending moments and the shearing forces at the same points in Table 1. In the case of TBIE the plate was discretized into 16 boundary elements. As can be seen in Tables 1 and 2, the results are in a good agreement with the exact solutions.

Table 1. Circular clamped plate results (generalized displacements)

$\frac{r}{a}$	$\frac{64D}{qa^4}w$		$\frac{16D}{qa^3}\phi_r$	
	Exact	TBIE	Exact	TBIE
0.00	1.1829	1.1839	0.0000	0.0000
0.25	1.0503	1.0501	0.2344	0.2342
0.50	0.6996	0.6995	0.3750	0.3750
0.75	0.2714	0.2714	0.3281	0.3281
1.00	0.0000	0.0000	0.0000	0.0000

Table 2. Circular clamped plate results (stress resultants)

$\frac{r}{a}$	$\frac{16}{qa^2}M_{\theta\theta}$		$\frac{16}{qa^2}M_{rr}$		$\frac{2}{qa}Q_r$	
	Exact	TBIE	Exact	TBIE	Exact	TBIE
0.00	1.3274	1.3329	1.3274	1.3329	0.00	0.00
0.25	1.2087	1.2085	1.1212	1.1210	-0.25	-0.25
0.50	0.8524	0.8523	0.5024	0.5023	-0.50	-0.50
0.75	0.2587	0.2586	-0.5288	-0.5289	-0.75	-0.75
1.00	-0.5726	-0.5629	-1.9726	-1.9694	-1.00	-1.00

9.2 Simply supported thin square plate

Consider a square plate of 4m side length and simply supported from all sides as shown in Figure 4. A uniform load of -0.64tf/m^2 is applied over the plate domain. Seven internal points are considered (see Figure 4). Due to the problem symmetry, only one quarter of the plate was modeled in the analysis for the points 1 to 6. The results for point 7 which is located at the center of the plate were obtained by modeling the complete plate.

Tables 3, 4 and 5 present the displacements, bending moments and shear

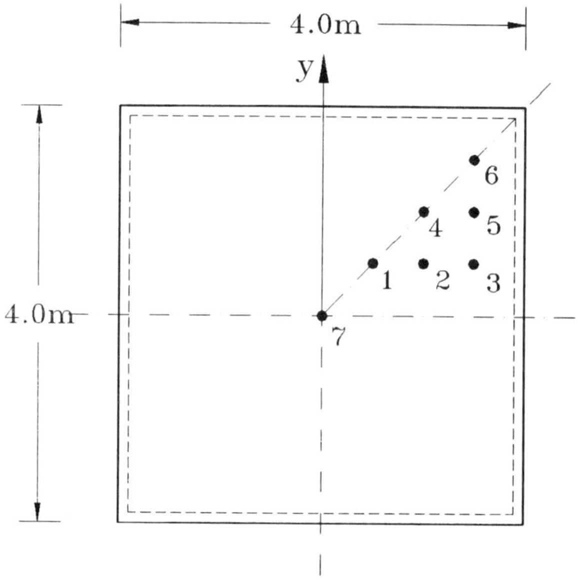

Figure 4. Simply supported thin square plate.

stresses at the internal points shown in Figure 4. The exact results for this problem can be found in Ref. [20] and the corresponding numerical results for both the DBIE and the TBIE are based on a model with 64 boundary elements. Referring to Ref. [21], accurate results for the DBIE can be achieved using only 8 continuous elements; but it was found that by employing 32 elements the results for the TBIE has an error of 2% to 7%; whereas, with 64 elements (as shown in the tables) the error is within 0.6%.

The direct evaluation of stress resultants on the boundary is one of the important applications of the hypersingular integral equations. To demonstrate this application, the same plate is considered and analyzed twice to compare the stress resultants on a boundary line (y-axes) for a quarter of the plate with the stress resultants when this line is considered as an internal line for the full plate analysis. In the first analysis, a quarter of the plate was considered by employing 16 boundary elements per side. The boundary stress resultants along the boundary along y-axis (see Figure 4) were computed using the SFD method and using the hypersingular integral equation (SIE) in eqns (31) and (44). In the second analysis, the complete plate was reanalyzed by using 16 continuous

Table 3. Thin square plate, displacements at internal points

Pt	$\phi_x \times -10^2$(rad)			$\phi_y \times -10^2$(rad)			$w \times -10^2$(m)		
	Exact	DBIE	TBIE	Exact	DBIE	TBIE	Exact	DBIE	TBIE
1	0.1844	0.1844	0.1855	0.1844	0.1844	0.1854	0.6134	0.6145	0.6176
2	0.3552	0.3551	0.3564	0.1424	0.1423	0.1433	0.4776	0.4785	0.4809
3	0.4902	0.4902	0.4916	0.0778	0.0778	0.0785	0.2641	0.2645	0.2664
4	0.2752	0.2752	0.2763	0.2752	0.2752	0.2763	0.3725	0.3733	0.3752
5	0.3822	0.3821	0.3833	0.1510	0.1510	0.1518	0.2066	0.2070	0.2084
6	0.2117	0.2116	0.2125	0.2117	0.2116	0.2125	0.1151	0.1154	0.1163
7	0.0000	0.0000	0.0000	0.0000	0.0000	0.0000	0.7094	0.7110	0.7119

Table 4. Thin square plate, bending moments at internal points

Pt	M_{xx}(tf.m/m)			M_{xy}(tf.m/m)			M_{yy}(tf.m/m)		
	Exact	DBIE	TBIE	Exact	DBIE	TBIE	Exact	DBIE	TBIE
1	-0.4396	-0.4396	-0.4402	0.0378	0.0377	0.0378	-0.4396	-0.4396	-0.4402
2	-0.3744	-0.3745	-0.3749	0.0715	0.0715	0.0717	-0.3504	-0.3505	-0.3509
3	-0.2410	-0.2410	-0.2414	0.0960	0.0960	0.0964	-0.2027	-0.2028	-0.2030
4	-0.3014	-0.3015	-0.3018	0.1367	0.1367	0.1370	-0.3014	-0.3015	-0.3018
5	-0.1982	-0.1983	-0.1985	0.1857	0.1856	0.1861	-0.1769	-0.1770	-0.1772
6	-0.1212	-0.1212	-0.1213	0.2577	0.2576	0.2580	-0.1212	-0.1212	-0.1213
7	-0.4905	-0.4904	-0.4904	0.0000	0.0000	0.0000	-0.4905	-0.4904	-0.4904

Table 5. Thin square plate, shear forces at internal points

Pt	Q_x(tf/m)			Q_y(tf/m)		
	Exact	DBIE	TBIE	Exact	DBIE	TBIE
1	0.1529	0.1527	0.1531	0.1529	0.1527	0.1531
2	0.3278	0.3275	0.3279	0.1200	0.1199	0.1203
3	0.5471	0.5467	0.5474	0.0668	0.0667	0.0671
4	0.2613	0.2610	0.2614	0.2613	0.2610	0.2614
5	0.4505	0.4502	0.4506	0.1485	0.1482	0.1485
6	0.2713	0.2710	0.2713	0.2713	0.2710	0.2713
7	0.0000	0.0000	0.0000	0.0000	0.0000	0.0000

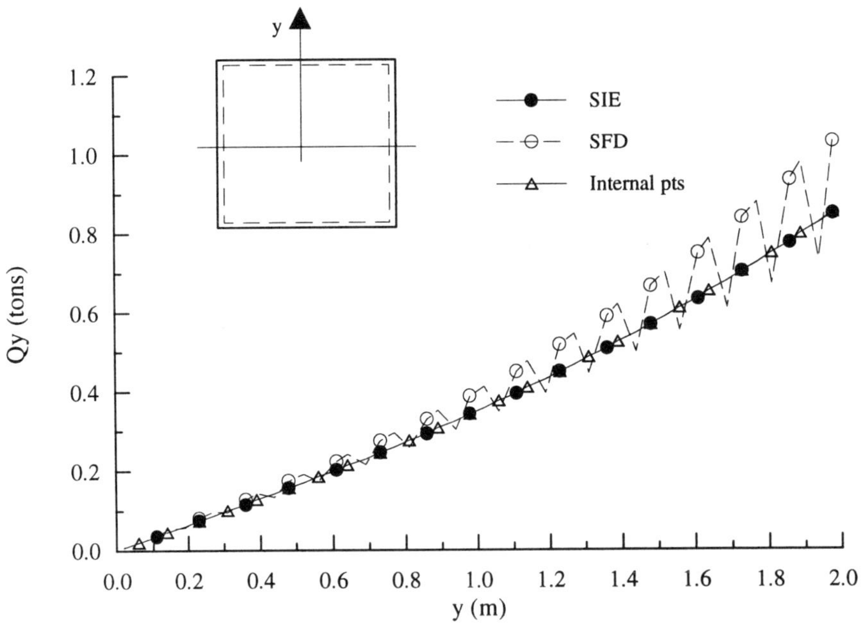

Figure 5. Shear stress resultant along the y-axis for the simply supported plate.

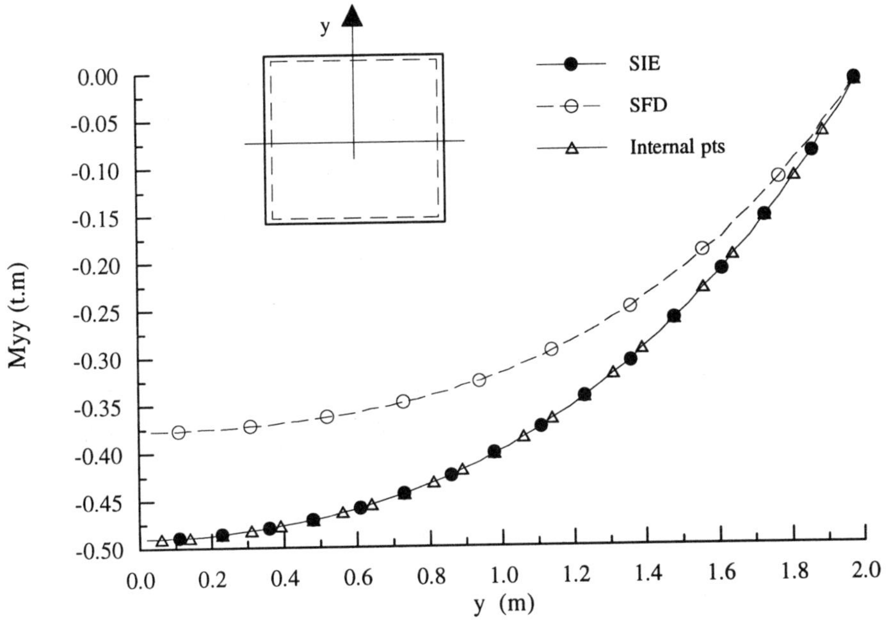

Figure 6. Bending stress resultant along the y-axis for the simply supported plate.

boundary elements per side. The stress resultants were computed at the internal points, which are in the same place as the boundary points in the first analysis. Figures 5 and 6 show the shear and the bending stress resultants respectively, along the considered line. It can be seen that the SFD method shows oscillations in the shear stress resultant. In addition, the SFD results are not accurate for the bending stress resultant, whereas, the SIE results have the same accuracy as those of the internal point computation.

9.3 Timoshenko beam

To demonstrate the capability of the proposed model in analyzing beams, a Timoshenko beam of length $10m$ is considered. The beam has a cross section of 3m depth $\times$ 1m width. The following material properties are considered $\nu = 0.2$ and $E = 2 \times 10^6 \mathrm{t/m^2}$. The beam was fixed from one end and left free as a cantilever. A concentrated load of $P = 1$ ton is applied at the free end of the beam. A mesh of 20 elements along the beam length (for each side) $\times$ 4 elements along the beam width (per side) is used to model the beam mid-plane. The analytical solution for that problem considering shear deformation can be found in Ref. [22]. The

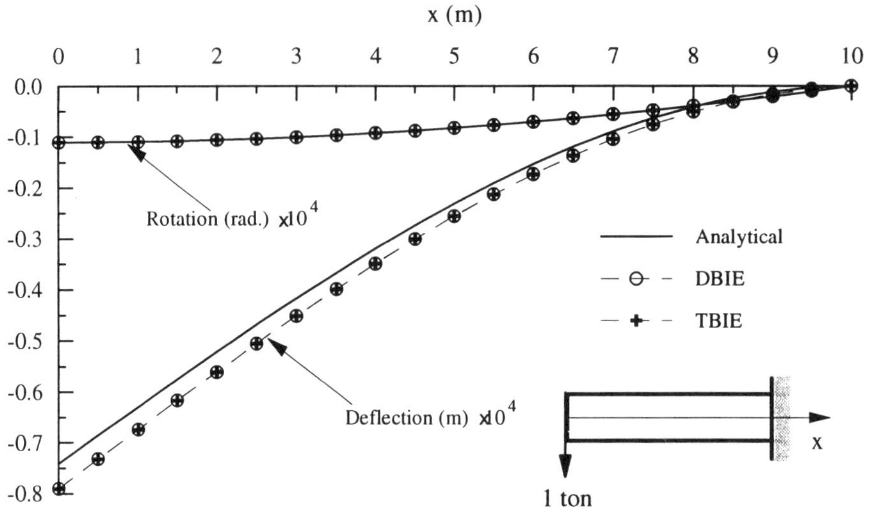

Figure 7. Rotation and deflection results for the Timoshenko beam.

results for the rotation in the x direction and the transverse deflection along the beam centre line are plotted together with the analytical results in Figure 7. The bending moment and shearing force digrams along the same line are also plotted in Figure 8. As can be seen from these figures the results of the TBIE are in good agreement with those of the DBIE. A difference (6%) between the BIE and the analytical solution for the maximum deflection of the beam is obtained. It has to be noted that there are small inaccuracies in the bending moment and the shearing force near the fixed edge. These are mainly due to the poor discretization of the beam width.

10 Conclusions

In this chapter, the stress resultant based integral equations were derived for the Reissner plate theory. The boundary stress resultants are computed directly and compared to the conventional way of evaluating them using the derivatives of the shape functions. Several examples were solved and the following conclusions can be drawn from them:

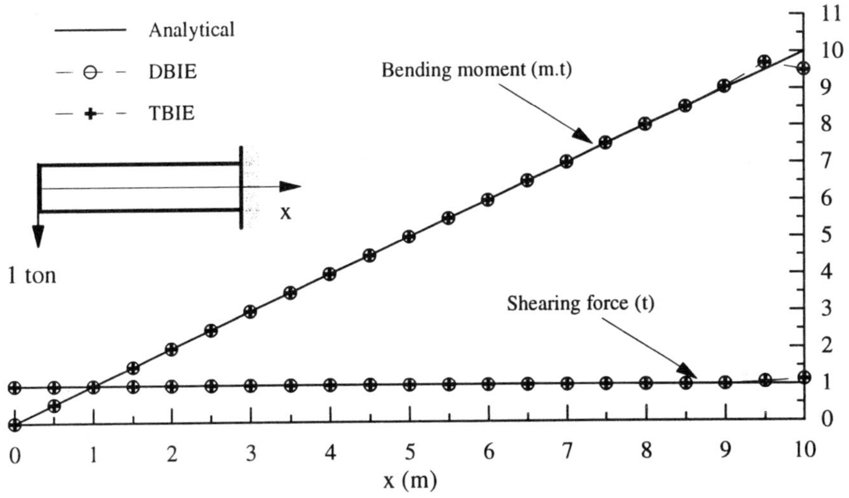

Figure 8. Bending moment and shearing force results for the Timoshenko beam.

1. The traction boundary integral equations require a finer discretization to obtain the same level of accuracy as in the displacement boundary integral equations.

2. The system of equations resulting from the displacement integral equation has small determinant value. Conversely, the determinant of the system of equations resulting from the traction integral equations is big and scaling is desirable when both of them are used together.

3. The stress resultant based integral equations can be used to evaluate the boundary stress resultants for any boundary condition type, unlike the evaluation of the boundary stress resultants via numerical differentiations of the local variables.

References

[1] Portela, A., Aliabadi, M.H. & Rooke, D.P. Dual boundary element incremental analysis of crack propagation, *Computers & Structures*, **46**(2), 237-247, 1993.

[2] Mi, Y. & Aliabadi, M.H. Three-Dimensional crack growth simulation using BEM , *Computers & Structures*, **52**(5), 871-878, 1994.

[3] Huber, O., Dallner, R., Partheymuller, P., & Kuhn, G. Evaluation of the stress tensor in 3-D elastoplasticity by direct solving of hypersingular integrals. *Int. J. Num. Methods Engineering*, **39**, 2555-2573, 1996.

[4] Paulino, G.H., Gray, L.J., & Zarikian, V. Hypersingular residuals: A new approach for error estimation in the boundary element method. *Int. J. Num. Methods Engineering*, **39**, 2005-2029, 1996.

[5] Bialecki, R., Dallner, R. & Kuhn, G. New application of hypersingular equations in the boundary element method. *Comp. Meth. in Applied Mechanics and Engineering*, **103**, 399-416, 1993.

[6] Bézine, G. Boundary integral formulation for plate flexure with arbitrary boundary conditions. *Mech. Res. Comm.*, **5**(4), 197-206, 1978.

[7] Stern, M. A general boundary integral formulation for the numerical solution of plate bending problems. *Int. J. Solids Structures*, **15**, 769-782, 1979.

[8] Knöpke, B. The hypersingular integral equation for the bending moments m_{xx}, m_{xy} and m_{yy} of the Kirchhoff plate. *Computational Mechanics*, **15**, 19-30, 1994.

[9] Reissner, E. On bending of elastic plates, *Quart. Appl. Math.*, **5**, 55-68, 1947.

[10] Vander Weeën, F. Application of the boundary integral equation method to Reissner's plate model, *Int. J. Num. Methods Engineering*, **18**, 1-10, 1982.

[11] Rashed, Y.F., Aliabadi, M.H. & Brebbia, C.A. Hypersingular boundary element formulation for Reissner plates. *Int. J. Solids Structures* (accepted).

[12] Rashed, Y.F., Aliabadi, M.H. & Brebbia, C.A. On the evaluation of the stresses in the BEM for Reissner plate bending problems. *Appl. Math. Modeling*, **21**, 155-163, 1997.

[13] Rashed, Y.F., Aliabadi, M.H. & Brebbia, C.A. Stress and displacement boundary integral formulations for shear-deformable plate bending analysis, *BEM XVIII*, eds C.A. Brebbia, J.B. Martins, M.H. Aliabadi and N. Haie, CMP, Southampton, pp. 493-502, 1996.

[14] Hörmander, L. *Linear Partial Differential Operators* Springer Verlag, Berlin, 1963.

[15] Abramowitz, M. & Stegun, I.A. *Handbook of Mathematical Functions* Dover, New York, 1965.

[16] Brebbia, C.A., Telles, J.C.F. & Wrobel, L.C. it Boundary Element Techniques: Theory and Applications in Engineering Springer-Verlag, Berlin, Heidelberg, 1984.

[17] Telles, J.C.F. A self-adaptive coordinate transformation for efficient numerical evaluation of general boundary element integrals, *Int. J. Num. Meth. Engng*, **24**, 959-973, 1987.

[18] Aliabadi, M.H., Hall, W.S. & Phemister, T.G. Taylor expansion for singular kernels in the boundary element method, *Int. J. Num. Methods Engineering*, **21**, 2221-2236, 1985.

[19] Aliabadi, M.H. & Rooke, D.P. *Numerical Fracture Mechanics* CMP, Kluwer Academic, 1992.

[20] Timoshenko, S. & Woinowsky-Krieger, S. it Theory of Plates and Shells, McGraw-Hill, New York, 1959.

[21] Karam, V.J. & Telles, J.C.F. On boundary elements for Reissner's plate theory, *Engineering Analysis*, **5**(1), 21-27, 1985.

[22] Timoshenko, S. & Goodier, J.N. it Theory of Elasticity McGraw-Hill, New York, 1934.

Appendix: Expressions for U^*_{ijk}, P^*_{ijk} and $W^*_{i\beta}$

$$\begin{aligned}
U^*_{\alpha\beta\gamma} &= \frac{1}{4\pi r}[(4A(z)+2zK_1(z)+1-\nu)(\delta_{\beta\gamma}r_{,\alpha}+\delta_{\alpha\gamma}r_{,\beta}) \\
&- 2(8A(z)+2zK_1(z)+1-\nu)r_{,\alpha}r_{,\beta}r_{,\gamma}+(4A(z)+1+\nu)\delta_{\alpha\beta}r_{,\gamma}] \\
U^*_{\alpha\beta 3} &= \frac{-(1-\nu)}{8\pi}\left[\left(2\frac{(1+\nu)}{(1-\nu)}\ln z-1\right)\delta_{\alpha\beta}+2r_{,\alpha}r_{,\beta}\right] \\
U^*_{3\beta\gamma} &= \frac{\lambda^2}{2\pi}[B(z)\delta_{\gamma\beta}-A(z)r_{,\gamma}r_{,\beta}] \\
U^*_{3\beta 3} &= \frac{1}{2\pi r}r_{,\beta} \qquad\qquad \text{(A1)}
\end{aligned}$$

$$
\begin{aligned}
P^*_{\alpha\beta\gamma} &= \frac{D(1-\nu)}{4\pi r^2}\{(4A(z)+2zK_1(z)+1-\nu)(\delta_{\gamma\alpha}n_\beta+\delta_{\gamma\beta}n_\alpha) \\
&+ (4A(z)+1+3\nu)\delta_{\alpha\beta}n_\gamma - (16A(z)+6zK_1(z)+z^2K_0(z)+2-2\nu) \\
&\times [(n_\alpha r_{,\beta}+n_\beta r_{,\alpha})r_{,\gamma}+(\delta_{\gamma\alpha}r_{,\beta}+\delta_{\gamma\beta}r_{,\alpha})r_{,n}] \\
&- 2(8A(z)+2zK_1(z)+1+\nu)(\delta_{\alpha\beta}r_{,\gamma}r_{,n}+n_\gamma r_{,\alpha}r_{,\beta}) \\
&+ 4(24A(z)+8zK_1(z)+z^2K_0(z)+2-2\nu)r_{,\alpha}r_{,\beta}r_{,\gamma}r_{,n}\} \\
P^*_{\alpha\beta 3} &= \frac{D(1-\nu)\lambda^2}{4\pi r}[(2A(z)+zK_1(z))(r_{,\beta}n_\alpha+r_{,\alpha}n_\beta) \\
&- 2(4A(z)+zK_1(z))r_{,\alpha}r_{,\beta}r_{,n}+2A(z)\delta_{\alpha\beta}r_{,n}] \\
P^*_{3\beta\gamma} &= \frac{-D(1-\nu)\lambda^2}{4\pi r}[(2A(z)+zK_1(z))(\delta_{\gamma\beta}r_{,n}+r_{,\gamma}n_\beta) \\
&+ 2A(z)n_\gamma r_{,\beta} - 2(4A(z)+zK_1(z))r_{,\gamma}r_{,\beta}r_{,n}] \\
P^*_{3\beta 3} &= \frac{D(1-\nu)\lambda^2}{4\pi r^2}[(z^2B(z)+1)n_\beta-(z^2A(z)+2)r_{,\beta}r_{,n}]
\end{aligned}
\tag{A2}
$$

$$
\begin{aligned}
W^*_{\alpha\beta} &= \frac{-r}{64\pi}\{(4\ln z-3)[(1-\nu)(r_{,\beta}n_\alpha+r_{,\alpha}n_\beta)+(1+3\nu)\delta_{\alpha\beta}r_{,n}] \\
&+ 4[(1-\nu)r_{,\alpha}r_{,\beta}+\nu\delta_{\alpha\beta}]r_{,n}\} - \frac{\nu}{(1-\nu)\lambda^2}U^*_{\alpha\beta\gamma}n_\gamma \\
W^*_{3\beta} &= \frac{1}{8\pi}[(2\ln z-1)n_\beta+2r_{,\beta}r_{,n}] - \frac{\nu}{(1-\nu)\lambda^2}U^*_{3\beta\gamma}n_\gamma
\end{aligned}
\tag{A3}
$$

Chapter 6

Fracture analysis of plate bending problems using the boundary element method

J.L.Wearing & S.Y. Ahmadi-Brooghani
Department of Mechanical Engineering, University of Sheffield, UK

Abstract

This paper discusses the application of the Boundary Element Method for the fracture analysis of plate bending problems. Results of K_I and K_{II} and K_{III} stress intensity factors for linear elastic fracture mechanics are presented for a number of case studies. Two approaches depending on the boundary conditions and loading conditions are used in the analysis. In the first approach use is made of the symmetry of the plates' boundary conditions and in the second approach, where symmetry cannot be exploited, the Dual Boundary Element Method is used. The *J*-integral method and the displacement extrapolation method have been used to determine the stress intensity factors. In all cases the boundary element results have been compared with analytical results or finite element results and good agreement has been achieved.

1 Introduction

The calculation of the stress distribution in the vicinity of cracks in engineering components is of vital importance to the engineer, as the presence of a crack in a component can significantly reduce its strength. The consequences of a crack in a component must therefore be considered when assessing the structural integrity of the component. Theoretical work relating to the fracture analysis of components has been well documented[1,2] and predicts a singular stress condition at the crack tip. The prediction of the stress distribution in the region of a crack depends on the stress intensity factor for the crack under consideration and there is a considerable amount of handbook data[3] available for the calculation of the stress intensity factors of a wide range of component geometries, loading conditions and crack configurations. Despite being extensive, there are many occasions when the handbook data does not fully satisfy the requirements of the component being analysed. As components, their loading conditions and crack configurations become more complex, stress analysts have sought to extend the available techniques for the determination of stress intensity factors in cracked components. Consequently, considerable attention has been focused in recent years on the use of numerical methods for the determination of stress intensity factors, where existing published data does not provide satisfactory answers. The two principal numerical methods which have been used for the analysis of cracked structures are the finite element method (FEM) and the boundary element method (BEM). In the field of linear elastic fracture mechanics (LEFM), the BEM is particularly attractive compared to the FEM, as the BEM requires considerably fewer degrees of freedom than the FEM to model the component being analysed. Considerable attention has therefore been given to the application of the BEM for the fracture analysis of engineering components and there are numerous examples where the BEM has been used in the field of linear elastic fracture mechanics.[4,5] There is, however, one major disadvantage when using the BEM for fracture analysis if symmetry of the component is unable to be exploited and both crack faces are modelled using displacement boundary element equations. The coincidence of the crack surfaces, in these cases, yields a system of singular algebraic equations. This problem can be overcome by dividing the problem into subregions, with opposite faces of the crack being in different subregions. Alternatively the dual boundary element method (DBEM) can be used, with the problem being analysed as a single region and one face of the crack being modelled using the displacement boundary integral equation and the other face of the crack being modelled using the traction boundary integral equations. The DBEM therefore eliminates the singular algebraic equations and the need for subregioning, and it has been used in two and three dimensional fracture analysis as illustrated by Portela[6] and Mi.[7] Although the BEM has been used extensively for the fracture analysis of two and three dimensional engineering problems there are considerable fewer illustrations of its use for the fracture analysis of plates. Wearing[8] and Ahmadi-Brooghani[9] have recently illustrated that the BEM can be used effectively for the determination of the stress intensity factors in plate bending problems. This chapter discusses the background theory relating to the fracture analysis of plate bending problems using the BEM, and presents a range of problems to demonstrate the effectiveness of the Boundary Element Method for the linear elastic fracture mechanics analysis of plate bending problems.

2 The boundary element method for plate bending

The plate bending boundary element equations which are used in this paper relate to Reissner's[10] plate bending equation which may be expressed as:

$$\Delta^*_{ij} u_j + b_i = 0, \tag{1}$$

where the Navier operator Δ^*_{ij} in the above equation is given by,

$$\Delta^*_{\alpha\beta} = D\frac{(1-\upsilon)}{2}\left[\left(\nabla^2 - \tau^2\right)\delta_{\alpha\beta} + \frac{(1+\upsilon)}{(1-\upsilon)}\frac{\partial^2}{\partial x_\alpha \partial x_\beta}\right], \tag{2a}$$

$$\Delta^*_{\alpha 3} = -\Delta^*_{3\alpha} = -D\frac{(1-\upsilon)}{2}\lambda^2\frac{\partial}{\partial x_\alpha}, \tag{2b}$$

$$\Delta^*_{33} = -D\frac{(1-\upsilon)}{2}\lambda^2\nabla^2, \tag{2c}$$

in which ∇^2 is the Laplace operator and b_i represents the generalised body force components acting on the plate as given by,

$$b_\alpha = \frac{\upsilon q_{1\alpha}}{\lambda^2(1-\upsilon)}, \tag{2d}$$

$$b_3 = q\,. \tag{2e}$$

The boundary element equations relating to Reissner's plate equation are given by Karam and Telles.[11] The displacements at an internal point on the plate are obtained from the following expression:

$$u_i(\xi) + \int_\Gamma T^*_{ij}(\xi,x)u_j(x)d\Gamma(x) = \int_\Gamma U^*_{ij}(\xi,x)t_j(x)d\Gamma(x) + \int_\Omega\left(U^*_{i3}(\xi,x) - \frac{\upsilon}{(1-\upsilon)\lambda^2}U^*_{i\alpha,\alpha}(\xi,x)\right)q(x)d\Gamma(x), \tag{3}$$

in which x is the field point, ξ is the source point and U^*_{ij} (ξ, x) and T^*_{ij} (ξ, x) are the fundamental solutions for displacement and traction, respectively. When the distance, r, between ξ and x is not zero, the integrals in eqn (3) are regular. The integrals become singular, however, when ξ is moved to the boundary of the plate and ξ coincides with x and in these circumstances eqn (3) becomes:

$$c_{ij}(\xi)u_j(\xi) + (CPV)\int_\Gamma T^*_{ij}(\xi,x)u_j(x)d\Gamma(x) = \int_\Gamma U^*_{ij}(\xi,x)t_j(x)d\Gamma(x) + \int_\Omega\left(U^*_{i3}(\xi,x) - \frac{\upsilon}{(1-\upsilon)\lambda^2}U^*_{i\alpha,\alpha}(\xi,x)\right)q(x)d\Gamma(x), \tag{4}$$

where $(CPV)\int$ denotes a Cauchy principal value integral.

Stresses at internal points are obtained from the integral equations for bending moment and shear force, which are derived from eqns (3) and (4) as:

$$M_{\alpha\beta}(\xi) = \int_\Gamma D^*_{\alpha\beta k}(\xi,x)t_k(x)d\Gamma(x) - \int_\Gamma S^*_{\alpha\beta k}(\xi,x)u_k(x)d\Gamma(x) + q\int_\Omega W^*_{\alpha\beta}(\xi,x)d\Gamma(x) + \frac{\upsilon}{(1-\upsilon)\lambda^2}q\delta_{\alpha\beta}, \quad (5)$$

$$Q_\beta(\xi) = \int_\Gamma D^*_{3\beta k}(\xi,x)t_k(x)d\Gamma(x) - \int_\Gamma S^*_{3\beta k}(\xi,x)u_k(x)d\Gamma(x) + q\int_\Omega W^*_{3\beta}(\xi,x)d\Gamma(x). \quad (6)$$

Equations (5) and (6) also become singular when ξ is moved to the boundary and ξ and x coincide. In these circumstances, provided the boundary is smooth, these equations become:

$$\frac{1}{2}M_{\alpha\beta}(\xi) = (CPV)\int_\Gamma D^*_{\alpha\beta k}(\xi,x)t_k(x)d\Gamma(x) - (HPV)\int_\Gamma S^*_{\alpha\beta k}(\xi,x)u_k(x)d\Gamma(x) + \left[(CPV)\int_\Gamma W^*_{\alpha\beta}(\xi,x)d\Gamma(x) + \frac{\upsilon}{(1-\upsilon)\lambda^2}\delta_{\alpha\beta}\right]q, \quad (7)$$

$$\frac{1}{2}Q_\beta(\xi) = (CPV)\int_\Gamma D^*_{3\beta k}(\xi,x)t_k(x)d\Gamma(x) - (HPV)\int_\Gamma S^*_{3\beta k}(\xi,x)u_k(x)d\Gamma(x) + q\int_\Gamma W^*_{3\beta}(\xi,x)d\Gamma(x), \quad (8)$$

in which $(HPV)\int$ represents a Hadamard principal value integral.

The components of traction, t_i, are obtained from:

$$t_\alpha = M_{\alpha\beta}\, n_\beta \text{ and } t_3 = Q_\alpha\, n_\alpha, \quad (9)$$

in which n represents the outward normal.

Equations (7) - (9) can therefore be used to obtain the traction components on a smooth boundary as:

$$\frac{1}{2}t_\alpha(\xi) = n_\beta(\xi)(CPV)\int_\Gamma D^*_{\alpha\beta k}(\xi,x)t_k(x)d\Gamma(x) - n_\beta(\xi)(HPV)\int_\Gamma S^*_{\alpha\beta k}(\xi,x)u_k(x)d\Gamma(x) + \left[n_\beta(\xi)(CPV)\int_\Gamma W^*_{\alpha\beta}(\xi,x)d\Gamma(x) + \frac{\upsilon}{(1-\upsilon)\lambda^2}n_\alpha(\xi)\right]q, \quad (10)$$

$$\frac{1}{2}t_3(\xi) = n_\beta(\xi)(CPV)\int_\Gamma D^*_{3\beta k}(\xi,x)t_k(x)d\Gamma(x)$$
$$- n_\beta(\xi)(HPV)\int_\Gamma S^*_{3\beta k}(\xi,x)u_k(x)d\Gamma(x) \tag{11}$$
$$+ qn_\beta(\xi)\int_\Gamma W^*_{3\beta}(\xi,x)d\Gamma(x),$$

In the case studies considered in this paper the domain integrals in eqns (3) and (4) resulting from a uniformly distributed load are transferred to boundary integrals using the following equation:

$$\int_\Omega \left[U^*_{13}(\xi,x) - \frac{\nu}{(1-\nu)\lambda^2} U^*_{i\alpha,\alpha}(\xi,x) \right] d\Omega =$$
$$\int_\Gamma \left[v^*_{i,\alpha}(\xi,x) - \frac{\nu}{(1-\nu)\lambda^2} v^*_{i\alpha}(\xi,x) n_\alpha(x) \right] d\Gamma, \tag{12}$$

where v^*_i satisfies the equation $v^*_{i,\alpha\alpha} = v^*_{ie}$ and is given by,

$$v^*_\alpha = \frac{1}{128\pi D\lambda^2} r_\alpha rz^2 (4\ln z - 5), \tag{13}$$

$$v^*_3 = \frac{-z^2}{256\pi D\lambda^4(1-\nu)} \left[64(\ln z - 1) - z^2(1-\nu)(2\ln z - 3) \right], \tag{14}$$

In eqns (1) - (14) the Roman subscripts *i*, *j* and *k* vary from 1 to 3 and the Greek subscripts α and β have values of 1 and 2.

3 Fracture mechanics in plate bending

The basic theory of fracture mechanics predicts that the stresses at the crack tip are infinite. In practice, however, these stresses are not infinite and, at the crack tip, the material yields, creating a plastic zone around the crack tip. The method of dealing with the stress fields in the vicinity of the crack tip depends on the size of the plastic zone. When the plastic zone is small compared to other crack dimensions the theory of linear elastic fracture mechanics (LEFM) applies. If, however, the size of the plastic zone is significant, elasto-plastic fracture mechanics (EPFM) applies. In this work only LEFM is considered and the expressions which are used to determine the stress intensity factors K_I, K_{II} and K_{III} must satisfy Reissner's sixth order plate bending equation (eqn (1)). The stress intensity factors at a point on the plate near the crack are related to the bending moments M_{11} and M_{22}, the twisting moments M_{12}

and M_{21} and the shear forces Q_1 and Q_2 at the point under consideration. The convention for the bending moments, twisting moments and shear forces for a small element of the plate are shown in Fig. 1.

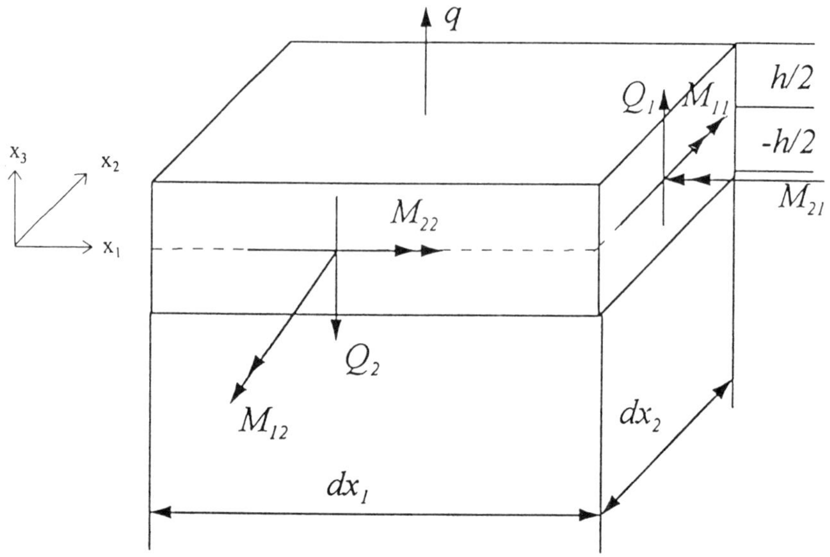

Fig. 1. Element of plate.

As the crack surfaces are assumed to be free from tractions, the bending moments, twisting moments and shear forces on the crack faces must be zero. Hence if a local set of axes is positioned on the crack tip as shown in Fig. 2, the following conditions apply on the crack faces.

$$M_{22} = 0, M_{12} = 0, Q_2 = 0 \text{ } for \text{ } x_1 \leq 0. \tag{15}$$

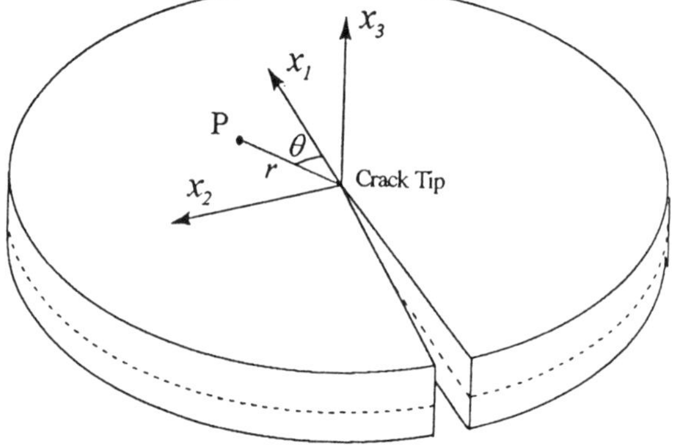

Fig. 2. Local co-ordinates at crack tip.

Reissner's plate theory is used to satisfy eqn (15) and yields the following expressions for the bending moment, twisting moment and shear force in the vicinity of the crack tip.

$$M_{11} = \frac{1}{\sqrt{2r}}\left[K_1 \cos\frac{\theta}{2}\left(1 - \sin\frac{\theta}{2}\sin\frac{3\theta}{2}\right) - K_2 \sin\frac{\theta}{2}\left(2 + \cos\frac{\theta}{2}\cos\frac{3\theta}{2}\right)\right], \quad (16a)$$

$$M_{22} = \frac{1}{\sqrt{2r}}\left[K_1 \cos\frac{\theta}{2}\left(1 + \sin\frac{\theta}{2}\sin\frac{3\theta}{2}\right) + K_2 \sin\frac{\theta}{2}\cos\frac{\theta}{2}\cos\frac{\theta}{2}\right], \quad (16b)$$

$$M_{12} = \frac{1}{\sqrt{2r}}\left[K_1 \sin\frac{\theta}{2}\cos\frac{\theta}{2}\cos\frac{3\theta}{2} + K_2 \cos\frac{3\theta}{2}\left(1 - \sin\frac{\theta}{2}\sin\frac{3\theta}{2}\right)\right], \quad (16c)$$

$$Q_1 = -\frac{1}{\sqrt{2r}} K_3 \sin\frac{\theta}{2}, \quad (16d)$$

$$Q_2 = -\frac{1}{\sqrt{2r}} K_3 \cos\frac{\theta}{2}. \quad (16e)$$

Reissner's equation also yields the following expressions for the rotations ϕ_1 and ϕ_2 about the x_1 and x_2 axes, respectively and the displacement w in the x_3 direction in the region of the crack tip.

$$\varphi_1 = \frac{1+\nu}{E}\left(\frac{12}{h^3}\right)\sqrt{\frac{r}{2}}\left[-K_1 \sin\frac{\theta}{2}\left(\frac{3-\upsilon}{1+\upsilon} - \cos\theta\right) - K_2 \cos\frac{\theta}{2}\left(\frac{1-3\upsilon}{1+\upsilon} + \cos\theta\right)\right], \quad (17a)$$

$$\varphi_2 = \frac{1+\nu}{E}\left(\frac{12}{h^3}\right)\sqrt{\frac{r}{2}}\left[-K_1 \cos\frac{\theta}{2}\left(\frac{3-\upsilon}{1+\upsilon} - \cos\theta\right) + K_2 \sin\frac{\theta}{2}\left(\frac{5+\upsilon}{1+\upsilon} + \cos\theta\right)\right], \quad (17b)$$

$$w = \frac{24(1+\nu)}{5Eh}\sqrt{\frac{r}{2}} K_3 \sin\frac{\theta}{2}. \quad (17c)$$

In eqns (16) and (17), r is the radial distance from the crack tip to a typical point P, θ is the angle of rotation, as shown in Fig. 2, and K_1 and K_2 are defined as bending moment and twisting moment intensity factors, respectively, and K_3 is the shear force intensity factor. The moment and shear force intensity factors are related to the stress intensity factors K_{I}, K_{II} and K_{III} in the region of the crack tip by the following equations:

$$K_{I} = \frac{12z}{h^3} K_1, \tag{18a}$$

$$K_{II} = \frac{12z}{h^3} K_2, \tag{18b}$$

$$K_{III} = \frac{3}{2h}\left[1 - \left\{\frac{2z}{h}\right\}^2\right] K_3, \tag{18c}$$

in which h is the plate thickness and z is the distance in the x_3 direction from the central neutral axis (i.e. the x_1, x_2 plane in Fig. 1).

4 Determination of stress intensity factors

The stress analyst, when faced with the problem of assessing the effect of a crack in a component, must be able to determine the stress and displacement fields near the crack tip for the evaluation of the stress intensity factors at the crack tip. Due to the complexity of most engineering components the stress and displacement fields cannot be obtained analytically, and numerical methods must therefore be used for their determination. In this work two approaches, using boundary element results, have been used for the evaluation of stress intensity factors at the crack tip in plate bending problems. These are the displacement extrapolation method and the *J*-integral method.

4.1 Displacement extrapolation method

The displacement extrapolation method for the determination of stress intensity factors uses the crack opening displacement between the two opposite faces of the crack. These crack opening displacements can be obtained for two opposite faces of the crack from eqn (17) as:

$$\begin{Bmatrix} \Delta\phi_1 \\ \Delta\phi_2 \\ \Delta w \end{Bmatrix} = \begin{Bmatrix} \phi_1 \\ \phi_2 \\ w \end{Bmatrix}_{\phi=180} - \begin{Bmatrix} \phi_1 \\ \phi_2 \\ w \end{Bmatrix}_{\phi=-180} = \sqrt{2r}\begin{bmatrix} \frac{48}{Eh^3} & 0 & 0 \\ 0 & \frac{48}{Eh^3} & 0 \\ 0 & 0 & \frac{24(1+\upsilon)}{5Eh} \end{bmatrix} \begin{Bmatrix} K_1 \\ K_2 \\ K_3 \end{Bmatrix}, \tag{19}$$

in which E is the modulus of elasticity and ν is Poisson's ratio.

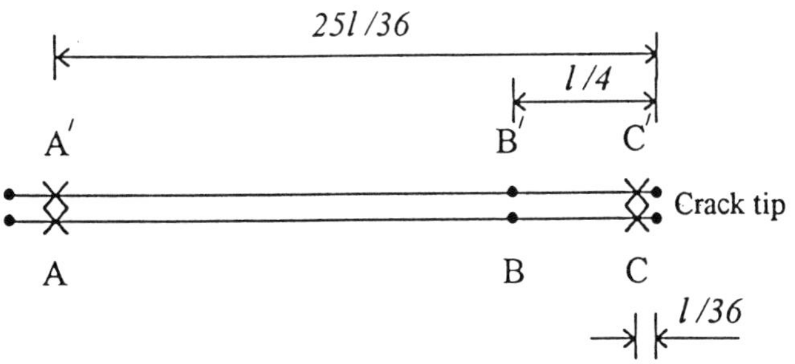

Fig. 3 Crack tip element.

The rotations ϕ_1 and ϕ_2 and the displacement w at nodes on opposite faces of the crack are obtained from the boundary element calculations and consequently K_1, K_2 and K_3 can be calculated for each pair of opposite nodes. In the displacement extrapolation method the stress intensity factors are calculated for the crack opening displacement at the nodes on the elements on the crack face adjacent to the crack tip. The crack tip elements, as shown in Fig. 3, are quadratic, discontinuous elements of length ℓ with the mid side node B, located at a distance of $\ell/4$ from the crack tip and the other two nodes positioned as indicated in Fig. 3.

The moment and shear force intensity factors are calculated at the nodes $A\text{-}A^1$, $B\text{-}B^1$ and $C\text{-}C^1$ from the displacements at the nodes obtained from the boundary element calculations. Hence from eqn (19):

$$\begin{Bmatrix} K_1 \\ K_2 \\ K_3 \end{Bmatrix}^{AA^1} = \frac{6}{5\sqrt{2\ell}} \begin{bmatrix} \frac{Eh^3}{48} & 0 & 0 \\ 0 & \frac{Eh^3}{48} & 0 \\ 0 & 0 & \frac{5Eh}{24(1+\upsilon)} \end{bmatrix} \begin{Bmatrix} \Delta\phi_1 \\ \Delta\phi_2 \\ \Delta w \end{Bmatrix}^{AA^1}, \tag{20a}$$

$$\begin{Bmatrix} K_1 \\ K_2 \\ K_3 \end{Bmatrix}^{BB^1} = \frac{2}{\sqrt{2\ell}} \begin{bmatrix} \frac{Eh^3}{48} & 0 & 0 \\ 0 & \frac{Eh^3}{48} & 0 \\ 0 & 0 & \frac{5Eh}{24(1+\upsilon)} \end{bmatrix} \begin{Bmatrix} \Delta\phi_1 \\ \Delta\phi_2 \\ \Delta w \end{Bmatrix}^{BB^1}, \tag{20b}$$

$$\begin{Bmatrix} K_1 \\ K_2 \\ K_3 \end{Bmatrix}^{CC^1} = \frac{6}{\sqrt{2\ell}} \begin{bmatrix} \frac{Eh^3}{48} & 0 & 0 \\ 0 & \frac{Eh^3}{48} & 0 \\ 0 & 0 & \frac{5Eh}{24(1+\upsilon)} \end{bmatrix} \begin{Bmatrix} \Delta\phi_1 \\ \Delta\phi_2 \\ \Delta w \end{Bmatrix}^{CC^1}, \tag{20c}$$

Equation (20) gives the moment and shear force intensity factors at points AA^1, BB^1 and CC^1 and these values can be extrapolated, using the following expressions, to give the intensity factors at the crack tip.

$$\begin{Bmatrix} K_1 \\ K_2 \\ K_3 \end{Bmatrix}^{AA^1-BB^1} = \frac{1}{8\sqrt{(2\ell)}} \begin{bmatrix} \frac{Eh^3}{48} & 0 & 0 \\ 0 & \frac{Eh^3}{48} & 0 \\ 0 & 0 & \frac{5Eh}{24(1+\upsilon)} \end{bmatrix} \times \left[25\begin{Bmatrix} \Delta\phi_1 \\ \Delta\phi_2 \\ \Delta w \end{Bmatrix}^{BB^1} - {}^{27}\!/_{5} \begin{Bmatrix} \Delta\phi_1 \\ \Delta\phi_2 \\ \Delta w \end{Bmatrix}^{AA^1} \right], \tag{21a}$$

$$\begin{Bmatrix} K_1 \\ K_2 \\ K_3 \end{Bmatrix}^{BB^1-CC^1} = \frac{1}{4\sqrt{(2\ell)}} \begin{bmatrix} \frac{Eh^3}{48} & 0 & 0 \\ 0 & \frac{Eh^3}{48} & 0 \\ 0 & 0 & \frac{5Eh}{24(1+\upsilon)} \end{bmatrix} \times \left[27\begin{Bmatrix} \Delta\phi_1 \\ \Delta\phi_2 \\ \Delta w \end{Bmatrix}^{CC^1} - {}^{27}\!/_{5} \begin{Bmatrix} \Delta\phi_1 \\ \Delta\phi_2 \\ \Delta w \end{Bmatrix}^{BB^1} \right] \tag{21b}$$

Equation (21a) and (21b) are therefore the extrapolation from points $AA^1 - BB^1$ and $BB^1 - CC^1$, respectively, to the crack tip to give the values of the bending moment, twisting moment and shear force intensity factors. The values obtained from either eqn (21a) or (21b) can then be substituted into eqn (18) to determine the stress intensity factors at the crack tip.

4.2 J-integral method

The stress intensity factor, as discussed initially by Rice[12], can be related to the integral of the path Γ as shown in Fig. 4. The line integral, which is known as the *J*-integral, does not depend on the position of the path, provided that the starting and finishing points of the path are on opposite sides of the crack and the crack tip is contained within the contour of the path.

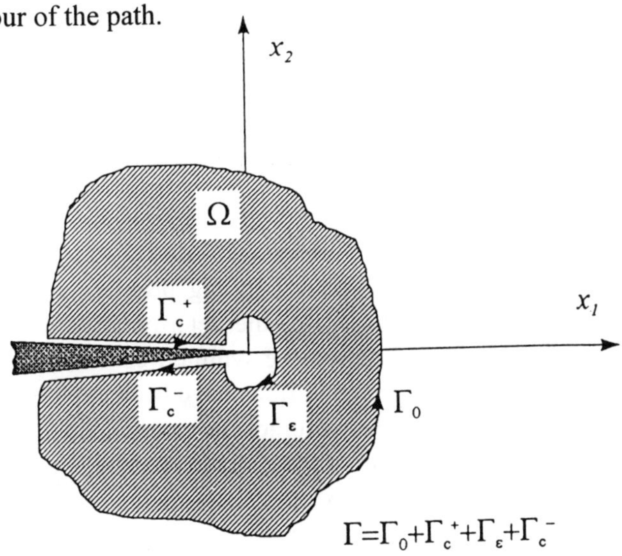

Fig. 4 *J*-integral path.

The *J*-integral for plate bending is given by the equation:

$$J_\delta = \oint_\Gamma \left\{ (W - pw)n_\delta - \left[M_{11}\varphi_{1,\delta} + M_{12}\varphi_{2,\delta} + Q_1 w_{,\delta} \right] n_1 - \left[M_{12}\varphi_{1,\delta} + M_{22}\varphi_{2,\delta} + Q_2 w_{,\delta} \right] n_2 \right\} d\Gamma, \tag{22}$$

where $\delta = 1, 2$, n_1 and n_2 are the outward normals to the path in the directions 1 and 2, respectively, and p is the uniform pressure load acting on the plate.

The strain energy density, W, in equation (22) for Reissner's plate theory is given by the following expression:

$$W = \frac{1}{2}\left[M_{11}\varphi_{1,1} + M_{12}\left(\varphi_{2,1} + \varphi_{1,2}\right) + M_{22}\varphi_{2,2} + Q_1\left(\varphi_1 + w_{,1}\right) + Q_2\left(\varphi_2 + w_{,2}\right)\right]. \tag{23}$$

The expressions for the moments, shear forces, rotations and displacement from eqns (16) and (17) can be substituted into eqn (22) to give the following expressions for the *J*-integrals, J_1 and J_2, in terms of the moment and shear force intensity factors when eqn (21) is integrated along the contour Γ.

$$J_1 = \frac{12\pi}{Eh^3}\left[K_1^2 + K_2^2 + \frac{h^2}{10}(1+\upsilon)K_3^2\right], \tag{24a}$$

$$J_2 = -\frac{24\pi}{Eh^3} K_1 K_2 + \int_{\Gamma_c}[W - pw]dx_1, \tag{24b}$$

where Γ_c denotes the path along the upper and lower crack faces.

5 Case studies

A number of case studies are presented to demonstrate the effectiveness of the BEM for the fracture analysis of plate bending problems. The case studies have been subdivided into two categories for the determination of the stress intensity factors. In the first group of case studies, the symmetrical nature of the plates' loading, crack configuration and boundary conditions have been exploited to overcome the problem of singular equations when two crack faces coincide. In the second group of case studies, there is no symmetry so the DBEM has been used to overcome the problem of singular equations when two crack faces coincide.

Eight case studies are presented. Four of these demonstrate the use of symmetry to determine the stress intensity factors and in the other four case studies the DBEM has been used to determine the stress intensity factors.

5.1 Case Study 1 (Fig. 5)

A rectangular plate, with a central crack, acted on by a uniformly distributed load p and simply supported on all boundaries.

5.2 Case Study 2 (Fig. 6)

A rectangular plate acted on by a uniformly distributed load, p simply supported on two opposite boundaries, with the other two boundaries free and having two equal opposite perpendicular cracks on the free edges.

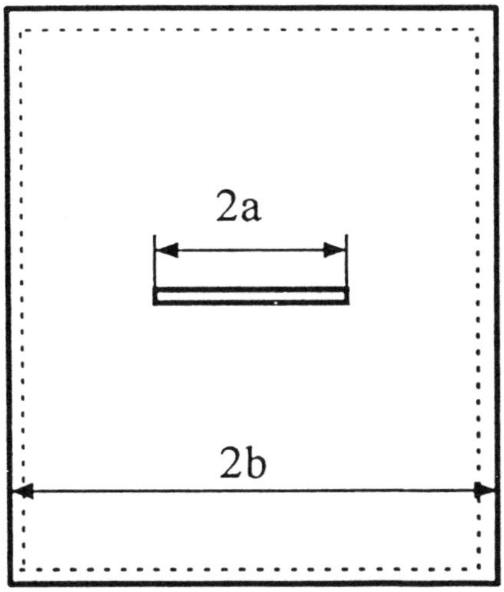

Fig. 5 Case study one

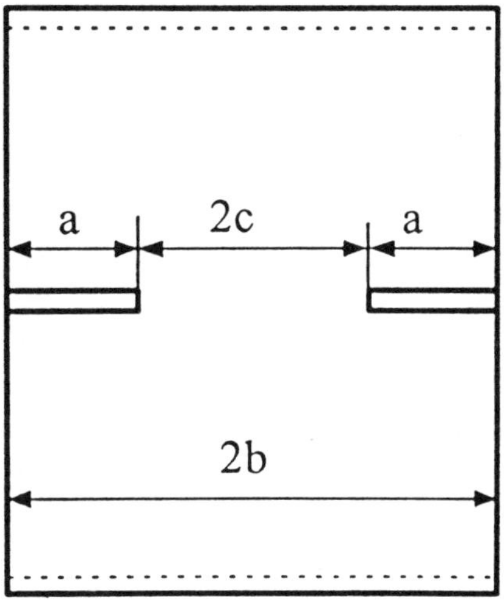

Fig. 6 Case study two

5.3 Case Study 3 (Fig. 7)

A rectangular plate loaded by two equal uniformly distributed moments M_o on opposite edges and the other two boundaries free, with two equal opposite edge cracks perpendicular to the free boundaries.

5.4 Case Study 4 (Fig. 8)

A rectangular plate with a central crack, loaded by two equal uniformly distributed moments M_o on opposite boundaries parallel to the crack, with the other two boundaries being free.

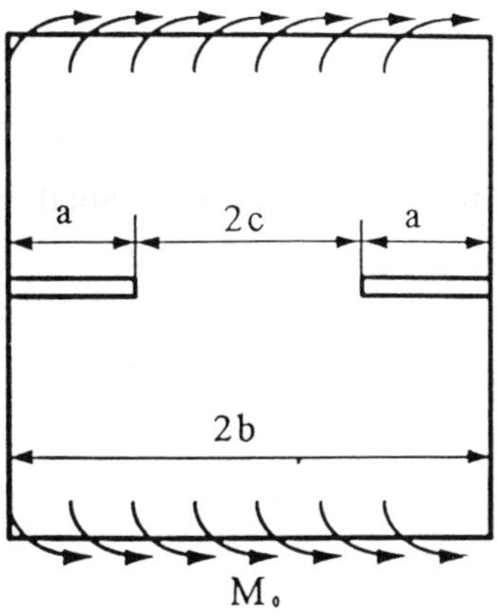

Fig. 7 Case study three

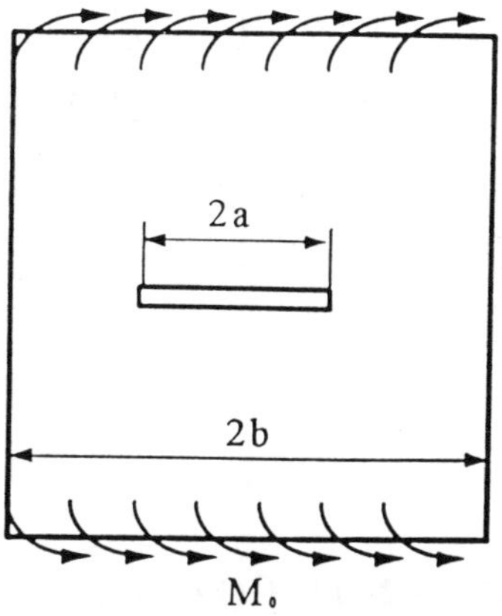

Fig. 8 Case study four

The plates in the above case studies were analysed using quadratic elements and resulting from the configuration of the loading, boundary conditions and cracks only a quarter of the plate needs to be analysed. In each case study semi-continuous elements were used at the junction of the external boundary of the quarter-plate model and the crack, the crack face was modelled using discontinuous elements, with quarter-point elements at the crack tip, and the remainder of the boundary was modelled using continuous elements. A convergence study, using five, ten, fifteen, twenty and twenty two elements, was undertaken for each case study. When five, ten, fifteen and twentry elements were used there was an equal number of elements on the crack face, the boundary collinear with the crack face and on each of the other boundaries. When twenty two elements were used, each uncracked boundary was modelled using four elements and the boundary containing the crack was modelled using ten elements with the number of elements being used to model the crack and the uncracked portion of that boundary depending on the crack length.

5.5 Case Study 5 (Fig. 9)

This case study comprises a rectangular plate with a central crack loaded by twisting moments M_o as shown. The boundaries parallel to the crack are acted on by two equal anti-clockwise twisting moments and the boundaries perpendicular to the crack face are loaded by two equal clockwise twisting moments. Twenty-four quadratic elements were used to model the plate. Each external boundary and each crack face was modelled using four elements. The crack faces were modelled using discontinuous quadratic elements, with quarter-point elements being used adjacent to the crack tip, and the boundaries of the plate were modelled using continuous quadratic elements.

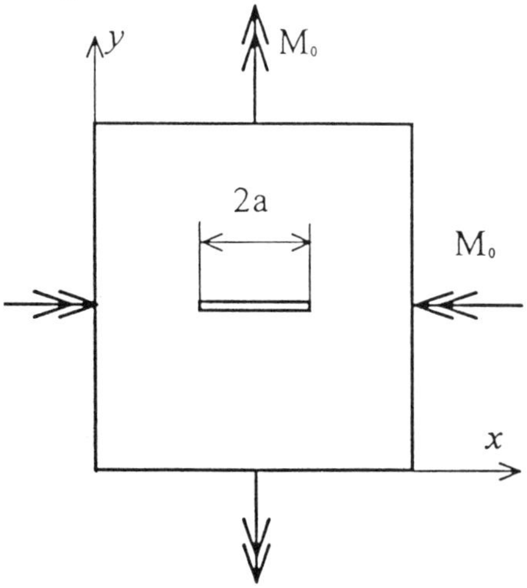

Fig. 9 Case study five

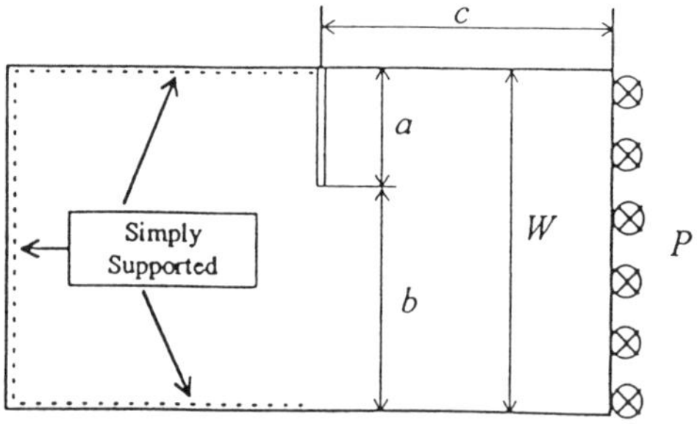

Fig. 10 Case study six

5.6 Case Study 6 (Fig. 10)

In this case study a rectangular plate with mixed boundary conditions is analysed. One short boundary and half of each long boundary is simply supported and the remainder of the boundary is free. A uniform edge load p acts on the short free boundary as shown in Fig. 10. The plate has a single edge crack perpendicular to the boundary at the junction of the simply supported and the free portions of one of the long boundaries as shown in Fig. 10. The boundary element model, used in the analysis, consisted of thirty two quadratic elements. Each long external boundary was modelled using eight elements, each short external boundary was modelled using four elements and each crack face was modelled using four elements. Discontinuous elements were used to model the crack face, semi-continuous elements were used at the junction of the crack face on the external boundary and continuous elements were used to model the remainder of the boundary. Quarter-point elements were used at the crack tip.

5.7 Case Study 7 (Fig. 11)

A cantilevered rectangular plate acted on by a uniformly distributed moment M_o on the free edge opposite the built-in edge. The plate has an interior crack, parallel to the built-in edge, and an interior circular hole as shown in Fig. 11. Quadratic elements were used in the analysis. The circumference of the hole was modelled using twelve continuous elements. Each crack face was modelled using eight discontinuous elements, with the elements adjacent to the crack tips being quarter-point elements. The external boundaries of the plate were each modelled using four continuous elements.

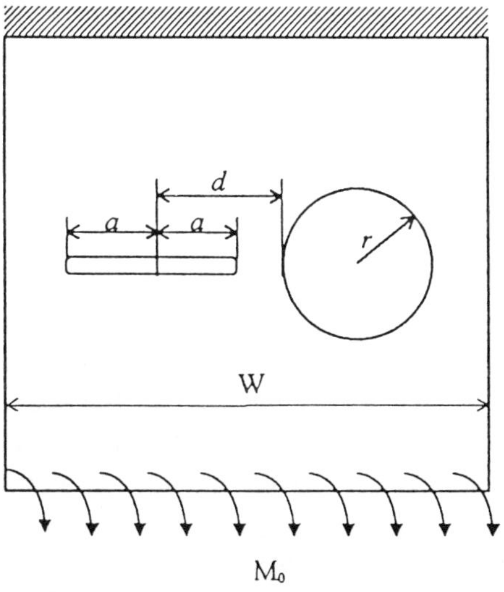

Fig. 11 Case study seven

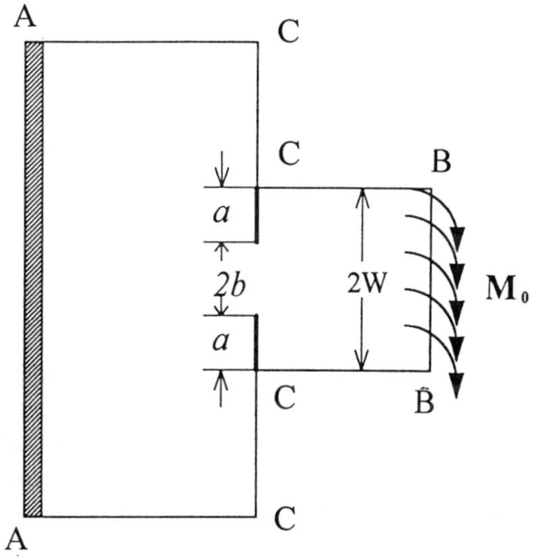

Fig.12 Case study eight

5.8 Case Study 8 (Fig. 12)

In this case study a tee shaped cantilevered plate was analysed. The plate is acted on by a uniformly distributed moment M_o acting on the free boundary opposite the built-in boundary and the plate has two equal opposite edge cracks collinear with the boundaries CC as shown in Fig. 12. The plate was modelled using fifty two quadratic elements. Boundary AA had eight elements, boundary BB had four elements and boundaries AC, CC and CB each had four elements. Each side of the crack face was modelled using four elements. Discontinuous elements were used to model the crack faces, semi-continuous elements were used at the junction of the crack face on the external boundary and the remainder of the boundary was modelled using continuous elements. Quarter-point elements were used at the crack tip.

In case studies five, six, seven and eight the DBEM was used in the analysis. Equation (4) for boundary displacements was used to model one of the crack faces and eqns (10) and (11) for boundary tractions were used to model the other crack face.

6 Results

In all case studies considered, the stress intensity factors at the crack tip were determined for LEFM using the J-integral method and the displacement extrapolation method. Two techniques, depending on the boundary conditions, loading conditions and crack configurations, have been used in the analysis to overcome the problem of singular equations resulting from the coincidence of nodes on opposite faces of the crack. In case studies one, two, three and four, because of the symmetrical nature of the problems, only a quarter of the plate was analysed in each case, thus avoiding the problem of singular algebraic equations. As there was no possibility of exploiting symmetry in case studies five, six, seven and eight, the dual boundary element method was used in the analyses to overcome the problem of singular algebraic equations.

Convergence studies ranging from five to twenty two elements were undertaken for case studies one, two, three and four and for case studies five, six, seven and eight a fixed number of elements, depending of the plate geometry and crack configuration was used for each case study.

Results for the K_{I} moment intensity factors for case studies one, two and three are shown in Figs 13-15. Square plates having breadth to depth ratios of two and ten and a range of crack lengths were analysed for each case study. The boundary element results, in each case, are for the twenty two element model. The J-integral results are compared with results previously obtained by Wearing and Ahmadi-Brooghani[8] who used the BEM and the displacement extrapolation method to determine the K_{I} stress intensity factors. The J-integral BEM results are also compared with finite element results obtained by Sosa and Eischen,[13] who used the J-integral method to determine the stress intensity factors. In case studies one and two the plate is acted on by a uniformly distributed load, p. Figure 13 gives details of the results for the variation of normalised stress intensity factor at the crack tip with the ratio of crack length to plate breadth as shown in Fig. 5. For the second case study (Fig. 6) results are shown in Fig. 14 for the variation of normalised stress intensity factor at the crack tip with the ratio of the distance between the crack tips to plate breadth. The plate in case study three was acted by two opposite, equally distributed bending moments on opposite boundaries parallel to the cracks (Fig. 7). The results for the third case study are shown in Fig. 15 for the variation of normalised stress intensity factor at the crack tip with the ratio of the distance between the crack tips to plate breadth.

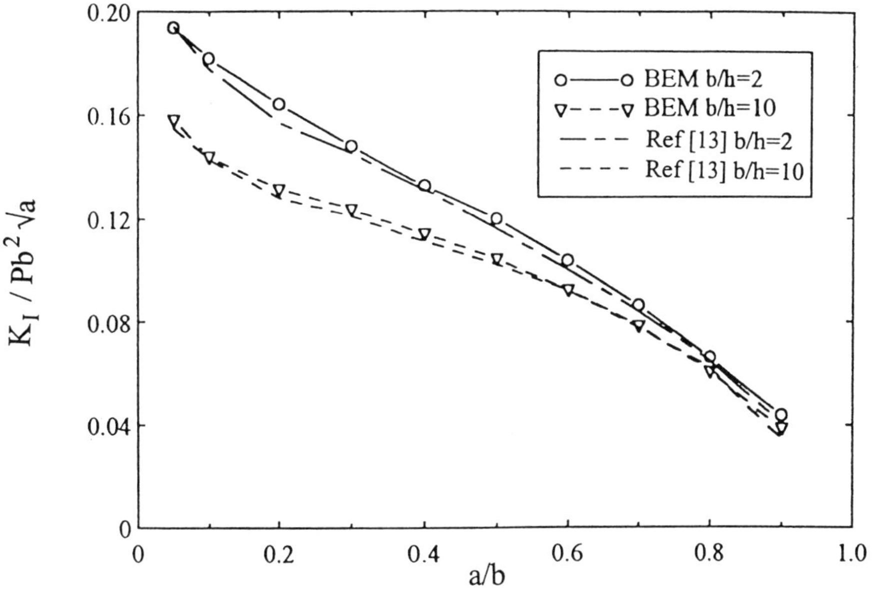

Fig. 13 Variation of bending moment intensity factor with crack length for case study one.

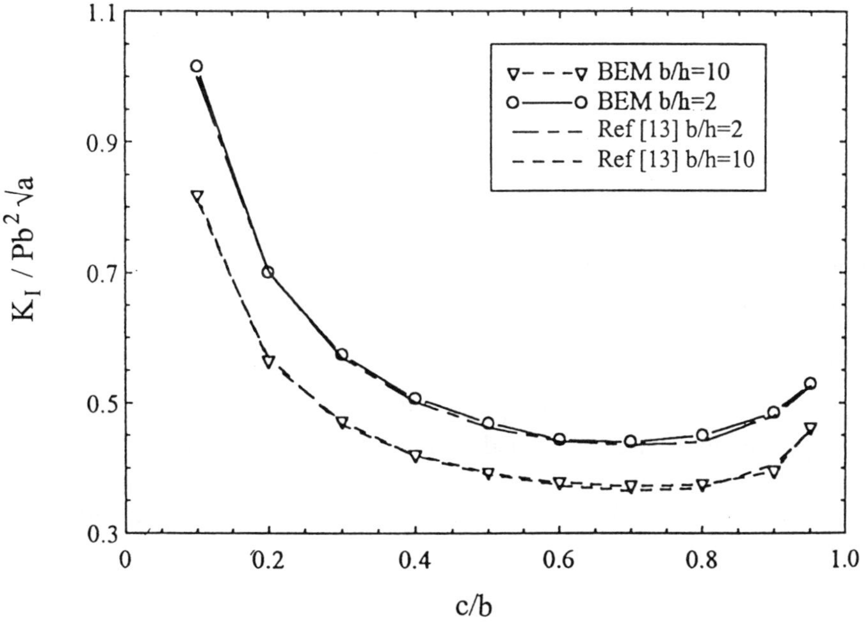

Fig. 14 Variation of bending moment intensity factor with crack length for case study two.

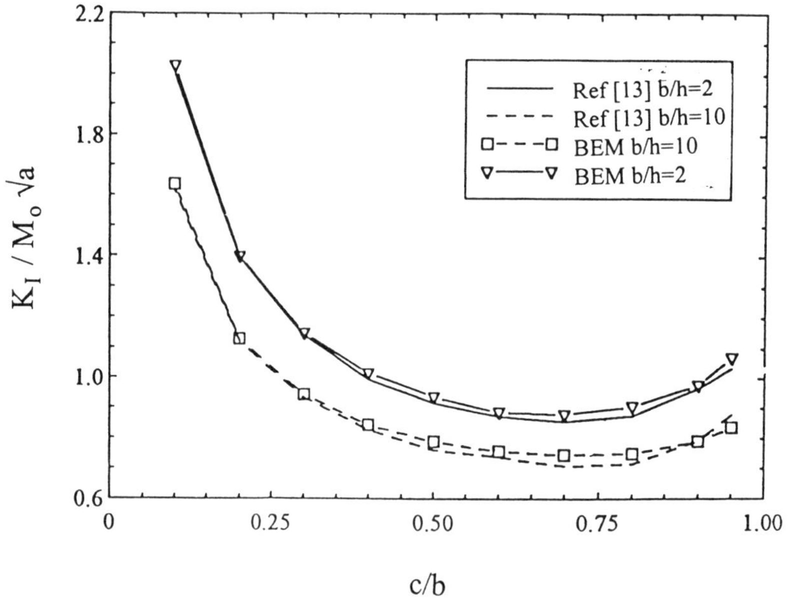

Fig. 15 Variation of bending moment intensity factor with crack length for case study three.

Results for the K_I moment intensity factors for case study four (Fig. 8) are shown in Fig. 16 and for the K_2 and K_3 moment and shear force intensity factors for case study five (Fig. 9) are shown in Fig. 17. Results for the normalised moment intensity factor $K_I / M_o\sqrt{a}$ (K_I is given by equation (18a)) for case study four are shown in Fig. 16 for a plate having a thickness to crack length ratio of $h / a\sqrt{10} = 0.1$. The *J*-integral results for the ratio of crack length to plate breadth varying from 0.05 to 0.9 are shown in Fig. 6 and these results are compared with Murakami's[14] analytical results and Ahmadi-Brooghani and Wearing's[9] previously published results where the DBEM was used in conjunction with the displacement extrapolation method to determine K_I. Figure 17 shows the normalised twisting moment intensity factor $K_2 / M_o\sqrt{a}$ and the normalised shear force intensity factor $K_3(1+\upsilon) / M_o\sqrt{10a}$ for case study five (Fig. 9) where K_2 and K_3 are obtained from eqn 18b and 18c. Results are presented for the ratio $h / \sqrt{10a}$ varying from 0.1 to 1.5 for a plate of thickness *h* and semi crack length *a*. The plate length in this case study was kept constant at one tenth of the breadth of the plate. Sih's[15] analytical results and Ahmadi-Brooghani and Wearing's[9] boundary element displacement results are also shown in Fig.17.

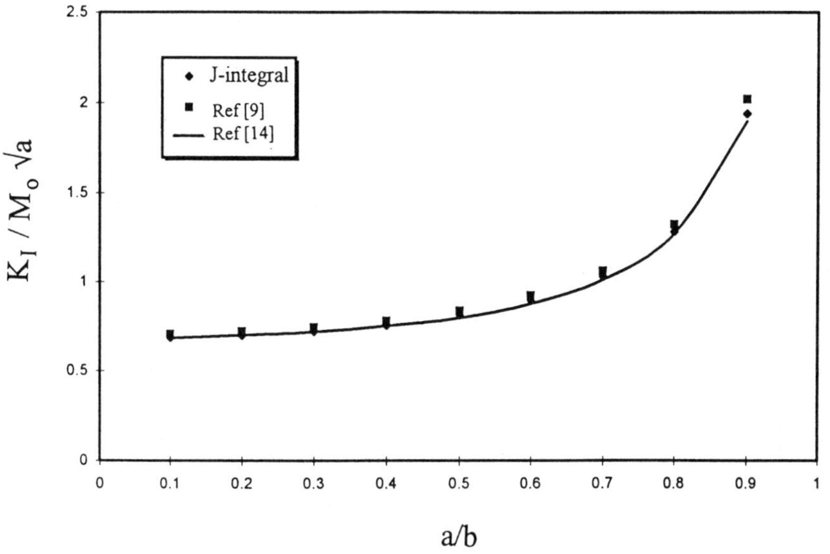

ig. 16 Variation of bending moment intensity factor with crack length for case study four.

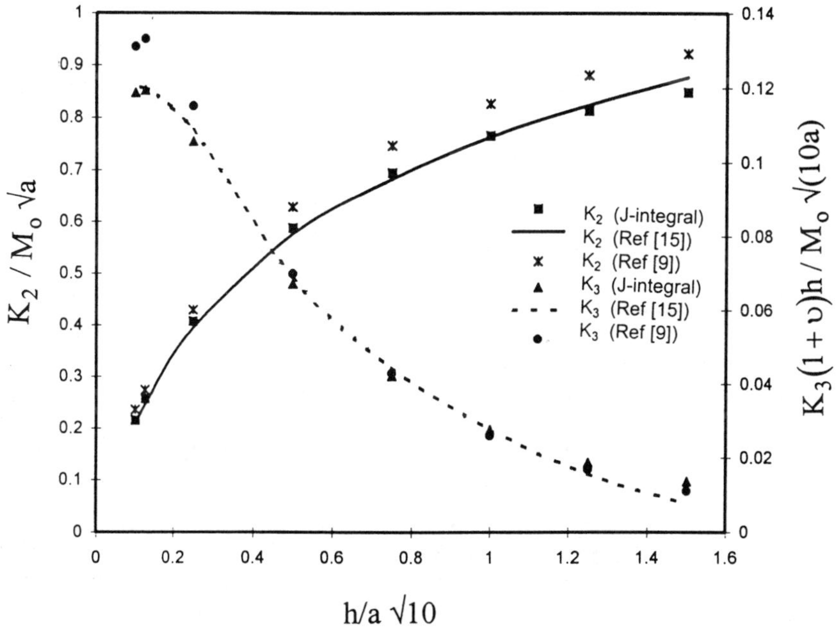

Fig. 17 Variation of twisting moment and shear force intensity factors with crack length for case study five.

Results for the normalised stress intensity factors F_I, F_{II} and F_{III} for case studies six, seven and eight using the *J*-integral approach and the displacement extrapolation method are shown in Figs 18, 19 and 20. The results for case study six are shown in Fig.18. The normalised stress intensity factors F_I, F_{II} and F_{III} have been calculated for a range of values of a/W from 0.1 to 0.9 where a is the crack length and W is the width of the plate as shown in Fig.10. The stress intensity factors have been obtained for a value of h/W of 0.5 where h is the plate thickness. The results for the normalised stress intensity factor F_I for case studies seven and eight (Figs 11 and 12) are shown in Figs 19 and 20, respectively. For case study seven the distance from the centre of the crack to the circumference of the hole equals the radius of the hole. The normalised stress intensity factors were calculated for a range of values of a/d from 0.1 to 0.9, where a is the semi-crack length and d is the distance from the centre of the crack to the circumference of the hole as shown in Fig. 11. Values of the variation of stress intensity factor are given in Fig. 19 for values of $h/d\sqrt{10}$ of 0.1 and 1.0 where h is the plate thickness. Finite element results using the *J*-integral approach are also shown on Fig. 19 for the range of crack lengths and plate thicknesses, for which DBEM results have been obtained. For case study eight (Fig. 12) results are presented in Fig. 20 for a range of ratios of crack length a to distance b between the crack tips.

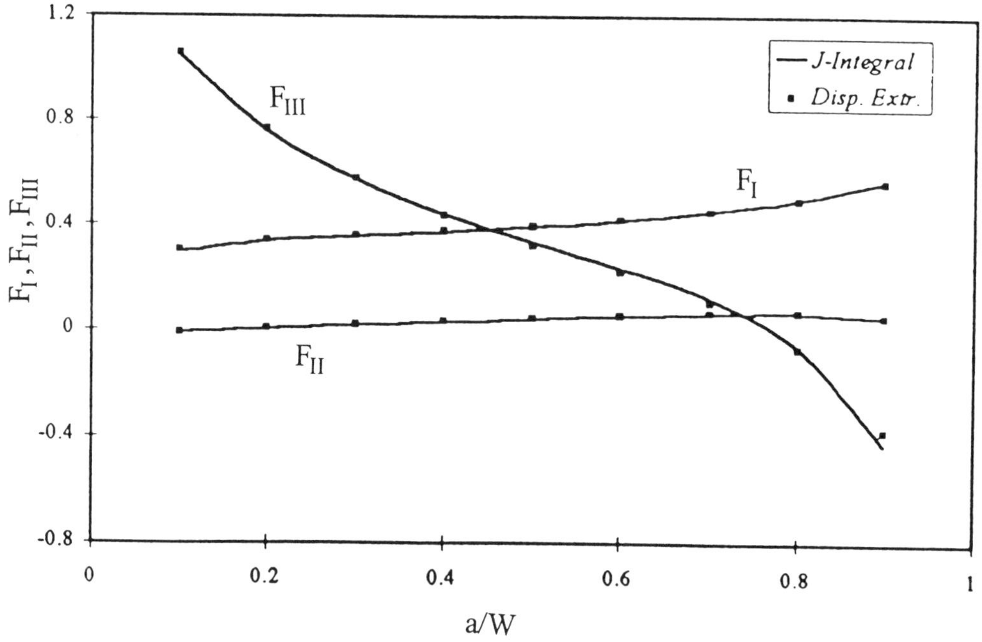

Fig. 18 Variation of normalised stress intensity factors with crack length for case study six.

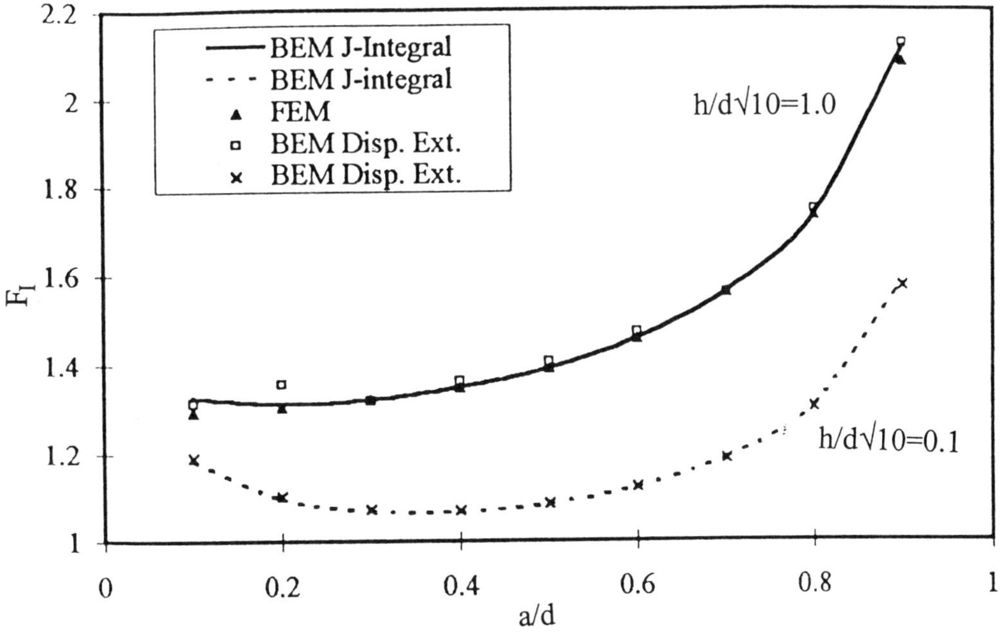

Fig. 19 Variation of normalised stress intensity factors with crack length for case study seven.

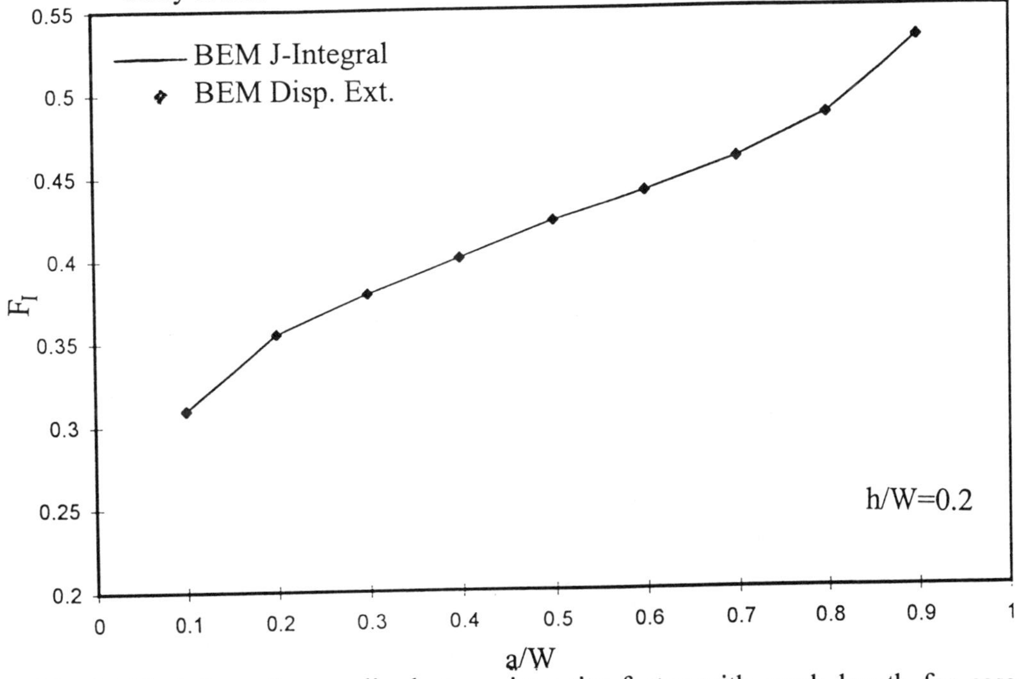

Fig. 20 Variation of normalised stress intensity factor with crack length for case study eight.

In case study six the normalised stress intensity factors were obtained from the following expressions as given by Hasabe[16]:

$$K_I = \frac{6PcW}{h^2\sqrt{b}} F_I, \tag{25a}$$

$$K_{II} = \frac{6PcW}{h^2\sqrt{b}} F_{II}, \tag{25b}$$

$$K_{III} = \frac{6PcW}{h^2\sqrt{b}} F_{III}. \tag{25c}$$

In case study seven the results for the variation of F_{I} have been obtained from the following expression as given by Murakami[14]:

$$K_I = \frac{6M_o\sqrt{a}}{h^2} F_I. \tag{25d}$$

The values of F_{I} for case study eight were obtained from the following expression as given by Hasabe[16]:

$$K_I = \frac{6M_oW}{h^2\sqrt{b}} F_I. \tag{25e}$$

In eqn (25) the stress intensity factors K_{I}, K_{II} and K_{III} are obtained from eqn (18). The factors F_{I}, F_{II} and F_{III} in eqn (25a) - (25d) relate the ratio of crack length a to plate width W and in eqn (25e) the F_{I} is related to the ratio of crack length a to distance b between the crack tips. In eqn (25a) - (25c) c is the length of the free portion of the long boundary in case study six.

7 Discussion of results

The results presented in Figs 13 - 20 for the plates shown in Figs 5 - 12 clearly indicate the effectiveness of the BEM for the determination of the stress intensity factors in the linear elastic fracture mechanics analyses of plate bending problems. Plates with a number of boundary conditions, loading conditions and crack configurations have been analysed and two different approaches have been used in the analyses to determine the stress intensity factors at the crack tip. In case studies one, two, three and four the symmetry of the plates was exploited and the displacement boundary integral equations were used to model the boundary of the quarter-plate models. Case studies five, six, seven and eight did not have the advantage of symmetry and the DBEM was used, in the analyses, eliminating the requirement for subregioning. In all case studies the *J*-integral method and the displacement extrapolation approach were used in conjunction with the BEM to determine the stress intensity factors. The boundary element results for the various, case studies were also compared with finite element results and with analytical solutions and in all cases the results from alternative sources compared well with the boundary element results. Quadratic elements were used to model the plates in all case studies, with quarter-point elements being used to model the crack tips.

Figures 13 - 15 give the results for the normalised K_1 moment intensity factors for case studies one, two and three. For each case study, plates with breadth to thickness ratios of two and ten, both of which can be accommodated within Reissner's plate theory, have been considered. In each of these case studies twenty two element models have been analysed and these showed rapid convergence from the five element models. The results from the twenty two element boundary element models are in close agreement with Sosa and Eischen's[13] finite element results. However, the size of the quarter plate boundary element models, with its twenty two elements and forty four nodes, is in sharp contrast to the one hundred and forty element quarter plate finite element models with four hundred and sixty nine nodes.

Figure 16 gives the results of the K_1 bending moment intensity factors for case study four for a range of crack lengths, and Fig. 17 gives the results of the K_2 twisting moment intensity factor and the K_3 shear force intensity factor for case study five. In each case study plates with a range of crack lengths were analysed. The boundary element J-integral and displacement extrapolation results for case study four are compared with Murakami's[14] analytical results and the boundary element results from case study five are compared with Sih's[15] analytical results. The boundary element results for both case studies are in close agreement with the analytical results.

Results for the normalised stress intensity factors, F_{I}, F_{II} and F_{III}, for case studies six, seven and eight are shown in Figs 18 - 20. Figure 18 gives the F_{I}, F_{II} and F_{III} normalised stress intensity factors for case study six, using both the J-integral and the displacement extrapolation method within the DBEM for a range of crack lengths. Figures 19 and 20 give the F_{I} normalised stress intensity factor for case studies seven and eight. For both these case studies the DBEM has been used to calculate the stress intensity factors using both the J-integral and displacement extrapolation methods. Finite element results are also shown in Fig. 19 for case study seven. All sets of results for case studies six, seven and eight are in close agreement. In case study seven both sets of DBEM results have been achieved with considerably fewer elements and degrees of freedom than the finite element results, with the boundary element model comprising forty four elements and the finite element model comprising one thousand and five hundred elements.

8 Conclusions

The results presented in this chapter confirm that the BEM can be used successfully for the determination of stress intensity factors in plate bending problems when LEFM conditions apply. Boundary element results have been presented for stress intensity factors using both the *J*-integral approach and the displacement extrapolation approach. The BEM results have been compared with finite element results and with analytical results. In all cases the BEM results are in close agreement with the results from other sources. The BEM has also been shown to have a clear advantage over the FEM as results of comparable accuracy have been achieved using the BEM with considerably fewer elements and degrees of freedom, compared to the FEM. The work discussed in the chapter also indicates that, where appropriate, the DBEM can be used for the fracture analysis of plate bending problems.

Acknowledgement

Mr Ahmadi-Brooghani would like to thank the Iranian Research organisation for Science and Technology, Mashad Centre, for providing the financial support to enable him to undertake the work discussed in this chapter.

References

[1] Broek, D., *Elementary Engineering Fracture Mechanics*, Sythoff and Noordhoff, Amsterdam, 1988.

[2] Ewalds, H.L. and Wanhill, R.J.H., *Fracture Mechanics*, Edward Arnold, London, 1984.

[3] Rooke, D.P. and Cartwright, D.J., *A Compendium of Stress Intensity Factors,* Her Majesty's Stationery Office, London, 1976.

[4] Aliabadi, M.H., Cartwright, D.J., Naehring, D.W. and Daney, W.A., Stress intensity factor weight functions for cracks at holes and half plane (ed. C.A. Brebbia, and J.J. Connor), pp. 83-98, *Proceedings of the 11th International Conference on Boundary Element Methods,* Vol. 3, Boston, U.S.A., 1989, Springer, Berlin, 1989.

[5] Stok, B. and Bukovec, B., Stress intensity factor analysis in cracked elastic bars under torsion by the boundary element method (ed. C.A. Brebbia and J.J. Connor), pp. 111-122, *Proceedings of the 11th International Conference on Boundary Element Methods,* Vol. 3, Boston, U.S.A., 1989, Springer, Berlin, 1989.

[6] Portela, A., Aliabadi, M.H. and Rooke, D.P., The dual boundary element method: effective implementation for crack problems, *International Journal for Numerical Methods in Engineering*, 1992, **33**, 1269-1287.

[7] Mi, Y., and Aliabadi, M.H., Dual boundary element method for three dimensional crack growth analysis (ed. C.A. Brebbia and J.J. Rencis), pp. 249-260, *Proceedings of the 15th International Conference on Boundary Elements,*

Vol. 2, Worcester, U.S.A., 1993, Computational Mechanics Publications, Southampton, 1993.

[8] Wearing, J.L. and Ahmadi-Brooghani, S.Y., Quarter point boundary elements for the analysis of crack problems in plate bending (ed. C.A. Brebbia), *Proceedings of the 18th International Conference on Boundary Element Methods,* Braga, Portugal, 1996, Computational Mechanics Publications, Southampton, 1996.

[9] Ahmadi-Brooghani, S.Y. and Wearing, J.L., The application of the dual boundary element method in linear elastic fracture mechanics (ed. C.A. Brebbia), *Proceedings of the 18th International Conference on Boundary Element Methods,* Braga, Portugal, 1996, Computational Mechanics Publications, Southampton 1996.

[10] Reissner, E., On the bending of elastic plates, *Quarterly of Applied Mathematics*, 1947, **5**, 55-68.

[11] Karam, V.J. and Telles, J.C.F., On boundary elements for Reissner's plate theory, *Engineering Analysis with Boundary Elements,* 1985, **5**, 21-27.

[12] Rice, J.R., A path independent integral and the approximate analysis of strain concentration by notches and cracks, *Transactions of the American Society of Mechanical Engineering, Journal of Applied Mechanics*, 1968, **35**, 379-386.

[13] Sosa, H.A. and Eischen, J.W., Computation of stress intensity factors for plate bending via a path dependent integral, *Engineering Fracture Mechanics,* 1986, **25**, 451-462.

[14] Murakami, Y., *Stress Intensity Factors Handbook*, Pergamon Press, Oxford 1987.

[15] Sih, G.C., Strain energy density theory applied to plate bending problems, Mechanics of fracture 3, *Plates and shells with cracks, (*ed. G.C. Sih), pp. XVII-XLVIII, Noordhoff, Amsterdam, 1977.

[16] Hasabe, N., Miwa, M. and Nakamuri, T., A mixed boundary value problem of a strip with a crack under concentrated bending and torsional moments, *Transactions of the Japanese Society of Civil Engineering,* 1990, **416**, 395-401.

Chapter 7

Adaptive boundary element formulations for plate bending analysis

Y. Sawaki[a] & N. Kamiya[b]
[a]*Department of Mechanical Engineering, Mie University, Kamihamacho, Tsu 514, Japan*
[b]*Department of Informatics and Natural Science, School of Informatics and Sciences, Nagoya University, Nagoya 464-01, Japan*

Abstract

Some recent studies on adaptive boundary elements are reviewed briefly at the beginning of this chapter. In the present chapter, an adaptive strategy for the boundary element method, based on an error estimate using a sample point error analysis, is applied to a bending problem for a thin elastic plate. According to the idea of the sample point error analysis, the errors on each boundary element are estimated on the basis of the magnitude of solution discrepancy except at the boundary nodes employed for the initial analysis. A boundary element refinement, the *h*-version adaptive scheme, is implemented by employing the non-dimensional error indicator composed of two different kinds of indicators (extended error indicators), of which dimensions are of deflection and of slope. The properties of the non-dimensional error indicators are examined through some numerical examples using uniform meshes. Effectiveness of the proposed adaptive strategy for the boundary element plate bending analysis is demonstrated through some numerical applications of plates with various edge conditions.

1 Introduction

Numerical methods such as the finite element method (FEM) and the boundary element method (BEM) play an important role in the process of the computer aided engineering (CAE). In the CAE system, the accuracy of the solution computed by such numerical techniques depends on several factors, e.g. the reliability of solver (software) and/or the users' computational skill. The former is affected by the theory, the mathematical modelling and the computer implementation of a given problem, while the latter depends on the choice of the order of interpolation function and/or a mesh discretization.

Such user-dependent factors in the CAE system may be expected to be removed, for example, by introducing the automated adaptive scheme (in other words, the automated mesh generation system). In order to obtain the optimal mesh discretization and the corresponding solutions within a desired accuracy in an efficient way, many adaptive mesh generation schemes have been proposed for FEM (e.g. Babuska & Rank[1]) and the monographs edited by Brebbia & Aliabadi[2] and Babuska *et al.*[3] have reviewed the adaptive techniques for the FEM and the BEM.

Nowadays, some of the adaptive techniques for the FEM have been installed in several commercial softwares. On the other hand, the development of the adaptive methods in the research fields of the BEM is more recent than those of the FEM. However, some progress has been achieved in the adaptive procedure for the BEM in recent years. Some of the recent studies on error estimation and adaptive mesh refinement in the BEM field have been reviewed by Kita & Kamiya,[4] Liapis[5] and a special issue with respect to the above topics is edited by Kamiya.[6] In the present chapter, an adaptive boundary element technique based on an appropriate idea for the collocation-type direct BEM is applied to a bending analysis of a thin elastic plate.

2 Error estimations for adaptive boundary elements

Several concepts are proposed on the adaptive BEM scheme. The adaptive scheme is based on the error estimation processes. However, some of the error estimation for the BEM are affected by those developed for the FEM, which may always inapplicable to the BEM directly. Therefore, the techniques intrinsic to the BEM should be developed for the precise error estimation.

The error estimation schemes proposed for the BEM are classified into several types. Alarcon *et al.*[7,8] employed the error estimation based on the residual of the descretized boundary integral equation. Rank[9] and Parreira & Dong[10] estimated the boundary element errors on the basis of the residual of the boundary integral equation on the collocation points excepting the initial nodes. Abe[11] and Sun & Zamani[12, 13] have shown a relationship between the solution error and the residual obtained from the boundary integral equation. The error estimation method by using the difference between the initial numerical solution and the smoothed solution obtained by the higher order interpolation functions has been proposed by Rencis *et al.*[14,15] Guiggiani[16,17] introduced the error estimate based on the sensitivity of the numerical solution by shifting the collocation point along the boundary. A local error estimation of residual type, originating from hypersingular integral equations, was proposed by Guiggiani,[17] Ingber & Mitra[18] and Paulino *et al.*,[19] independently. The idea of local reanalysis was adopted in the error estimation by Charafi *et al.*[20] In this method, the local error is estimated through the subdivision of a boundary element under consideration. Yuuki *et al.*[21] defined the error by using the difference between the solutions obtained both by the singular BEM (ordinary BEM) and by the regular (nonsingular) BEM (source point is taken outside). Kamiya *et al.*[22~26] proposed an error estimation scheme exploiting a character of the collocation-type direct BEM, which was called a sample point error analysis. The scheme was extended to various applications for the potential problem,[22] the elastostatics,[23] the multi domain problem,[24] the thermo-elastics[25] and the elastic problem with body force.[26]

The above-indicated error estimation schemes have been combined with h-, p- and r-version and their hybrid type adaptive schemes.

In the present chapter, an h-version adaptive scheme based on the sample point error analysis is extended to a boundary element bending analysis of thin elastic plates.[27,28] The errors on each boundary element are estimated by the magnitude of discrepancy between the solution interpolated by the initial nodal solution and the solution evaluated at the arbitrary boundary source point which does not coincide with the initial boundary node.

As mentioned above, the adaptive scheme based on the sample point error analysis has been applied to the potential problems and the elastic problems by one of the authors.[22~26] In case of the potential and elastic problems, the governing equations are expressed by the second order differential equations, while the governing equation for the plate bending is formulated by the forth order differential equation (biharmonic differential equation). And then the additional equation for the slope on the boundary is employed. In case of the plate bending problem, we must deal with four kinds of variables and the corresponding errors of different dimension simultaneously. In order to unify the dimensions of errors, the extended error indicators defined as the multiple of each error by the corresponding fundamental solution were exploited in the previous paper.[22~26] However, even if the extended error indicators are introduced, there still exist two different kinds of error indicator, of which dimensions are of deflection and of slope. In order to treat them in a unified manner, an alternative error indicator composed of the non-dimensional extended error indicators is introduced in the present scheme. The property of the new error indicator proposed here is investigated in advance through a simple bending problem of plate analyzed by using uniform boundary elements. Boundary element refinement, h-version adaptive scheme, is implemented based on the magnitude of the error indicator defined in this chapter. The usefulness of the proposed adaptive solution procedure is demonstrated through the results of some analyses of plates with various edge conditions.

3 Sample point error analysis for plate bending problems

3.1 Boundary integral formulations

The governing equation for the bending of thin elastic plates under lateral distributed load is expressed by the following biharmonic differential equation:

$$\nabla^4 w = \frac{\bar{p}}{D} \tag{1}$$

where w, D and $\bar{p}$ mean the deflection, the bending rigidity and the lateral load, respectively. According to the direct boundary element procedure, the boundary integral equation for the plate bending is formulated in terms of four boundary quantities, i.e. the deflection w, the normal slope $w_{,n}$, the normal bending moment M_n and the equivalent shear force K_n , as follows:

$$cw(p_i) = \frac{1}{D}\int_\Gamma [K_n^*(p_i,p_j)w(p_j) - M_n^*(p_i,p_j)w_{,n}(p_j)$$

$$+w_{,n}^*(p_i,p_j)M_n(p_j) - w^*(p_i,p_j)K_n(p_j)]\mathrm{d}\Gamma - \frac{1}{D}\int_\Omega w^*(p_i,p_k)\bar{p}(p_k)\mathrm{d}\Omega \tag{2}$$

and further, another boundary integral equation for the normal derivative of the deflection $w_{,n_0}$ at the boundary source point p_i written as

$$cw_{,n_0}(p_i) = \frac{1}{D}\int_\Gamma [K_{n,n_0}^*(p_i,p_j)w(p_j) - M_{n,n_0}^*(p_i,p_j)w_{,n}(p_j)$$

$$+w_{,nn_0}^*(p_i,p_j)M_n(p_j) - w_{,n_0}^*(p_i,p_j)K_n(p_j)]\mathrm{d}\Gamma$$

$$-\frac{1}{D}\int_\Omega w_{,n_0}^*(p_i,p_k)\bar{p}(p_k)\mathrm{d}\Omega \tag{3}$$

is required for the analysis.[29,30] In the above integral equations, w^* is the fundamental solution of the two-dimensional biharmonic equation and the other functions with an asterisk, $w^*_{,n}$, M_n^*, and K_n^*, are the associated kernels with the fundamental solution w^*, and the coefficient c is the free term. Ω and Γ are the domain and smooth boundary of plate under consideration, and p_j and p_k are points on the boundary Γ and inside the domain Ω, respectively.

3.2 Sample point error analysis

The sample point error analysis is applied to the integral equations (2) and (3) for the plate bending in the following way. According to the idea of a collocation method, the following integral equations hold for the approximate solutions, $\hat{w}$, $\hat{w}_{,n}$, $\hat{M}_n$ and $\hat{K}_n$ obtained for the initial boundary source points p_i:

$$c\hat{w}(p_i) = \frac{1}{D}\int_\Gamma [K_n^*\hat{w} - M_n^*\hat{w}_{,n} + w_{,n}^*\hat{M}_n - w^*\hat{K}_n]\mathrm{d}\Gamma - \frac{1}{D}\int_\Omega w^*\bar{p}\,\mathrm{d}\Omega \tag{4}$$

and

$$c\hat{w}_{,n_0}(p_i)=\frac{1}{D}\int_\Gamma [K^*_{n,n_0}\hat{w}-M^*_{n,n_0}\hat{w}_{,n}+w^*_{,nn_0}\hat{M}_n-w^*_{,n_0}\hat{K}_n]\mathrm{d}\Gamma -\frac{1}{D}\int_\Omega w^*_{,n_0}\bar{p}\mathrm{d}\Omega \qquad (5)$$

If the source point is displaced to any other boundary point p_i' (referred to a sample point) on the element from the initial node p_i, the equality of eqns (4) and (5) is disturbed due to the property of the collocation method. It must be noted that the sample point is on the element where the initial node is located. In order to maintain the equality of eqns (4) and (5), appropriate modifications e_w, e_{w_n}, e_{M_n} and e_{K_n} should be added to the corresponding quantities on each boundary element. The magnitudes of such modifications are considered to predict the actual errors on each element, and assumed to be ignored for the value-prescribed boundary (boundary condition specified) if the numerical interpolation is implemented properly. By adding the modifications, e_w, e_{w_n}, e_{M_n} and e_{K_n} to the relevant quantities, the following equations are formulated for the sample point p_i'

$$c\hat{w}(p_i')+ce_w(p_i')=\frac{1}{D}\int_\Gamma [K^*_n(\hat{w}+e_w)-M^*_n(\hat{w}_{,n}+e_{w_n}) +w^*_{,n}(\hat{M}_n+e_{M_n})-w^*(\hat{K}_n+e_{K_n})]\mathrm{d}\Gamma-\frac{1}{D}\int_\Omega w^*\bar{p}\mathrm{d}\Omega \qquad (6)$$

and

$$c\hat{w}_{,n_0}(p_i')+ce_{w_n}(p_i')=\frac{1}{D}\int_\Gamma [K^*_{n,n_0}(\hat{w}+e_w)-M^*_{n,n_0}(\hat{w}_{,n}+e_{w_n}) +w^*_{,nn_0}(\hat{M}_n+e_{M_n})-w^*_{,n_0}(\hat{K}_n+e_{K_n})]\mathrm{d}\Gamma-\frac{1}{D}\int_\Omega w^*_{,n_0}\bar{p}\mathrm{d}\Omega \qquad (7)$$

where $\hat{w}(p_i')$ and $\hat{w}_{,n_0}(p_i')$ are evaluated from the interpolation of the approximate solutions obtained at the initial node p_i. By defining the values of the deflection $w'(p_i')$ and normal slope $w'_{,n_0}(p_i')$ at the sample point p_i' as follows:

$$cw'(p_i')\equiv\frac{1}{D}\int_\Gamma [K^*_n\hat{w}-M^*_n\hat{w}_{,n}+w^*_{,n}\hat{M}_n-w^*\hat{K}_n]\mathrm{d}\Gamma-\frac{1}{D}\int_\Omega w^*\bar{p}\mathrm{d}\Omega \qquad (8)$$

and

$$cw'_{,n_0}(p_i')\equiv\frac{1}{D}\int_\Gamma [K^*_{n,n_0}\hat{w}-M^*_{n,n_0}\hat{w}_{,n}+w^*_{,nn_0}\hat{M}_n-w^*_{,n_0}\hat{K}_n]\mathrm{d}\Gamma -\frac{1}{D}\int_\Omega w^*_{,n_0}\bar{p}\mathrm{d}\Omega \qquad (9)$$

we obtain the relations between the errors e_w, e_{w_n}, e_{M_n} and e_{K_n} on the sample point and those on the observation point as follows:

$$c\hat{w}(p_i') - cw'(p_i') + ce_w(p_i') = \frac{1}{D}\int_\Gamma [K_n^* e_w - M_n^* e_{w_n} + w_{,n}^* e_{M_n} - w^* e_{K_n}]d\Gamma \quad (10)$$

and

$$c\hat{w}_{,n_0}(p_i') - cw'_{,n_0}(p_i') + ce_{w_n}(p_i') = \frac{1}{D}\int_\Gamma [K_{n,n_0}^* e_w - M_{n,n_0}^* e_{w_n} + w_{,nn_0}^* e_{M_n} - w_{,n_0}^* e_{K_n}]d\Gamma \quad (11)$$

where $w'(p_i')$ and $w'_{,n_0}(p_i')$ defined by eqns (8) and (9) were shown to predict the actual behavior of the exact solution. By considering the difference defined as

$$d(p_i') = c\hat{w}(p'_i) - cw'(p_i') \quad (12)$$

and

$$d_n(p_i') = c\hat{w}_{,n_0}(p'_i) - cw'_{,n_0}(p_i') \quad (13)$$

to be the residues of the deflection and normal slope at the sample point, equations (10) and (11) are rewritten as follows:

$$d(p_i') = \frac{1}{D}\int_\Gamma [K_n^* e_w - M_n^* e_{w_n} + w_{,n}^* e_{M_n} - w^* e_{K_n}]d\Gamma - ce_w(p_i') \quad (14)$$

and

$$d_n(p_i') = \frac{1}{D}\int_\Gamma [K_{n,n_0}^* e_w - M_{n,n_0}^* e_{w_n} + w_{,nn_0}^* e_{M_n} - w_{,n_0}^* e_{K_n}]d\Gamma - ce_{w_n}(p_i') \quad (15)$$

The location of the sample point p_i' is generally arbitrary except the one of the initial node, but there should be a location where the differences between the sample point solutions $w'(p_i')$, $w'_{,n_0}(p_i')$ and the interpolated approximate solutions $\hat{w}(p_i')$, $\hat{w}_{,n_0}(p_i')$ are relatively large. Once the residues d and d_n, the left-hand side of the above equations, are evaluated approximately in any way, the errors can be obtained by solving eqns (14) and (15) simultaneously.

3.3 Discretization

In order to evaluate the above defined errors on each boundary element, eqns (14) and (15) are discretized by N constant boundary elements as follows:

$$d\,(p_i') = \frac{1}{D}\sum_{j=1}^{N}[\,e_{wj}\,(\int_{\Gamma_j} K_n^* \mathrm{d}\Gamma - c\delta_{ij}) - e_{w_n j}\int_{\Gamma_j} M_n^* \mathrm{d}\Gamma + e_{M_n j}\int_{\Gamma_j} w_{,n}^* \mathrm{d}\Gamma - e_{K_n j}\int_{\Gamma_j} w^* \mathrm{d}\Gamma] \tag{16}$$

and

$$d_n(p_i') = \frac{1}{D}\sum_{j=1}^{N}[\,e_{wj}\int_{\Gamma_j} K^*_{n,n_0} \mathrm{d}\Gamma - e_{w_n j}(\int_{\Gamma_j} M^*_{n,n_0} \mathrm{d}\Gamma + c\delta_{ij})$$

$$+ e_{M_n j}\int_{\Gamma_j} w^*_{,nn_0} \mathrm{d}\Gamma - e_{K_n j}\int_{\Gamma_j} w^*_{,n_0} \mathrm{d}\Gamma] \tag{17}$$

where δ_{ij} is the Kronecker's delta. Considering here the case of fully clamped supported plate for simplicity, the deflection w and the normal slope $w_{,n}$ on the boundary are specified to be null. In this case, the errors of these values are thought to vanish from the assumption:

$$e_w = 0, \quad e_{w_n} = 0$$

By substituting the above equations, equations (16) and (17) are reduced to

$$d\,(p_i') = \frac{1}{D}\sum_{j=1}^{N}[\,e_{M_n j}\int_{\Gamma_j} w_{,n}^* \mathrm{d}\Gamma - e_{K_n j}\int_{\Gamma_j} w^* \mathrm{d}\Gamma] \tag{18}$$

and

$$d_n(p_i') = \frac{1}{D}\sum_{j=1}^{N}[\,e_{M_n j}\int_{\Gamma_j} w^*_{,nn_0} \mathrm{d}\Gamma - e_{K_n j}\int_{\Gamma_j} w^*_{,n_0} \mathrm{d}\Gamma] \tag{19}$$

The above equations can be expressed formally in the matrix form

$$\mathbf{d} = \mathbf{A}\mathbf{e} \tag{20}$$

where the vectors **d**, **e** and the matrix **A** are expressed as

$$\mathbf{d}=\begin{pmatrix} d(p_i') \\ d_n(p_i') \end{pmatrix}, \qquad \mathbf{e}=\begin{pmatrix} e_{M_n} \\ e_{K_n} \end{pmatrix}$$

$$\mathbf{A}=\frac{1}{D}\begin{bmatrix} \int_\Gamma w^*_{,n}\mathrm{d}\Gamma & -\int_\Gamma w^*\mathrm{d}\Gamma \\ \int_\Gamma w^*_{,nn_0}\mathrm{d}\Gamma & -\int_\Gamma w^*_{,n_0}\mathrm{d}\Gamma \end{bmatrix} \tag{21}$$

After estimating the residue **d** on each sample point, we can obtain the required modifications by solving eqn (20) as the simultaneous linear algebraic equation. The modifications for the problem with other boundary conditions can be evaluated by employing the procedure similar to the above. In the case of fully supported plate, for example, the vector **e** and matrix **A** corresponding to eqn (21) are replaced by

$$\mathbf{e}=\begin{pmatrix} e_{w_n} \\ e_{K_n} \end{pmatrix}, \qquad \mathbf{A}=\frac{1}{D}\begin{bmatrix} -\int_\Gamma M^*_n\mathrm{d}\Gamma & -\int_\Gamma w^*\mathrm{d}\Gamma \\ -\int_\Gamma M^*_{n,n_0}\mathrm{d}\Gamma - c\delta_{ij} & -\int_\Gamma w^*_{,n_0}\mathrm{d}\Gamma \end{bmatrix} \tag{22}$$

4 Boundary element refinement

4.1 Non-dimensional error indicator

The dimensions of the errors (or modifications) e_w, e_{w_n}, e_{M_n} and e_K estimated by the above mentioned sample point error analysis are different from each other, i.e. the deflection, the slope, the bending moment and the equivalent shear force, respectively. Consequently it seems difficult to compare them directly. In order to reduce the number of error varieties, we introduce, instead of the error itself, the following extended error indicator defined as the multiple of the error on each element and the corresponding fundamental solutions:

$$\tilde{e}_w=\frac{1}{D}\int_\Gamma K^*_n e_w\mathrm{d}\Gamma, \quad \tilde{e}_{w_n}=\frac{1}{D}\int_\Gamma M^*_n e_{w_n}\mathrm{d}\Gamma$$
$$\tilde{e}_{M_n}=\frac{1}{D}\int_\Gamma w^*_{,n} e_{M_n}\mathrm{d}\Gamma, \quad \tilde{e}_{K_n}=\frac{1}{D}\int_\Gamma w^* e_{K_n}\mathrm{d}\Gamma \tag{23}$$

$$\tilde{e}_{w,n_0}=\frac{1}{D}\int_\Gamma K^*_{n,n_0} e_w\mathrm{d}\Gamma, \quad \tilde{e}_{w_n,n_0}=\frac{1}{D}\int_\Gamma M^*_{n,n_0} e_{w_n}\mathrm{d}\Gamma$$
$$\tilde{e}_{M_n,n_0}=\frac{1}{D}\int_\Gamma w^*_{,nn_0} e_{M_n}\mathrm{d}\Gamma, \quad \tilde{e}_{K_n,n_0}=\frac{1}{D}\int_\Gamma w^*_{,n_0} e_{K_n}\mathrm{d}\Gamma \tag{24}$$

Equations (23) and (24) have the dimensions of deflection and slope, respectively. It must be noticed here that in order to evaluate eqns (23) and (24) we may utilize the multiple of the components of matrix **A** [e.g. eqns (21) or (22)] and corresponding errors without direct numerical integration. Substitution of eqns (23) and (24) into eqns (14) and (15) leads to

$$d(p_i') = \tilde{e}_w + \tilde{e}_{w_n} + \tilde{e}_{M_n} + \tilde{e}_{K_n} \tag{25}$$

and

$$d_n(p_i') = \tilde{e}_{w,n_0} + \tilde{e}_{w_n,n_0} + \tilde{e}_{M_n,n_0} + \tilde{e}_{K_n,n_0} \tag{26}$$

Employing each component of the above extended errors

$$\tilde{e}_w(i,j),\quad \tilde{e}_{w_n}(i,j),\quad \tilde{e}_{M_n}(i,j),\quad \tilde{e}_{K_n}(i,j) \tag{27}$$

$$\tilde{e}_{w,n_0}(i,j),\quad \tilde{e}_{w_n,n_0}(i,j),\quad \tilde{e}_{M_n,n_0}(i,j),\quad \tilde{e}_{K_n,n_0}(i,j) \tag{28}$$

which are considered as "error influence" on the *i*-th sample point from the *j*-th element error, we introduce the following error indicators

$$\eta(i,j) = \left|\tilde{e}_w(i,j)\right| + \left|\tilde{e}_{w_n}(i,j)\right| + \left|\tilde{e}_{M_n}(i,j)\right| + \left|\tilde{e}_{K_n}(i,j)\right| \tag{29}$$

and

$$\eta_n(i,j) = \left|\tilde{e}_{w,n_0}(i,j)\right| + \left|\tilde{e}_{w_n,n_0}(i,j)\right| + \left|\tilde{e}_{M_n,n_0}(i,j)\right| + \left|\tilde{e}_{K_n,n_0}(i,j)\right| \tag{30}$$

where $\eta(i,j)$ and $\eta_n(i,j)$ are considered as the influence intensities of the *j*-th element error on the error on the sample point *i*, and have the dimensions of deflection and slope, respectively. It must be noted that any two components of the right-hand side of eqn (29) or (30) vanish since the errors of the specified boundary quantities are thought to be null. Therefore, equations (29) and (30) are reduced to the following expression depending on the boundary conditions:

$$\eta(i,j) = \left|\tilde{e}_{M_n}(i,j)\right| + \left|\tilde{e}_{K_n}(i,j)\right| \tag{31}$$

$$\eta_n(i,j) = \left|\tilde{e}_{M_n,n_0}(i,j)\right| + \left|\tilde{e}_{K_n,n_0}(i,j)\right| \tag{32}$$

for the clamped supported edge,

$$\eta(i,j) = \left|\tilde{e}_{w_n}(i,j)\right| + \left|\tilde{e}_{K_n}(i,j)\right| \tag{33}$$

$$\eta_n(i,j) = \left|\tilde{e}_{w_n,n_0}(i,j)\right| + \left|\tilde{e}_{K_n,n_0}(i,j)\right| \tag{34}$$

for the supported edge, and

$$\eta\,(i,j)=\left|\tilde{e}_{w}(i,j)\right|+\left|\tilde{e}_{w_n}(i,j)\right| \tag{35}$$

$$\eta_n\,(i,j)=\left|\tilde{e}_{w,n_0}(i,j)\right|+\left|\tilde{e}_{w_n,n_0}(i,j)\right| \tag{36}$$

for the free edge.

As already mentioned above, $\eta(i,j)$ and $\eta_n(i,j)$ are of the dimensions of deflection and slope, respectively. In order to treat them in the unified manner, the non-dimensional error indicator is defined here as follows:

$$\bar{\eta}\,(i,j)=\frac{D}{\bar{p}L^4}\eta\,(i,j)+\frac{D}{\bar{p}L^3}\eta_n(i,j) \tag{37}$$

for the distributed lateral load $\bar{p}$, and

$$\bar{\eta}\,(i,j)=\frac{D}{PL^2}\eta\,(i,j)+\frac{D}{PL}\eta_n(i,j) \tag{38}$$

for the concentrated lateral load P. In eqns (37) and (38), L is the reference length of a plate and regarded, for example, as the maximum length of a plate. When the plate is subjected to a concentrated moment or distributed moment, the non-dimensional error indicator is defined in a similar way without any difficulty.

4.2 Error estimation

In order to examine the property of non-dimensional error influence intensity defined by eqn (37) or (38), we consider a bending problem of a fully clamped thin elastic square plate subjected to uniform lateral load [edge length = 100mm, thickness = 5mm, Young's modulus = 206,000MPa, Poisson's ratio = 0.3, lateral load = 2.4MPa]. The plate boundary is discretized into uniform constant elements (40 elements) as shown in Fig. 1. The errors are evaluated at two sample points disposed on a boundary element symmetrically about the source point at the center. Each sample point is placed just in the middle between the source point and the extreme ends of the boundary element for the present analysis. A distribution of magnitude of non-dimensional error influence intensity $\bar{\eta}(i,j)$ calculated by uniform meshes is depicted stereographically in Fig. 2. From this figure, it is recognized that the magnitude of $\bar{\eta}(i,j)$ becomes extreme when the distance between the elements i and j is maximum. This result shows that the magnitude of $\bar{\eta}(i,j)$ depends more on the distribution of fundamental solution than on that of the error. By examining sufficiently the distribution of $\bar{\eta}(i,j)$, the following two expressions

$$\bar{\eta}_i=\sum_j \bar{\eta}\,(i,j)/N \tag{39}$$

and

$$\overline{\eta}_j = \sum_i \overline{\eta}\,(i,j)/\,N \tag{40}$$

are considered as the candidates of the indicators for adaptive mesh refinement. In the above equations, N is the total number of boundary elements. Figure 3 shows the distribution of the magnitude of above-defined indicators against each node. In this figure, the whole data along the boundary are plotted for completeness while the present problem has the symmetry. It is recognized from Fig. 3 that the distribution of the magnitude of $\overline{\eta}_j$ varies considerably, while those of the magnitude of $\overline{\eta}_i$ is relatively monotonous. Therefore $\overline{\eta}_j$ is thought to be preferable to $\overline{\eta}_i$ as the indicator for the adaptive mesh refinement. An averaged value of $\overline{\eta}(i,j)$ defined as

$$\eta_{ave} = \sum_{i,j} \overline{\eta}\,(i,j)/\,N^2 \tag{41}$$

is also depicted by the broken line in the same figure for reference.

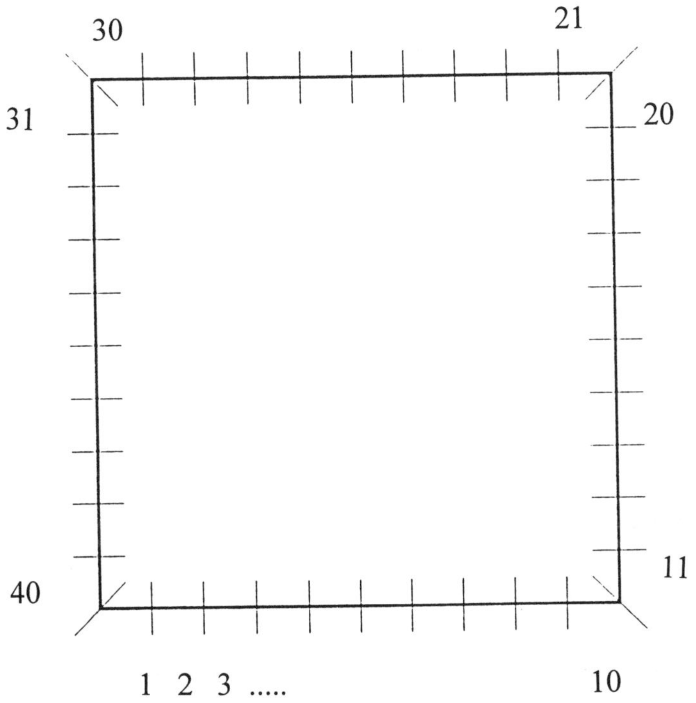

Figure 1. Uniform mesh and node numbering for benchmark problem.

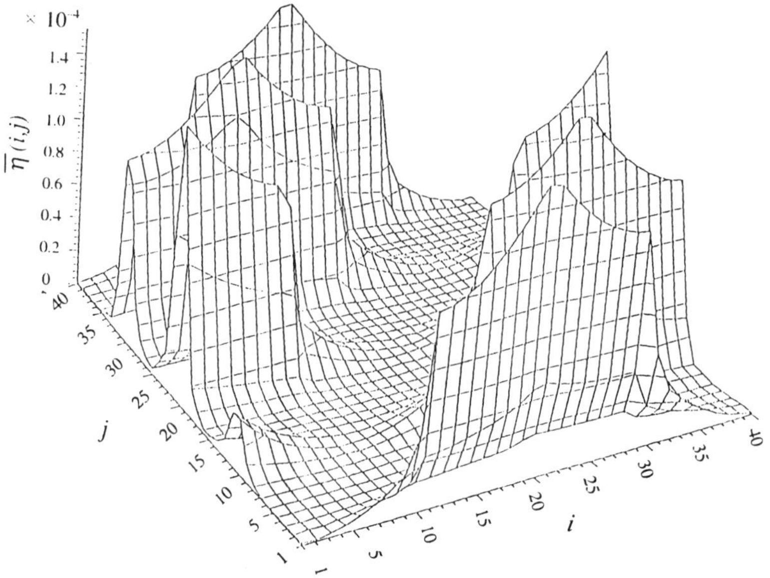

Figure 2. Stereographic distribution of non-dimensional error indicator $\eta(i,j)$ for uniform mesh.

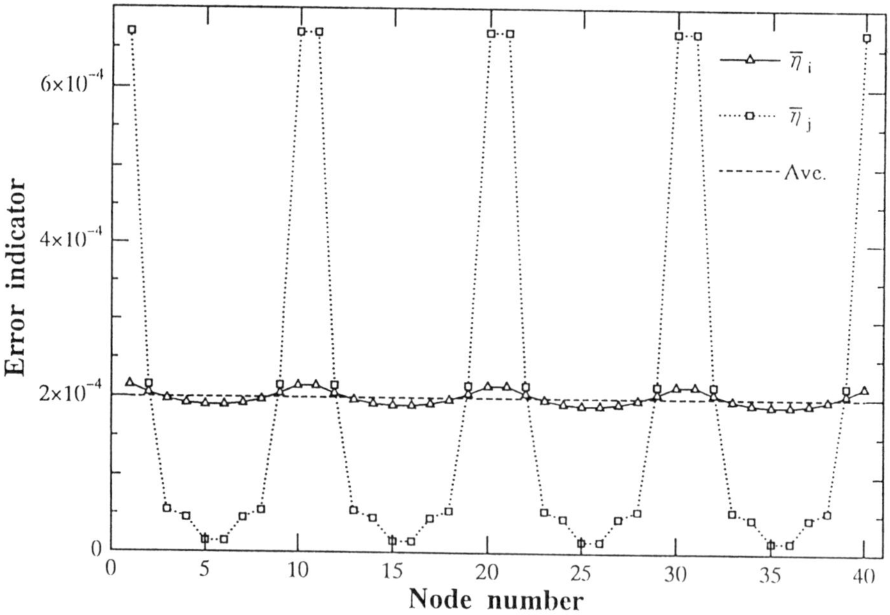

Figure 3. Distribution of indicators η_i and η_j for uniform mesh.

As a consequence, in this study, we will employ $\overline{\eta}_j$ as a indicator for the adaptive refinement and its averaged value $\overline{\eta}_{ave}$ as a criterion of mesh refinement. That is, the j-th boundary element is refined according to the following criterion:

$$\overline{\eta}_j \geq \overline{\eta}_{ave} \tag{42}$$

The convergence of the solution is judged by the k-th and k+1-st magnitudes of $\overline{\eta}_{ave}$ as follows:

$$\frac{\overline{\eta}_{ave}^{(k+1)} - \overline{\eta}_{ave}^{(k)}}{\overline{\eta}_{ave}^{(k)}} < \varepsilon \tag{43}$$

where ε is a specified tolerance. In the following analysis the magnitude of ε is taken around 10^{-2}.

5 Numerical applications

In order to examine the performance of the adaptive approach proposed here, some fundamental examples are shown for the bending problems of thin elastic plates with various edge conditions. It is possible to apply the present adaptive scheme to the general bending problems.

5.1 A fully clamped square plate subjected to uniformly distributed lateral load

We consider the example same as that of the previous section, i.e. a fully clamped square plate subjected to uniformly distributed lateral load. Since the boundary condition of this problem is uniform along a whole plate boundary, the only four boundary elements (N = 4) taken on the entire sides are required for the initial computation. Figure 4 shows the process of boundary element refinement for each iteration step. The distribution of the magnitude of normal bending moment and equivalent shear force calculated by each mesh is presented in Figs 5 and 6, respectively. In these figures, only numerical results on the half edge are plotted due to the symmetry of the problem. The BEM solutions of the first two iteration steps (N = 4, 8) are omitted because of their insufficient accuracy. The analytical solutions for each quantity are also inserted by solid line in the figures for comparison purpose. It is recognized from these results that the accuracy of the boundary element solutions is improved according to the mesh refinement near the corners of the plate.

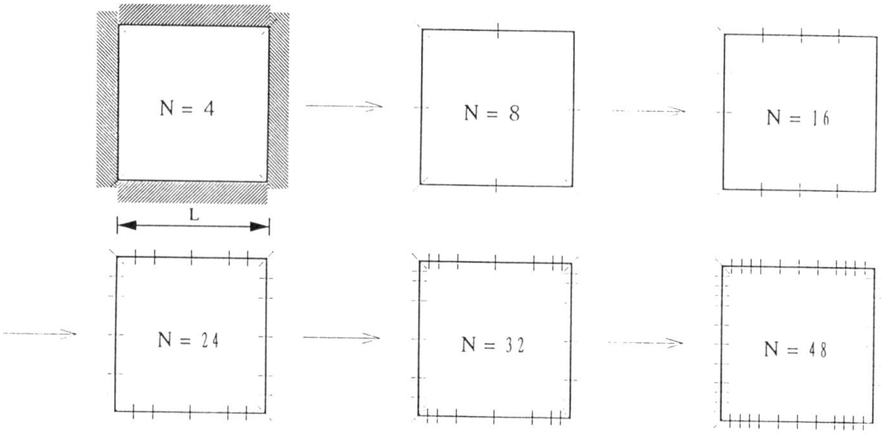

Figure 4. Mesh refinement process for clamped-supported square plate.

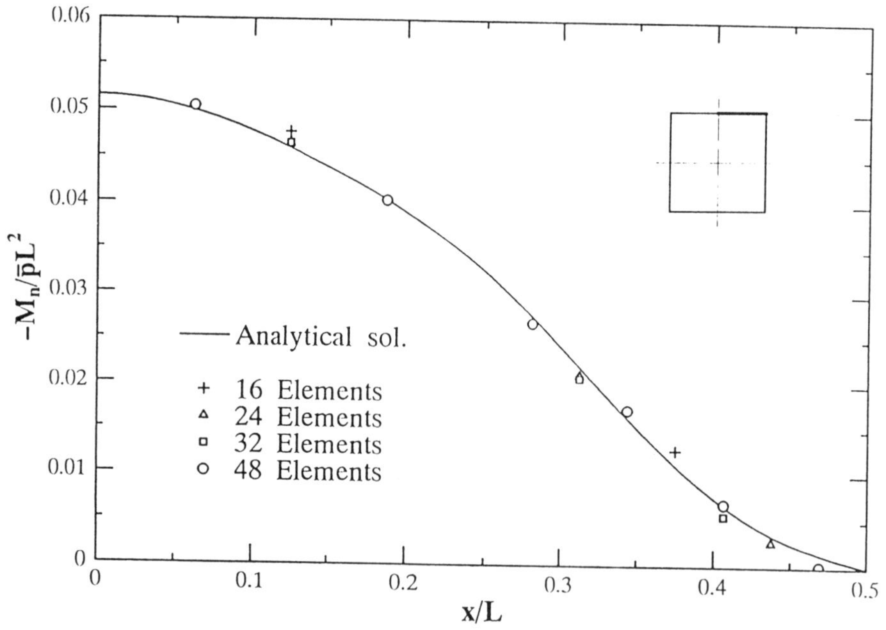

Figure 5. Normal bending moment of clamped-supported square plate.

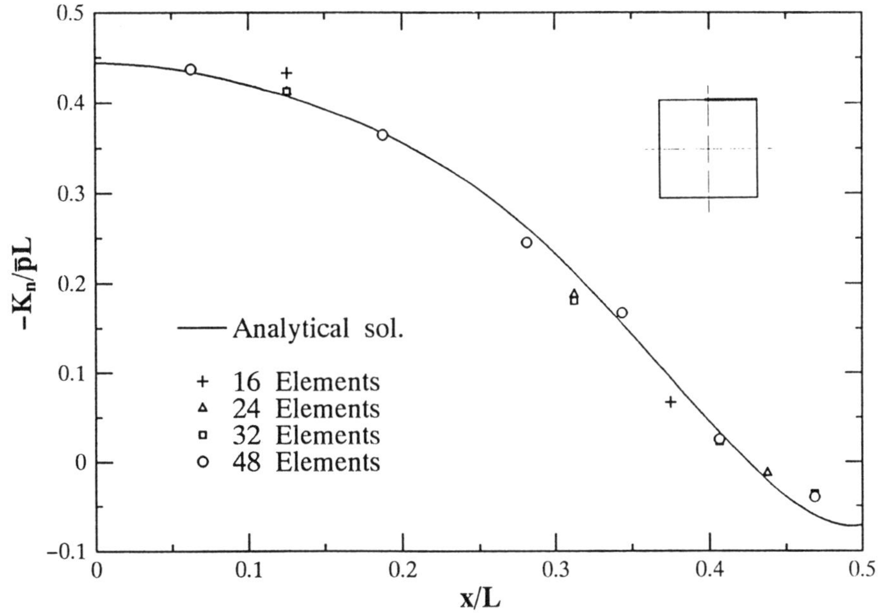

Figure 6. Equivalent shear force of clamped-supported square plate.

5.2 A fully simply-supported thin square plate subjected to uniform lateral load

A fully simply supported thin square plate subjected to a uniform lateral load is analyzed by using the proposed adaptive meshing scheme. The process of boundary element refinement for each iteration step is shown in Fig. 7. It can be seen from this figure that the remeshing process in the case of simply supported plate is a little different from that in the case of clamped supported plate. Figure 8 shows the distribution of normal slope on the supported edge. The boundary element solutions are modified appropriately depending on the adaptive addition of boundary elements.

5.3 A square plate under uniform load with two opposite edges simply supported and the other two edges clamped

A square plate under uniform load with two opposite edges simply supported and the other two edges clamped is considered. The element distribution at each mesh refinement stage is shown in Fig. 9. At the final stage, the mesh refinement proceeds more on the clamped edges compared with on the simply supported edges. In Figs 10 and 11 are shown the numerical results of the normal bending moment on the clamped edge and the normal slope on the simply supported edge, respectively. Both boundary element solutions on the final mesh refinement stage are sufficiently accurate compared with the corresponding analytical solutions.

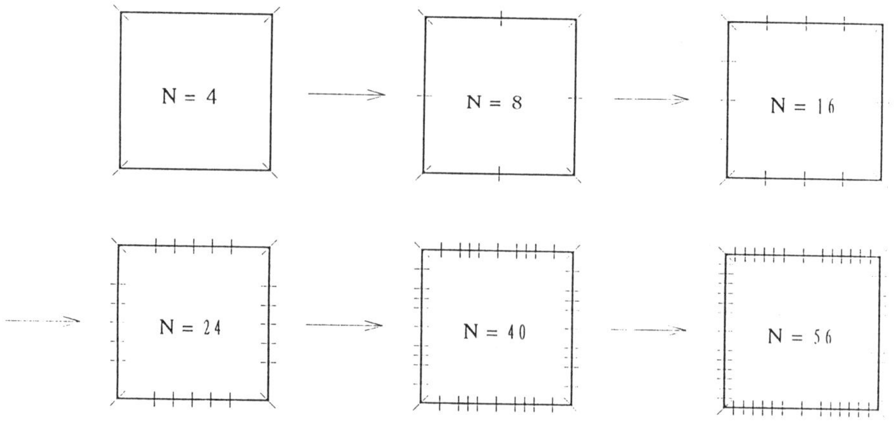

Figure 7. Mesh refinement process for simply-supported square plate.

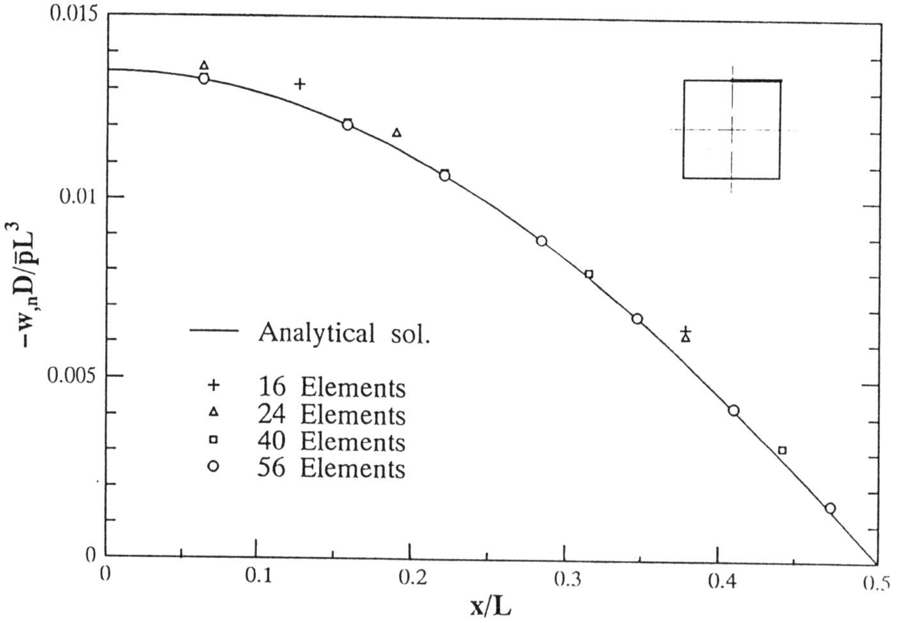

Figure 8. Normal slope of simply-supported square plate.

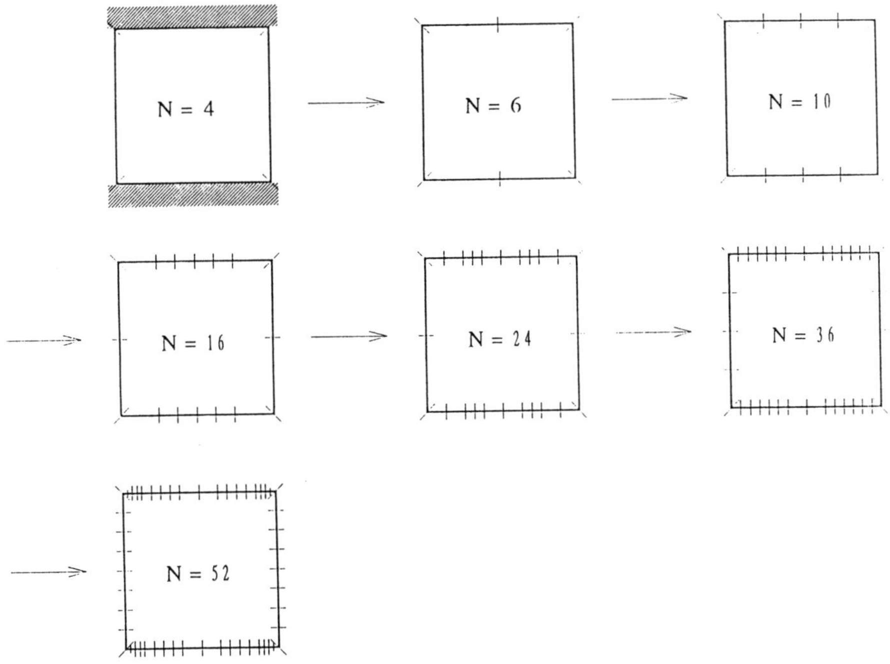

Figure 9. Mesh refinement process for square plate with two opposite edges simply supported and other two edges clamped.

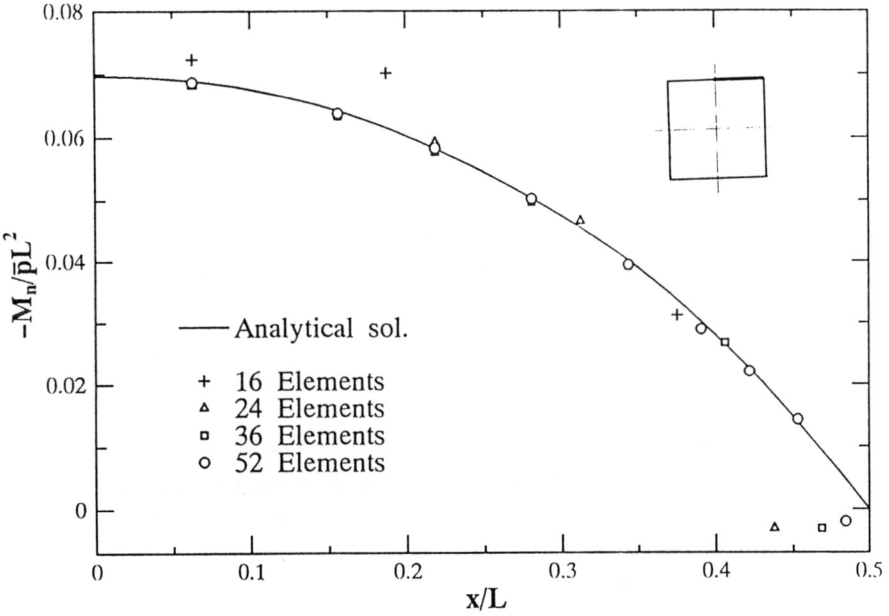

Figure 10. Normal bending moment on clamped-supported edge.

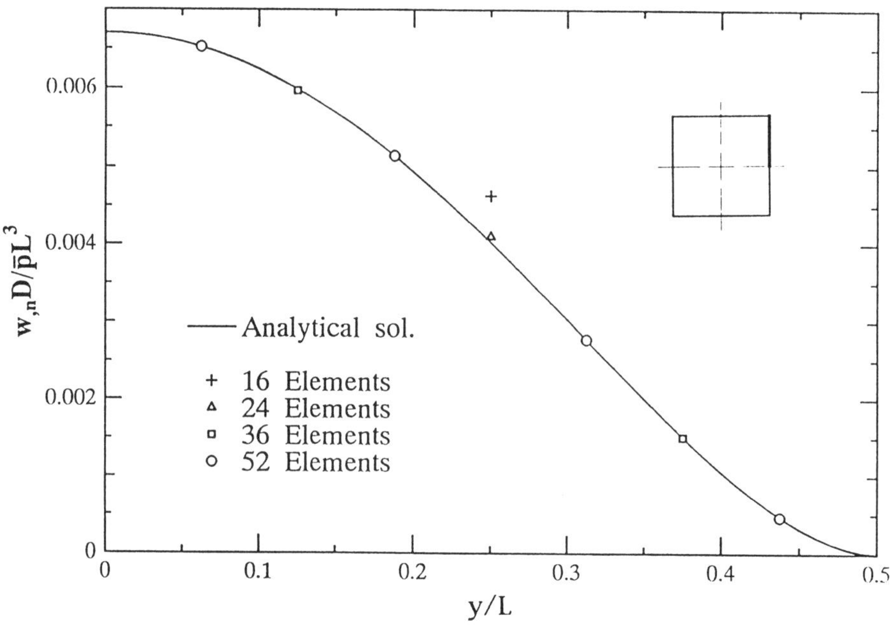

Figure 11. Normal slope on simply-supported edge.

5.4 A square plate with three edges simply supported and one edge clamped

A case of square plate with one edge clamped and three other edges simply supported is considered in this example. The plate is assumed to be under uniformly distributed load. Four boundary elements are also used for an initial discretization of the plate. The boundary element refinement process is shown in Fig. 12. The results on each edge by the present adaptive boundary elements are compared with those of the corresponding analytical solutions. The distribution of the normal bending moment on a fixed side and of the normal slope on the supported edges is depicted in Fig. 13 and Figs 14 and 15, respectively. It is noticed from these figures that the accuracy of the boundary element solutions is improved in consequence of the adaptation.

5.5 A simply-supported square plate under an eccentrically concentrated load

We consider lastly a simply-supported square plate under an eccentrically concentrated load. It can be seen from Fig. 16 that the mesh refinement proceeds more on the edges in the neighborhood of the nearest corner to the loaded point. Figure 17 shows the distribution of the normal slope on the upper edge of the plate. After the fourth iteration, we can obtain enough accurate solutions only by 24 boundary elements produced by the proposed adaptive scheme.

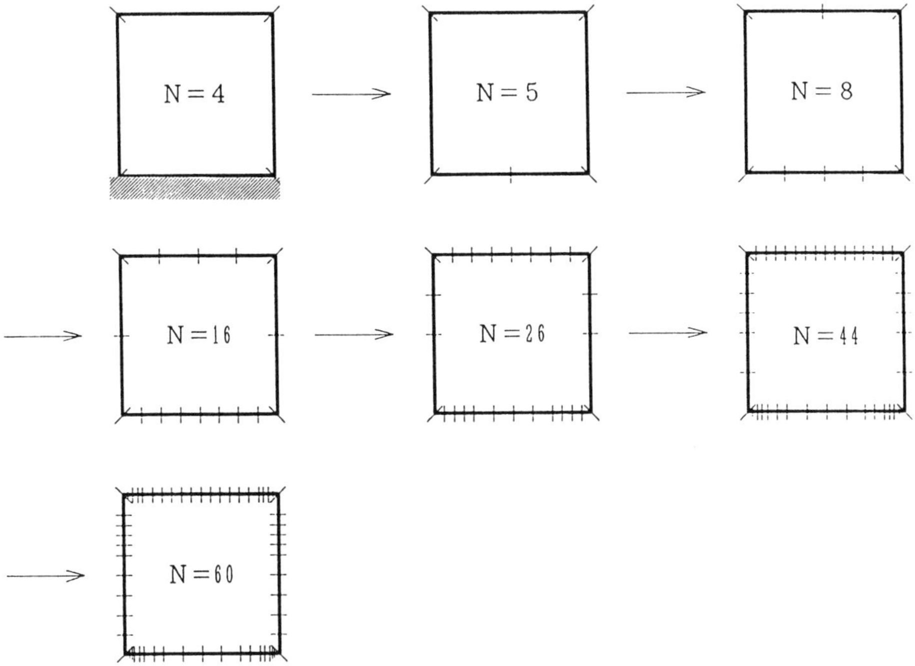

Figure 12. Mesh refinement process for square plate with three edges simply supported and one edge clamped.

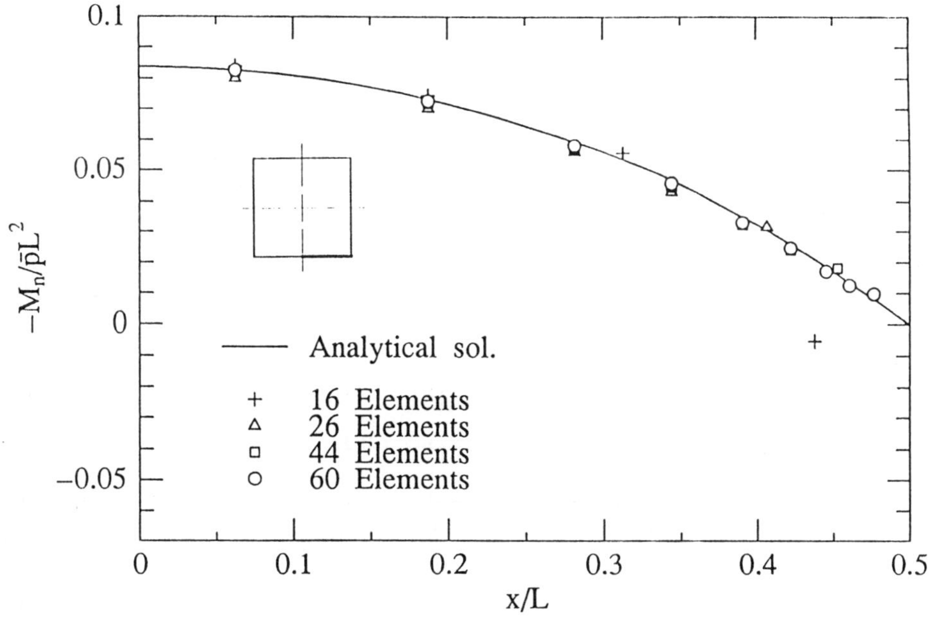

Figure 13. Normal bending moment on clamped-supported edge.

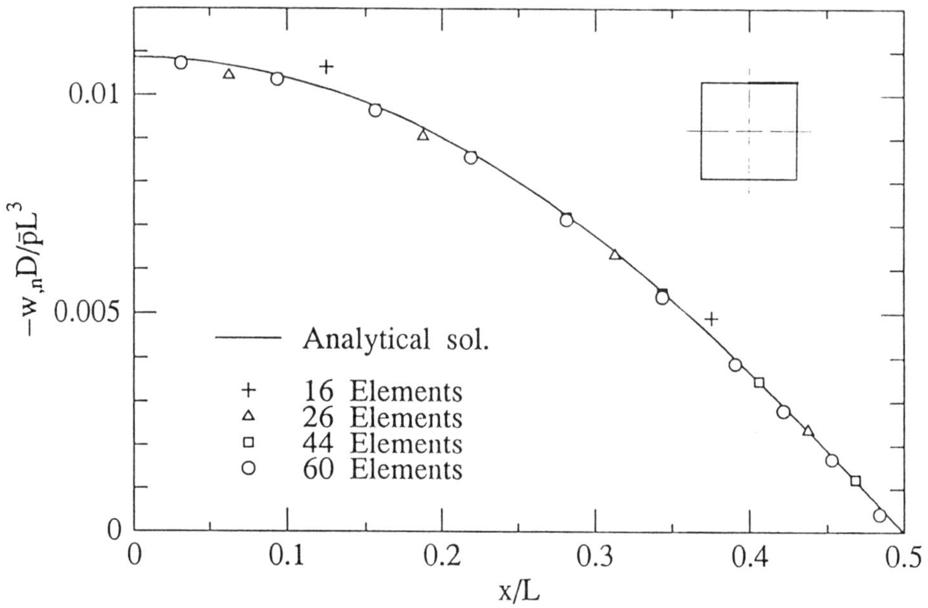

Figure 14. Normal slope on simply-supported edge opposite with clamped-supported edge.

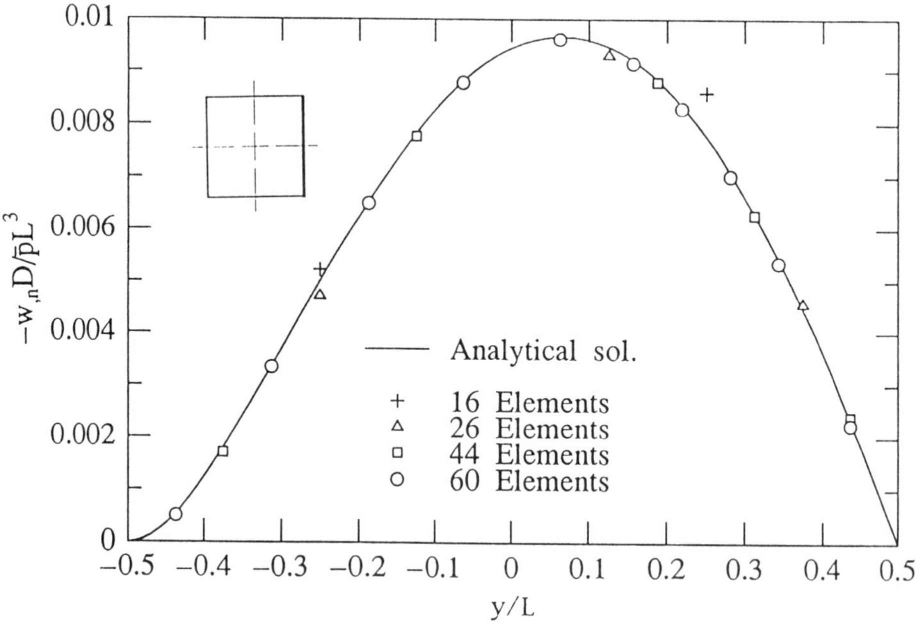

Figure 15. Normal slope on two opposite simply-supported edges.

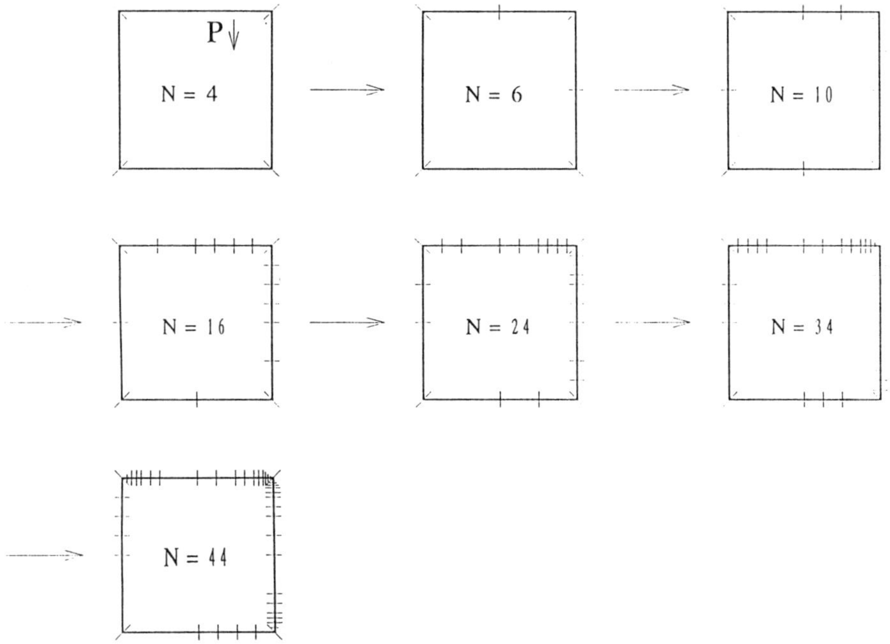

Figure 16. Mesh refinement process for simply-supported square plate under an eccentrically concentrated load.

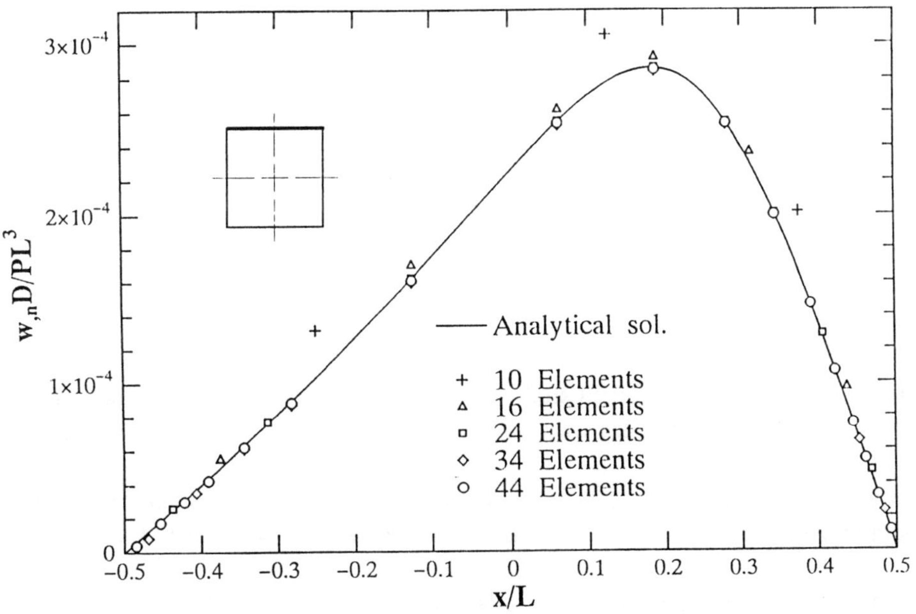

Figure 17. Normal slope of an eccentrically concentrated loaded plate.

6 Concluding remarks

An h-version adaptive boundary element scheme based on the sample point error analysis was applied to the bending problem of thin elastic plates. In the case of plate bending analysis by the sample point error estimate, there exists two kinds of error indicators which have different dimensions, i.e. those of the deflection and the slope. In order to treat them in the unified manner, the non-dimensional alternative error indicator was proposed in this paper. The effectiveness of the proposed indicator to an h-version adaptive boundary element scheme was proved by analyzing some numerical examples. The solution with high accuracy was obtained by the relatively small number of boundary elements produced by the proposed adaptive scheme.

Acknowledgements

The authors wish to express their thanks to Professor M. Tokuda of Mie University for his encouragement to the first author and also wish to thank Mr K. Tago, a graduated student of Mie University, for his great efforts in the numerical calculations.

References

[1] Babuska, I. & Rank, E. An expert system for optimal mesh design in the *h-p* version of the finite element method, *International Journal for Numerical Methods in Engineering*, **24**, 2087~2106, 1987.

[2] Brebbia, C.A. & Aliabadi, M.H. eds. *Adaptive Finite and Boundary Element Methods.* Computational Mechanics Publications, Southampton, Boston, 1993.

[3] Babuska, I., Zienkiewicz, O.C., Gago, J. & de A. Oliveira, E.R., eds. *Accuracy Estimates and Adaptive Refinements in Finite Element Computations*, John Wiley and Sons, 1986.

[4] Kita, E. & Kamiya, N., Recent studies on adaptive boundary element methods, *Advances in Engineering Software*, **19**, 21~32, 1994.

[5] Liapis, S., A review of error estimation and adaptivity in the boundary element method, *Engineering Analysis with Boundary Elements*, **14**, 315~323, 1994.

[6] Kamiya, N., ed., Error estimate and adaptive meshes for FEM/BEM, *Special Issue of Advances in Engineering Software*, **15**, 1992.

[7] Alarcon, E. & Reverter, A. *P*-adaptive boundary elements, *International Journal*

for Numerical Methods in Engineering, **23**, 801~829, 1986.

[8] Cerrolaza, M., Gomez-Lera, M.S., & Alarcon, E. Elastostatic *p*-adaptive boundary elements for micros, *Software forEngineering Workst*ations, **4**, 18~24, 1988.

[9] Rank, E. Adaptive *h*- *p*- and *hp*-versions for boundary integral element methods, *International Journal for Numerical Methods in Engineering*, **28**, 1335~1349, 1989.

[10] Parreira, P. & Dong, Y.F. Adaptive heirachical boundary elements, *Advances in Engineering Software*, **15**, 249~259, 1992.

[11] Abe K. A new residue and nodal error evaluation in h-adaptive boundary element method, *Advances in Engineering Software*, **15**, 231~247, 1992.

[12] Sun, W. & Zamani, N.G. Adaptive mesh refinement/redistribution for the equations of linear elasticity, boundary element formulation, *Computers and Structures*, **44**, 627~637, 1992.

[13] Zamani, N.G. & Sun, W. Adaptive *r* and *h-r* algorithms for boundary elements, *Advances in Engineering Software*, **15**, 241~247, 1992.

[14] Mullen, R.L. & Rencis, J.J. Adaptive mesh refinement technique for boundary element methods, *Proceedings on Advanced Topics in Boundary Element Analysis*. ASME, New York, pp. 235~255, 1985.

[15] Rencis, J.J. & Jong, K.Y. A self-adaptive *h*-refinement technique for the boundary element method, *Computer Methods in Applied Mechanics and Engineering*, **73**, 295~316, 1989.

[16] Guiggiani, M. Error indicators for adaptive mesh refinement in the boundary element methods - a new approach, *International Journal for Numerical Methods in Engineering*, **29**, 1247~1269, 1990.

[17] Guiggiani, M. Sensitivity analysis for boundary element error estimation and mesh refinement, *International Journal for Numerical Methods in Engineering*, **39**, 2907~2920, 1996.

[18] Ingber, M.J. & Mitra, A.K. Grid redistribution based on measureable error indicators for the direct boundary element method, *Engineering Analysis with Boundary Elements*, **9**, 13~19, 1992.

[19] Paulino G.H., Gray L.J. & Zarikian V. Hypersingular residuals - a new approach for error estimation in the boundary element method, *International Journal for*

Numerical Methods in Engineering, **39**, 2005~2029, 1996.

[20] Charafi, A., Neves, A.C. & Wrobel, L.C. *h*-Hierarchical adaptive boundary element method using local reanalysis, *International Journal for Numerical Methods in Engineering*, **38**, 2185~2207, 1995.

[21] Yuuki, R., Cao, G.Q. & Tamaki, M. Efficient error estimation and adaptive meshing method for boundary element analysis, *Advances in Engineering Software*, **15**, 279~287, 1992.

[22] Kawaguchi, K. & Kamiya, N. An adaptive BEM by sample point error analysis, *Engineering Analysis with Boundary Elements*, **9**, 225~262, 1992.

[23] Kamiya, N. & Kawaguchi, K. Error analysis and adaptive refinement of boundary elements in elastic problem, *Advances in Engineering Software*, **15**, 223~230, 1992.

[24] Kamiya, N. & Koide, M., Adaptive boundary element for multiple subregions, *Computational Mechanics*, **12**, 69~80, 1993.

[25] Kamiya, N., Aikawa, Y. & Kawaguchi, K. An adaptive boundary element scheme for steady thermoelastic analysis, *Computer Methods in Applied Mechanics and Engineering*, **119**, 311~324, 1994.

[26] Kamiya, N., Aikawa, Y. & Kawaguchi, K. Elasto-static analysis considering body force by *h*-adaptive boundary element method, *Proceedings of Sixth China-Japan Conferences on Boundary Element Methods*, 271~278, 1994.

[27] Sawaki, Y., Tanaka, M. & Kamiya, N. Adaptive remeshing for boundary element bending analysis, in *Boundary Elements XIV*, Vol. 1, 1992, 667~678.

[28] Sawaki, Y., Tago, K. & Kamiya, N. Adaptive boundary elements for plate bending analysis, *Engineering Analysis with Boundary Elements*, **14**, 219~227, 1994.

[29] Bezine, G. Boundary integral formulation for plate flexure with arbitrary boundary conditions, *Mechanics Research Communications*, **5**, 197~206, 1987.

[30] Stern, M. A general boundary integral formulation for the numerical solution of plate bending problems, *International Journal on Solids and Structures*, **15**, 769~782, 1979.

Chapter 8

Nonlinear analysis of plate bending by boundary element method

Qing Hua Qin
Department of Engineering Mechanics, Tsinghua University, Beijing 100084, People's Republic of China

Abstract

In this paper, the boundary integral formulation has been presented for nonlinear analysis of plate bending problems which include large deflection of thick plates and postbuckling problem of thin plates with or without elastic foundations. The thick plate may be of Reissner plate or sandwich plate. The fundamental solutions are derived by way of resolution method of differential operator, and then the boundary integral equations are established through use of the solutions and some newly constructed variational functionals. Numerical iteration calculations are carried out by means of discretized algebraic equations with piecewise constant boundary elements and linear domain cells. Some numerical results are presented to illustrate the application of the boundary element formulation.

1 Introduction

The boundary element method (BEM), which is based on an integral formulation of a given problem, has been so well developed during the last 20 years that it is considered to be a very popular computational tool. This method consists of formulating the problem in terms of an integral equation relating only boundary values and determing its solution numerically. Thus, it requires only a surface discretization and not a full-domain discretization with "domain-type" techniques, such as the finite difference method (FDM) and the finite element method (FEM). If interior domain values are needed, however, it can be subsequently calculated from boundary data. As a result, the dimensionality of the problem will be reduced by one and the system of equations resulting from the discretization of the boundary integral formulation is much smaller in size than the one encountered in either FDM or FEM, which are both based on a differential equation formulation. Remarkable advances in this area during the past 15 years can be found, for example, in Brebbia,[1,2] Beskos,[3] Aliabadi & Brebbia.[4] For the linear plate bending problems, the method has been widely used in small deflection of thin plates[5-9] and of moderately thick plates.[10-13] As far as the plate resting on an elastic foundation is concerned, Katsikadelis & Armenakas appear to address this direction firstly.[14] The same problem was analysed by Costa and Brebbia for various boundary conditions.[15] Puttonen and Varpasuo extended the BIE analysis to plates resting on two-parameter foundation.[16] As a further progressive step, various boundary integral formulations have been developed to treat large deflection of plates in the decades. Kamiya and Sawaki proposed a so-called direct formulation for analysing finite deflection of thin plates.[17,18] Consequently, they extended their procedure to the case of sandwich plates and shells.[19-21] The formulation proposed by Tanaka[22] is alternative to those of Kamiya *et al.*[17,18] He obtained a coupled boundary and inner domain integral equations in terns of stress and displacement functions. Later on, many researchers investigated the BEM for large deflection of plates with the so-called generalized Green identities,[23-25] the dual reciprocity process,[26] the weighted residual method[27] and the spline function method.[28] Postbuckling problems of thin plates with or without elastic foundation have been examined by Qin *et al.*[29,30] Further, geometrically nonlinear plates on elastic foundation have been considered by Katsikadelis.[31] More recently, He and Qin have developed a set of boundary integral equations for analysing large deflection of Reissner plates based on the variational approach as well as a newly derived fundamental solution.[32] Later, Qin extended the formulation to plates resting on elastic foundation.[33]

2 Reissner plates with or without elastic foundation

2.1 Basic equations

Consider a Reissner plate of uniform thickness t, occupying a two dimensional arbitrary shaped region Ω bounded by its boundary $\partial\Omega$ and resting on an elastic foundation. We use a cartesian coordinate system in which the x- and y-axes lie in the plate middle plane. Throughout this paper, repearted indices imply the summation convention of Einstein. The indices i, j and k take take values in the range $\{1,2\}$, m takes a value in the range $\{3,4,5\}$, and p, q take values in the range $\{1,2,3,4,5\}$. The nonlinear behaviour of the plate for moderately large deflection is governed by the following equations[33]:

(1) Equilibrium equations in Ω:

$$\begin{aligned} &N_{ij,j} = 0, && (i=1,2) \\ &M_{ij,j} - Q_i = 0, && (i=1,2) \\ &Q_{i,i} + N_{ij} w_{,ij} - q + \bar{k} w = 0 \end{aligned} \tag{1}$$

(2) Constitutive relationships in Ω:

$$N_{ij} = N_{ij}^{l} + N_{ij}^{n}$$

$$N_{ij}^{l} = Gt\left\{u_{i,j} + u_{j,i} + \frac{2\mu}{1-\mu}u_{k,k}\delta_{ij}\right\},$$

$$N_{ij}^{n} = Gt\left\{w_{,i}w_{,j} + \frac{\mu}{1-\mu}w_{,k}w_{,k}\delta_{ij}\right\}, \tag{2}$$

$$M_{ij} = \frac{1-\mu}{2}D\left(\psi_{i,j} + \psi_{j,i} + \frac{2\mu}{1-\mu}\psi_{k,k}\delta_{ij}\right),$$

$$Q_i = C(w_{,i} - \psi_i)$$

(3) Natural boundary conditions

$$\begin{aligned}
&N_n = N_{ij}n_i n_j = \overline{N}_n(\text{on } C_{N_n}), \quad N_{ns} = N_{ij}n_i s_j = \overline{N}_{ns}(\text{on } C_{N_{ns}}),\\
&M_n = M_{ij}n_i n_j - \alpha G_p w = \overline{M}_n(\text{on } C_{M_n}),\\
&M_{ns} = M_{ij}n_i s_j = \overline{M}_{ns}(\text{on } C_{M_{ns}}),\\
&R_n = Q_i n_i + N_n w_{,n} + N_{ns}w_{,s} = \overline{R}_n(\text{on } C_R)
\end{aligned} \tag{3}$$

(4) Essential boundary conditions

$$\begin{aligned}
&u_n = u_i n_i = \overline{u}_n(\text{on } C_{u_n}), \qquad u_s = u_i s_i = \overline{u}_s(\text{on } C'_{u_s}),\\
&\psi_n = \psi_i n_i = \overline{\psi}_n(\text{on } C_{\psi_n}), \quad \psi_s = \psi_i s_i = \overline{\psi}_s(\text{on } C_{\psi_s}),\\
&w = \overline{w}(\text{on } C_w)
\end{aligned} \tag{4}$$

$$(\partial\Omega = C_{u_n} \cup C_{N_n} = C_{u_s} \cup C_{N_{ns}} = C_{\psi_n} \cup C_{M_n} = C_{\psi_s} \cup C_{M_{ns}} = C_R \cup C_w) \tag{5}$$

where a comma followed by a subscript indicates partial differentiation with respect to that subscript, $\alpha=1$ for the Pasternak-type foundation,[34] $\alpha=0$ for others, and (for unstated symbols , see Appendix I):

$$\bar{k} = \begin{cases} k_w & \text{for Winkler-type foundation} \\ k_p - G_p\nabla^2 & \text{for Pasternak-type foundation} \\ 0 & \text{without foundation} \end{cases} \tag{6}$$

It should be pointed out that the main difference between Winkler-type and Pasternak-type foundations is whether the effect of shear interactions is included.

2.2 Fundamental solutions

The fundamental solution plays an important role in the derivation of the boundary integral equation. In this subsection, the construction of the fundamental solutions for Reissner plate on elastic foundation will be discussed in detail. To obtain the fundamental solution, we will first consider the linear deformation case, i.e. small deflection of plate theory. For this case, the nonlinear terms in equation (1) can be dropped out and the in-plane and out-of-plane deformations become uncoupled. The resulting field equations are[33]:

(1) Equations related to in-plane deformations

$$N^{I}_{ij,j} = 0 \qquad (i = 1,2) \tag{7}$$

(2) Equations related to bending deformations

$$\begin{aligned} &D[\psi_{x,xx} + 0.5(1-\mu)\psi_{x,yy} + 0.5(1+\mu)\psi_{y,xy}] + C(w_{,x} - \psi_x) = 0 \\ &D[\psi_{y,yy} + 0.5(1-\mu)\psi_{y,xx} + 0.5(1+\mu)\psi_{x,xy}] + C(w_{,y} - \psi_y) = 0 \\ &C(\nabla^2 w - \psi_{x,x} - \psi_{y,y}) + \bar{k}w - q = 0 \end{aligned} \tag{8}$$

where, for the sake of simplicity, vanishing distributed moment loads, $m_x = m_y = 0$, have been assumed.

The fundamental solutions to eqn (7) is obviously Kelvin's solution (for the plane stress case, see Reference 30, for example) as

$$u^{*}_{ij}(r) = \frac{1+\mu}{4\pi E}\left[(\mu - 3)\delta_{ij}\ln r + (1+\mu)r_{,i}r_{,j}\right] \tag{9}$$

In the following attention will be focused on finding the fundational solution to eqn (8). It can be seen from eqn (8) that the coupling of the variables w, ψ_x and ψ_y makes it difficult to generate the fundational solutions. To by-pass this problem, two auxiliary functions, g and f, are introduced such that[33]:

$$\psi_x = g_{,x} + f_{,y}, \quad \psi_{,y} = g_{,y} - f_{,x} \tag{10}$$

It should be pointed out that

$$g_{0,x} + f_{0,y} = 0 \text{ and } g_{0,y} - f_{0,x} = 0 \tag{11}$$

are Cauchy Riemann equations, the solution of which always exists. As a consequence, ψ_x and ψ_y remain unchanged if f and g in eqn (10) are replaced by $f+f_0$ and $g+g_0$. This property will play an important role in the subsequent part of this section.

The solution of eqns $(11)_{1,2}$ may conveniently be expressed in a complex form (with $i = \sqrt{-1}$) as

$$f_0 + ig_0 = \Phi(x + iy) \tag{12}$$

The substitution of eqn (10) into the first two differential equations in eqn (8) yields

$$\begin{aligned} &\frac{\partial}{\partial x}[D\nabla^2 g + C(w-g)] + \frac{\partial}{\partial y}[\frac{1}{2}D(1-\mu)\nabla^2 f - Cf] = 0 \\ &\frac{\partial}{\partial y}[D\nabla^2 g + C(w-g)] + \frac{\partial}{\partial x}[\frac{1}{2}D(1-\mu)\nabla^2 f - Cf] = 0 \end{aligned} \tag{13}$$

Now, if the contents of the two brackets are considered as independent generalized variables,

$$A=[\frac{1}{2}D(1-\mu)\nabla^2 f - Cf] \quad \text{and} \quad B=[D\nabla^2 g + C(w-g)] \tag{14}$$

we again get a set of Cauchy Riemann equations

$$B_{,x}+A_{,y}=0, \quad B_{,y}-A_{,x}=0 \tag{15}$$

and, in the same manner as in eqn (12), we can set

$$A+iB=[\frac{1}{2}D(1-\mu)\nabla^2 f - Cf]+i[D\nabla^2 g + C(w-g)] = F(x+iy) \tag{16}$$

This relation is a non-homogeneous equation with independent unknown function f, g and w. Its solution can be composed of a particular solution and a homogeneous solution. Since $F(x+iy)$ is a harmonic function, it is easy to see that the particular solution can be taken as

$$f+ig=-F(x+iy)/C \quad \text{and} \quad w=0 \tag{17}$$

It is obvious, see eqns (10) and (11), that this solution leads to $\psi_x=\psi_y=w=0$. Therefore, the particular solution may simply be omitted and we only need to consider the homogeneous part of eqn (16), namely

$$\frac{1}{2}D(1-\mu)\nabla^2 f - Cf = 0 \quad \text{and} \quad D\nabla^2 g + C(w-g)=0 \tag{18}$$

From the second equation of eqns (18), one sees

$$w = g - D\nabla^2 g/C \tag{19}$$

The substitution of this relation and of the expression (10) into eqn $(8)_3$, finally leads to

$$D\nabla^4 g + \frac{\bar{k}}{C}D\nabla^2 g - \bar{k}g + q = 0 \tag{20}$$

As a result, we obtain for g and f the following differential equations

$$D\nabla^4 g + \frac{\bar{k}}{C}D\nabla^2 g - \bar{k}g + q = 0$$

$$\nabla^2 f - \lambda^2 f = 0 \tag{21}$$

where

$$\lambda^2 = \begin{cases} 10/t^2 & \text{for a homogeneous plate} \\ \dfrac{4G_c(1+\mu)}{Et(h+t)} & \text{for a sandwich plate} \end{cases} \tag{22}$$

The corresponding displacements and rotations are obtained from eqns (10) and (19).

Equation (21) is the well-known modified Helmholtz equation and its fundational solution is

$$f^*(r) = \frac{\lambda^2}{2\pi C} K_0(\lambda r) \tag{23}$$

where $K_0(\ )$ is a modified Bessel function of zero order of the second kind.

The next step is to derive the fundamental solution of eqn (20). To this end, consider the homogeneous equation

$$D\nabla^4 g^* + D\bar{k}\nabla^2 g^* / C - \bar{k} g^* = D(\nabla^2 + C_1)(\nabla^2 - C_2) g^* = \delta(P,Q) \tag{24}$$

where $\delta(P,Q)$ is the Dirac delta function, P and Q stand for source point and field point, respectively, and

$$C_1 = \frac{k_w}{2C} + \sqrt{(\frac{k_w}{2C})^2 + \frac{k_w}{D}},$$
$$C_2 = -\frac{k_w}{2C} + \sqrt{(\frac{k_w}{2C})^2 + \frac{k_w}{D}}, \tag{25}$$

for a Winkler-type foundation, or

$$C_1 = \frac{\sqrt{b} + k_p / C + G_p / D}{2(1 - G_p / C)},$$
$$C_2 = \frac{\sqrt{b} - k_p / C - G_p / D}{2(1 - G_p / C)}, \tag{26}$$
$$b = (k_p / C + G_p / D)^2 + 4k_p(1 - G_p / C)/D$$

for a Pasternak-type foundation.

To find the solution of eqn (14), we set

$$(\nabla^2 - C_2) g^* = A \tag{27}$$

It follows from eqn (24) that

$$D(\nabla^2 + C_1) A = \delta(P,Q) \tag{28}$$

The fundamental solution of eqn (28) can be easily obtained as

$$A = Y_0(r\sqrt{C_1})/4D \tag{29}$$

where $Y_0(\ )$ is the Bessel function of zero order of the second kind.

In a similar way, let

$$(\nabla^2 + C_1) g^* = B \tag{30}$$

then we have

$$D(\nabla^2 - C_2) B = \delta(P,Q) \tag{31}$$

The fundamental solution of eqn (31) is, then, given as

$$B = \frac{1}{2\pi D} K_0(r\sqrt{C_2}) \tag{32}$$

Subtracting eqn (27) from eqn (30) and noting eqns (29) and (32), the fundamental solution g^* can be given in the form

$$g^*(r) = \frac{B-A}{C_1+C_2} = \frac{1}{C_1+C_2}\left[\frac{1}{2\pi D}K_0(r\sqrt{C_2}) - \frac{1}{4D}Y_0(r\sqrt{C_1})\right] \tag{33}$$

In the absence of elastic foundation ($C_1=C_2=0$), the fundamental solution (33) reduces to

$$g^*(r) = \frac{1}{8\pi D}r^2 \ln r \tag{34}$$

With the substitution of eqns (23) and (33) [or (34)] into eqns (10) and (19), we have

$$\begin{aligned} w^* &= g^* - D\nabla^2 g^* / C, \\ \psi_x^* &= \partial g^* / \partial x + \partial f^* / \partial y, \\ \psi_y^* &= \partial g^* / \partial y - \partial f^* / \partial x \end{aligned} \tag{35}$$

2.3 Boundary integral formulation

The boundary integral equation for nonlinear Reissner (or sandwich) plates on an elastic foundation can be established by the variational approach.[32] The approach is mainly based on a modified variational principle. Following the line of argument of He & Qin,[32] the related functional used for deriving boundary integral equation can be given in the form[33]:

$$\begin{aligned} \Pi^m = \Pi_1 &+ \int_{C_{u_n}} (\bar{u}_n - u_n)N_n \mathrm{d}c + \int_{C_{u_s}} (\bar{u}_s - u_s)N_{ns}\mathrm{d}c + \int_{C_{\psi n}} (\bar{\psi}_n - \psi_n)M_n \mathrm{d}c \\ &+ \int_{C_{\psi s}} (\bar{\psi}_s - \psi_s)M_{ns}\mathrm{d}c + \int_{C_w} (\bar{w} - w)R_n \mathrm{d}c \end{aligned} \tag{36}$$

where

$$\begin{aligned} \Pi_1 = \iint_\Omega (U - wq)\mathrm{d}\Omega &- \int_{C_{N_n}} \bar{N}_n u_n \mathrm{d}c - \int_{C_{N_{ns}}} \bar{N}_{ns} u_s \mathrm{d}c - \int_{C_{M_n}} \bar{M}_n \psi_n \mathrm{d}c \\ &- \int_{C_{M_{ns}}} \bar{M}_{ns}\psi_s \mathrm{d}c - \int_{C_R} \bar{R}_n w \mathrm{d}c \end{aligned} \tag{37}$$

$$2U = N_{ij}(u_{i,j} + u_{j,i} + w_{,i}w_{,j})/2 + M_{ij}(\psi_{i,j} + \psi_{j,i})/2 + Q_i(w_{,i} - \psi_i) + U^F$$

with

$$U^F = \begin{cases} k_w w^2 & \text{for Winkler-type foundation} \\ (k_p w^2 + G_p w_{,i} w_{,i}) & \text{for Pasternak-type foundation} \\ 0 & \text{without foundation} \end{cases} \tag{38}$$

The terminology "modified variational principle" refers, here, to the use of conventional potential functional Π_1 and some modified terms for the construction of a special variational principle. Consequently, we will discuss some properties and their proof on the functional. They are:

(i) Modified variational principle

$$\delta\Pi^m = 0 \Rightarrow (1),\ (3) \text{ and } (4) \tag{39}$$

(ii) Theorem on the existence of extremum: if the expression

$$\begin{aligned}&\iint_\Omega \delta^2 U \mathrm{d}\Omega + \iint_\Omega N_{ij}\delta w_{,i}\delta w_{,j}\mathrm{d}\Omega - \int_{C_{u_n}} \delta N_n \delta u_n \mathrm{d}c - \int_{C_{u_s}} \delta N_{ns}\delta u_s \mathrm{d}c \\ &- \int_{C_{\psi_n}} \delta M_n \delta\psi_n \mathrm{d}c - \int_{C_{\psi_s}} \delta M_{ns}\delta\psi_s \mathrm{d}c - \int_{C_w} \delta R_n \delta w \mathrm{d}c\end{aligned} \tag{40}$$

is uniformly positive (or negative) at the neighborhood of $\mathbf{U}_0$ ($\mathbf{U}_0=\{u_{10}, u_{20}, \psi_{10}, \psi_{20}, w_0\}$), where $\mathbf{U}_0$ is such a value that $\Pi^m(\mathbf{U}_0) = (\Pi^m)_0$, and where $(\Pi^m)_0$ stands for stationary value of Π^m, we have

$$\Pi^m \geq (\Pi^m)_0 [\text{or } \Pi^m \leq (\Pi^m)_0] \tag{41}$$

Proof: from the first, we derive the stationary conditions of functional Π^m. To this end, taking vanishing variation of Π^m, one gets

$$\begin{aligned}\delta\Pi^m = &\iint_\Omega \{-N_{ij,j}\delta u_i - (M_{ij,j} - Q_i)\delta\psi_i - (Q_{i,i} + N_{ij}w_{,ij} + \bar{k}w - q)\delta w\} d\Omega \\ &+ \int_{C_{u_n}} (\bar{u}_n - u_n)\delta N_n dc + \int_{C_{u_s}} (\bar{u}_s - u_s)\delta N_{ns} dc + \int_{C_w} (\bar{w} - w)\delta R_n dc \\ &+ \int_{C_{\psi_n}} (\bar{\psi}_n - \psi_n)\delta M_n dc + \int_{C_{\psi_s}} (\bar{\psi}_s - \psi_s)\delta M_{ns} dc + \int_{C_{N_n}} (N_n - \bar{N}_n)\delta u_n dc \\ &+ \int_{C_{N_{ns}}} (N_{ns} - \bar{N}_{ns})\delta u_s dc + \int_{C_R} (R_n - \bar{R}_n)\delta w dc + \int_{C_{M_n}} (M_n - \bar{M}_n)\delta\psi_n dc \\ &+ \int_{C_{M_{ns}}} (M_{ns} - \bar{M}_{ns})\delta\psi_s dc = 0\end{aligned} \tag{42}$$

Therefore, the Euler equations for eqn (42) are eqns (1), (3) and (4). The principle (39) has, thus, been proved.

As for the proof of the theorem on the existence of extremum, we may complete it by way of the so-called second variational approach.[32,35] In doing this, taking variation of $\delta\Pi^m$, we see

$$\begin{aligned}\delta^2\Pi^m = &\iint_\Omega N_{ij}\delta w_{,i}\delta w_{,j}\mathrm{d}\Omega - \int_{C_{u_n}} \delta N_n\delta u_n \mathrm{d}c - \int_{C_{u_s}} \delta N_{ns}\delta u_s \mathrm{d}c - \int_{C_{\psi_n}} \delta M_n \delta\psi_n \mathrm{d}c \\ &- \int_{C_{\psi_s}} \delta M_{ns}\delta\psi_s \mathrm{d}c - \int_{C_w} \delta R_n \delta w \mathrm{d}c + \iint_\Omega \delta^2 U \mathrm{d}\Omega = \text{expression (40)}\end{aligned} \tag{43}$$

So the theorem has been proved from the sufficient condition on the existence of local extreme of a functional.[35]

What follows is to transform eqn (42) into a boundary integral equation. Noting that the displacement vector **U** in eqn (42) is not constrained by the essential boundary condition (4), the quantity $\delta\mathbf{U}$ can be arbitrarily assumed. Naturally let

$$\delta\mathbf{U} = \varepsilon\mathbf{U}_p^*(P,Q) \tag{44}$$

where ε is an infinitesimal, the asterisked symbol $\mathbf{U}_p^*(P,Q)$ represents the related fundamental solutions which have been obtained in subsection 2.2, the components $U_{pq}^*(P,Q)$ of $\mathbf{U}_p^*(P,Q)$ mean the in-plane displacements (for q=1 and 2) or the rotations (for q=3 and 4) or the deflection (for q=5) at the field point Q of an infinite plate when a unit point force (for p=1,2 and 5) or a unit point couple (for p=3 and 4) is applied at the source point P. $N_{ijp}^*(P,Q)$, $Q_{ip}^*(P,Q)$ and $M_{ijp}^*(P,Q)$ can be calculated from $U_{pq}^*(P,Q)$ by using the constitutive relationships in eqn (2). The fundamental solutions for out-of-plane deformation are listed in Appendix II, while the in-plane fundamental solutions has been given in eqn (9).

Moreover, the menbranous equation (7) and bending one (8) are independent of each other, which means that

$$U_{im}^* = U_{mi}^* = 0 \tag{45}$$

With the substitution of eqn (44) and noting eqn (45), eqn (42) can be transformed following two sets of boundary integral equations through a series of derivations

$$\begin{aligned}\alpha(P)U_k(P) &- \int_{C_{N_n}} [u_{nk}^*(P,Q)\overline{N}_n - N_{nk}^*(P,Q)u_n]\mathrm{d}c - \int_{C_{N_{ns}}} [u_{sk}^*(P,Q)\overline{N}_{ns} \\ &- N_{nsk}^*(P,Q)u_s]\mathrm{d}c - \int_{C_{u_n}} [u_{nk}^*(P,Q)N_n - N_{nk}^*(P,Q)\overline{u}_n]\mathrm{d}c \\ &- \int_{C_{u_s}} [u_{sk}^*(P,Q)N_{ns} - N_{nsk}^*(P,Q)\overline{u}_s]\mathrm{d}c = -\frac{1}{2}\iint_\Omega w_{,i}w_{,j}N_{ijk}^*(P,Q)\mathrm{d}\Omega\end{aligned} \tag{46}$$

$$\begin{aligned}\alpha(P)U_m(P) &- \int_{C_{M_n}} [\psi_{nm}^*(P,Q)\overline{M}_n - M_{nm}^*(P,Q)\psi_n]\mathrm{d}c \\ &- \int_{C_{M_{ns}}} [\psi_{sm}^*(P,Q)\overline{M}_{ns} - M_{nsm}^*(P,Q)\psi_s]\mathrm{d}c - \int_{C_R} [w_m^*(P,Q)\overline{R}_n \\ &- Q_{nm}^*(P,Q)w]\mathrm{d}c - \int_{C_{\psi_n}} [\psi_{nm}^*(P,Q)M_n - M_{nm}^*(P,Q)\overline{\psi}_n]\mathrm{d}c \\ &- \int_{C_{\psi_s}} [\psi_{sm}^*(P,Q)M_{ns} - M_{nsm}^*(P,Q)\overline{\psi}_s]\mathrm{d}c - \int_{C_w} [w_m^*(P,Q)R_n \\ &- Q_{nm}^*(P,Q)\overline{w}]\mathrm{d}c = \iint_\Omega [w_m^*(P,Q)q - N_{ij}w_{,i}w_{m,j}^*(P,Q)]\mathrm{d}c\end{aligned} \tag{47}$$

where $\alpha(P)$ is a conventional boundary shape coefficient, $\alpha(P)$=1 if $P \in \Omega$, $\alpha(P)$=1/2, if P is on the smooth boundary $\partial\Omega$.

2.4 Numerical implementation

The analytical solutions of eqns (46) and (47) are not, generally, possible to obtain and therefore a numerical procedure must be used to solve the equations.

To obtain the solution of eqns (46) and (47), as in the usual BEM, the boundary $\partial\Omega$ and the domain Ω of the plate are divided into a series of constant boundary elements and linear cells, respectively. For a particular constant element, the node is taken to be the centre of the element. After performing the discretization with boundary elements and internal cells, eqns (46) and (47) are reduced to a system of algebraic equations including 10 boundary vectors, $\mathbf{N}_n$, $\mathbf{N}_{ns}$, $\mathbf{u}_1$, $\mathbf{u}_2$, $\mathbf{M}_n$, $\mathbf{M}_{ns}$, $\mathbf{R}_n$, ψ_1, ψ_2, $\mathbf{w}$:

$$\begin{bmatrix} \mathbf{Q}_{11} & \mathbf{Q}_{12} \\ \mathbf{Q}_{21} & \mathbf{Q}_{22} \end{bmatrix} \begin{Bmatrix} \mathbf{N}_n \\ \mathbf{N}_s \end{Bmatrix} + \begin{bmatrix} \mathbf{S}_{11} & \mathbf{S}_{12} \\ \mathbf{S}_{21} & \mathbf{S}_{22} \end{bmatrix} \begin{Bmatrix} \mathbf{u}_1 \\ \mathbf{u}_2 \end{Bmatrix} = \begin{Bmatrix} \mathbf{F}_1 \\ \mathbf{F}_2 \end{Bmatrix}$$

$$\begin{bmatrix} \mathbf{H}_{11} & \mathbf{H}_{12} & \mathbf{H}_{13} \\ \mathbf{H}_{21} & \mathbf{H}_{22} & \mathbf{H}_{23} \\ \mathbf{H}_{31} & \mathbf{H}_{32} & \mathbf{H}_{33} \end{bmatrix} \begin{Bmatrix} \mathbf{M}_n \\ \mathbf{M}_{ns} \\ \mathbf{R}_n \end{Bmatrix} + \begin{bmatrix} \mathbf{G}_{11} & \mathbf{G}_{12} & \mathbf{G}_{13} \\ \mathbf{G}_{21} & \mathbf{G}_{22} & \mathbf{G}_{23} \\ \mathbf{G}_{31} & \mathbf{G}_{32} & \mathbf{G}_{33} \end{bmatrix} \begin{Bmatrix} \psi_1 \\ \psi_2 \\ \mathbf{w} \end{Bmatrix} = \begin{Bmatrix} \mathbf{F}_3 \\ \mathbf{F}_4 \\ \mathbf{F}_5 \end{Bmatrix} \tag{48}$$

Of the 10 quantities, five need to be prescribed on the boundary points and the remaining five are to be determined. These 10 vectors only contain boundary variables, while [**Q**], [**S**], [**H**], [**G**] and $\{\mathbf{F}_i\}$ denote, respectively, the coefficient matrices and pseudo-loading vectors, whose components are given by

$$(\mathbf{Q}_{k1})_{ab} = -\int_{\partial\Omega_b} u^*_{nk}(P_a, Q)\mathrm{d}c(Q),$$

$$(\mathbf{Q}_{k2})_{ab} = -\int_{\partial\Omega_b} u^*_{sk}(P_a, Q)\mathrm{d}c(Q),$$

$$(\mathbf{S}_{ki})_{ab} = n_{ai}(\mathbf{S}^*_{k1})_{ab} + s_{ai}(\mathbf{S}^*_{k2})_{ab} + \alpha(P_a)\delta_{ab}\delta_{ki} \quad (a \text{ not summed})$$

$$(\mathbf{S}^*_{k1})_{ab} = \int_{\partial\Omega_b} N^*_{nk}(P_a, Q)\mathrm{d}c(Q),$$

$$(\mathbf{S}^*_{k2})_{ab} = \int_{\partial\Omega_b} N^*_{nsk}(P_a, Q)\mathrm{d}c(Q), \tag{49a}$$

$$(\mathbf{F}_i)_a = -\frac{1}{2}\int_{\Omega} w_{,j} w_{,k} N^*_{jki}(P_a, Q)\mathrm{d}\Omega(Q),$$

$$(\mathbf{H}_{d1})_{ab} = -\int_{\partial\Omega_b} \psi^*_{nd}(P_a, Q)\mathrm{d}c(Q),$$

$$(\mathbf{H}_{d2})_{ab} = -\int_{\partial\Omega_b} \psi^*_{sd}(P_a, Q)\mathrm{d}c(Q),$$

$$(\mathbf{H}_{d3})_{ab} = -\int_{\partial\Omega_b} w^*_{d}(P_a, Q)\mathrm{d}c(Q),$$

$$(\mathbf{G}_{di})_{ab} = n_{ai}(\mathbf{G}^*_{d1})_{ab} + s_{ai}(\mathbf{G}^*_{d2})_{ab} + \alpha(P_a)\delta_{ab}\delta_{di} \quad (a \text{ not summed})$$

$$
\begin{aligned}
(\mathbf{G}^*_{d1})_{ab} &= \int_{\partial\Omega_b} M^*_{nd}(P_a, Q)\mathrm{d}c(Q), \\
(\mathbf{G}^*_{d2})_{ab} &= \int_{\partial\Omega_b} M^*_{nsd}(P_a, Q)\mathrm{d}c(Q), \\
(\mathbf{G}_{d3})_{ab} &= \int_{\partial\Omega_b} Q^*_{nd}(P_a, Q)\mathrm{d}c(Q) + \alpha(P_a)\delta_{ab}\delta_{d3}, \\
(\mathbf{F}_{d+2})_a &= \iint_{\Omega} [q w^*_d(P_a, Q) - N_{ij} w_{,i} w^*_{d,j}(P_a, Q)]\mathrm{d}\Omega(Q)
\end{aligned}
\tag{49b}
$$

where $\mathbf{n}_a = \{n_{a1}\ n_{a2}\}^{\mathrm{T}}$ and $\mathbf{s}_a = \{s_{a1}\ s_{a2}\}^{\mathrm{T}}$ are, respectively, the unit normal and unit tangent to the boundary $\partial\Omega$ at point a, the subscript d takes value in the range $\{1,2,3\}$.

Rearranging eqn (48) by collecting the unknown terms to the left-hand side and the known terms to the right-hand side, we get

$$
\begin{aligned}
[\mathbf{E}]\{\mathbf{X}\} &= \{\mathbf{R}_1\} \\
[\mathbf{F}]\{\mathbf{Y}\} &= \{\mathbf{R}_2\}
\end{aligned}
\tag{50}
$$

Note that, since eqns (50) are not, in general, suitable for numerical analysis, an incremental form of the equation should therefore be adopted. Denoting the incremental variable by the superim-posed dot and omitting the infinitesmal terms resulting from the product of incremental variables, one obtains

$$
\begin{aligned}
[\mathbf{E}]\{\dot{\mathbf{X}}\} &= \{\dot{\mathbf{R}}_1\} \\
[\mathbf{F}]\{\dot{\mathbf{Y}}\} &= \{\dot{\mathbf{R}}_2\}
\end{aligned}
\tag{51}
$$

It follows that eqn (51) is linear with respect to the incremental variables. However, the right-hand side vectors, $\{\dot{\mathbf{R}}_1\}$ and $\{\dot{\mathbf{R}}_2\}$, contain the domain unknown variables. To avoid solving these variables directly, an iterative procedure is required.

It is noted that $\{\dot{\mathbf{R}}_1\}$ depends only upon $\dot{w}$ [see eqn $(49)_6$], so only an initial value w^0 is required. As long as the value of $\dot{w}$ in Ω is known, we can calculate the pseudo loading vector $\{\dot{\mathbf{R}}_1\}$, and then, all of the unknown variables in eqn $(51)_1$ are at the boundary. We may solve it for $\dot{u}_1$ and $\dot{u}_2$. As a consequence, $\{\dot{\mathbf{R}}_1\}$ can be evaluated from the current values of $\dot{u}_1$, $\dot{u}_2$ and $\dot{w}$. An iterative scheme may be established according to the above analysis. Specifically, suppose that $\mathbf{U}^{(k)}$ stands for kth approximations, which can be obtained from the preceding cycle of iteration. The (k+1)th solution may be evaluated as follows

(a) Assume the initial value $\dot{w}^0$ in Ω;

(b) Enter the iterative cycle for i=1,2, …, calculate $\{\dot{\mathbf{R}}_1\}$ by way of eqn $(49)_6$, solve eqn $(51)_1$ for $\{\dot{\mathbf{X}}\}^{(i)}$, and then determine the values $\dot{u}_1^i$ and $\dot{u}_2^i$ in Ω;

(c) calculate $\{\dot{\mathbf{R}}_2\}$ using the current values of $\mathbf{U}$, and then solve eqn $(51)_2$ for $\{\mathbf{Y}\}^{(i)}$ and determine the value of $\dot{w}^i$ in Ω;

(d) if $\varepsilon_i = [(\mathbf{U}^{(i)})^T \mathbf{U}^{(i)} - (\mathbf{U}^{(i-1)})^T \mathbf{U}^{(i-1)}] / (\mathbf{U}^{(i-1)})^T \mathbf{U}^{(i-1)} \leq \varepsilon$ (ε is a convergence tolerance), proceed to the next loading step and calculate

$$\mathbf{U}^{(k+1)} = \mathbf{U}^{(k)} + \dot{\mathbf{U}}^{i} \tag{52a}$$

otherwise, let

$$\dot{w}^0 = \dot{w}^i \quad \text{and} \quad w^0 = w^{\mathrm{k}} + \dot{w}^i \tag{52b}$$

and go back to step (b).

Moreover, it is important to note that once the matrices [**Q**], [**S**], [**H**] and [**G**] in eqn (48) have been formed, they can be stored in the core and used in each cycle of iteration without any change. This is because the matrices depend only upon the geometric and material parameters of the plate and the foundation. Obviously it can save a large amount of computing time.

3 Postbuckling plate on elastic foundation

3.1 Basic equations

The basic equations govenring postbuckling behavior of thin plate on an elastic foundation, in this case, are given as[29]:

(1) Equilibrium equations in Ω:

$$\begin{aligned} &N_{ij,j} = 0, \qquad (i = 1,2) \\ &D\nabla^4 w + p_0 \nabla^2 w + \bar{k} w = N_{ij} w_{,ij} \end{aligned} \tag{53}$$

(2) Constitutive relationships in Ω:

$$\begin{aligned} &N_{ij} = N_{ij}^{l} + N_{ij}^{n} \\ &N_{ij}^{l} = Gt\left\{ u_{i,j} + u_{j,i} + \frac{2\mu}{1-\mu} u_{k,k}\delta_{ij} \right\}, \\ &N_{ij}^{n} = Gt\left\{ w_{,i} w_{,j} + \frac{\mu}{1-\mu} w_{,k} w_{,k}\delta_{ij} \right\}, \\ &M_{ij} = -D\big((1-\mu) w_{,ij} + \mu w_{,kk}\delta_{ij}\big) \end{aligned} \tag{54}$$

(3) Natural boundary conditions

$$\begin{aligned} &N_n = N_{ij} n_i n_j = \bar{N}_n (\text{on } C_{N_n}), \qquad N_{ns} = N_{ij} n_i s_j = \bar{N}_{ns} (\text{on } C_{N_{ns}}), \\ &M_n = M_{ij} n_i n_j - \alpha G_p w = \bar{M}_n (\text{on } C_{M_n}), \\ &R_n = M_{ij,j} n_i + M_{ns,s} + N_n w_{,n} + N_{ns} w_{,s} = \bar{R}_n (\text{on } C_R) \end{aligned} \tag{55}$$

(4) Essential boundary conditions

$$u_n = u_i n_i = \bar{u}_n \,(\text{on } C_{u_n}), \qquad u_s = u_i s_i = \bar{u}_s \,(\text{on } C_{u_s}),$$
$$w_{,n} = w_{,i} n_i = \bar{w}_{,n} \,(\text{on } C_{w_n}), \quad w = \bar{w} \,(\text{on } C_w) \tag{56}$$

$$(\partial\Omega = C_{u_n} \cup C_{N_n} = C_{u_s} \cup C_{N_{ns}} = C_{w_n} \cup C_{M_n} = C_w \cup C_R)) \tag{57}$$

where p_0 is the external radial uniform compressive load(see Fig.1), N_{ij}, here, represent the disturbed membrane force components, $M_{ns} = M_{ij} n_i s_j$.

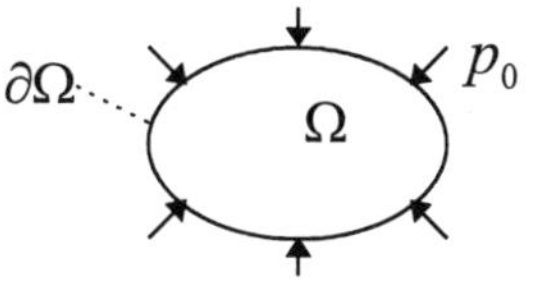

Fig. 1. Compressive load p_0

3.2 Fundamental solutions

The fundamental solution to the first two equations in eqns (53) has been given in eqn (9). What follows is to derive the fundamental solution of eqn $(53)_3$. In a similar way as previous section, consider

$$(D\nabla^4 + p_0\nabla^2 + \bar{k})w^* = D(\nabla^2 + C_3)(\nabla^2 + C_4)w^* = \delta(P,Q) \tag{58}$$

and then, let

$$(\nabla^2 + C_3)w^* = A, \qquad (\nabla^2 + C_4)w^* = B \tag{59}$$

we have

$$w^*(r) = \frac{1}{4D(C_4 - C_3)}[Y_0(r\sqrt{C_3}) - Y_0(r\sqrt{C_4})] \tag{60}$$

where

$$C_{3,4} = \frac{1}{2}\left(\frac{p_0}{D} \mp \sqrt{(\frac{p_0}{D})^2 - \frac{4k_w}{D}}\right) \tag{61}$$

for a Winkler-type foundation, or

$$C_{3,4} = \frac{1}{2}\left(\frac{p_0 - G_p}{D} \mp \sqrt{(\frac{p_0 - G_p}{D})^2 - \frac{4k_p}{D}}\right) \tag{62}$$

for a Pasternak-type foundation.

In the absence of elastic foundation ($C_3 = 0,\ C_4 = p_0/D$), the solution of eqn $(59)_1$ reduces to

$$A = (1/2\pi D)\ln r \tag{63}$$

Thus the fundamental solution for the postbuckling plate without foundation becomes

$$w^*(r) = [2\ln r - \pi Y_0(r\sqrt{C_4})]/4\pi DC_4 \tag{64}$$

3.3 Boundary integral equation

The related modified variational principle used for deriving boundary integral equation can, in this case, be stated as

$$\delta\Pi^{mp} = 0 \Rightarrow \text{eqns (53), (55) and (56)} \tag{65}$$

where

$$\begin{aligned}\Pi^{mp} &= \Pi_1^p + \int_{C_{u_n}} (\bar{u}_n - u_n)N_n \mathrm{d}c + \int_{C_{u_s}} (\bar{u}_s - u_s)N_{ns} \mathrm{d}c \\ &+ \int_{C_{w_n}} (\bar{w}_{,n} - w_{,n})M_n \mathrm{d}c + \int_{C_w} (\bar{w} - w)R_n \mathrm{d}c\end{aligned} \tag{66}$$

with

$$\begin{aligned}\Pi_1^p &= \iint_\Omega U^p \mathrm{d}\Omega - \int_{C_{N_n}} \bar{N}_n u_n dc - \int_{C_{N_{ns}}} \bar{N}_{ns} u_s \mathrm{d}c \\ &+ \int_{C_{M_n}} \bar{M}_n w_{,n} \mathrm{d}c - \int_{C_R} \bar{R}_n w \mathrm{d}c \\ 2U^p &= N_{ij}(u_{i,j} + u_{j,i} + w_{,i}w_{,j})/2 + D[(\nabla^2 w)^2 \\ &+ 2(1-\mu)(w_{,12}^2 - w_{,11}w_{,22})] + U^F\end{aligned} \tag{67}$$

As was done in the Subsection 2.3, we can prove the principle (65) and show how to transform it into a boundary integral equation. For the sake of brevity, however, we omit those details. The resulting formulation is, now, given as

$$\begin{aligned}&\alpha(P)u_k(P) - \int_{C_{N_n}} [u_{nk}^*(P,Q)\bar{N}_n - N_{nk}^*(P,Q)u_n]\mathrm{d}c - \int_{C_{N_{ns}}} [u_{sk}^*(P,Q)\bar{N}_{ns} \\ &- N_{nsk}^*(P,Q)u_s]\mathrm{d}c - \int_{C_{u_n}} [u_{nk}^*(P,Q)N_n - N_{nk}^*(P,Q)\bar{u}_n]\mathrm{d}c \\ &- \int_{C_{u_s}} [u_{sk}^*(P,Q)N_{ns} - N_{nsk}^*(P,Q)\bar{u}_s]\mathrm{d}c = -\frac{1}{2}\iint_\Omega w_{,i}w_{,j}N_{ijk}^*(P,Q)\mathrm{d}\Omega\end{aligned} \tag{68}$$

$$\begin{aligned}&\alpha(P)w(P) + \int_{C_{M_n}} [w_{,n}^*(P,Q)\bar{M}_n - M_{nm}^*(P,Q)w_{,n}]\mathrm{d}c - \int_{C_R} [w^*(P,Q)\bar{R}_n \\ &- R_n^*(P,Q)w]\mathrm{d}c - \int_{C_w} [w_m^*(P,Q)R_n - R_n^*(P,Q)\bar{w}]\mathrm{d}c \\ &+ \int_{C_{w_n}} [w_{,n}^*(P,Q)M_n - M_{nm}^*(P,Q)w_{,n}]\mathrm{d}c - \sum_{a=1}^{d}[\Delta M_{ns}^*(P,Q_a)w(Q_a) \\ &- \Delta M_{ns}(Q_a)w^*(P,Q_a)] = -\iint_\Omega N_{ij}w_{,i}w_{,j}^*(P,Q)\mathrm{d}c\end{aligned} \tag{69}$$

where $\Delta(\ ) = (\ )^+ - (\ )^-$ is discontinuity jump at the corner point, $(\ )^+$ and $(\ )^-$ stand for the values before and after the corner point, respectively, and d is the number of corners. According to the thin plate theory, M_n^* and R_n^* corresponding to the fundamental solution w^* can be calculated through the higher order derivatives of w^*.

The boundary integral equations (68) and (69) contain 8 independent variables, u_n, u_s, N_n, N_{ns}, M_n, $w_{,n}$, R_n and w. Of the 8 quantities, 4 are prescribed and the remaining 4 are to be determined. So an additional boundary integral equation is required to ensure the solution to be unique [it should be noted that the corner variables ΔM_{ns} and w_c in eqn (69) can be solved by using the corner condition, *a priori*, and therefore, they are dependent variables]. To this end, we differentiate eqn (69) with respect to n_0 (here n_0 is the outward normal at source point P). This yields

$$\begin{aligned}
&\alpha(P)w_{,n_0}(P)+\int_{C_{M_n}}[w^*_{,nn_0}(P,Q)\overline{M}_n-M^*_{n,n_0}(P,Q)w_{,n}]\mathrm{d}c\\
&-\int_{C_R}[w^*_{,n_0}(P,Q)\overline{R}_n-R^*_{n,n_0}(P,Q)w]\mathrm{d}c-\int_{C_w}[w^*_{,n_0}(P,Q)R_n\\
&-R^*_{n,n_0}(P,Q)\overline{w}]\mathrm{d}c+\int_{C_{w_n}}[w^*_{,nn_0}(P,Q)M_n-M^*_{n,n_0}(P,Q)w_{,n}]\mathrm{d}c \qquad (70)\\
&-\sum_{a=1}^{d}[\Delta M^*_{ns,n_0}(P,Q_a)w(Q_a)-\Delta M_{ns}(Q_a)w^*_{,n_0}(P,Q_a)]\\
&=-\iint_{\Omega}N_{ij}w_{,i}w^*_{,jn_0}(P,Q)\mathrm{d}c
\end{aligned}$$

The discretized form of eqns (68) to (70) follow an analogous pattern to that in eqn (48).

4 Numerical examples

As numerical illustration of the above-stated integral equations and calculation scheme, some results have been obtained and comparison is made with those known solutions. In all the calculations, the convergence tolerance is taken to be ε=0.0001. The examples are described as follows.

4.1 Nonlinear Reissner plates

In this subsection, five benchmark problems are considered. To study the convergence properties of the proposed calculation scheme, several meshes for the solution domain are used in each example.

Example 1: *a square plate*. Consider a square plate with two opposite edges clamped and the others simply supported under a uniformly lateral load q (Q=16qa^4/Et^4) and with thick-ness/span ratio $t/2a$=0.05, where $2a$ is the side-length of the square plate. The boundary conditions are:

$$x = \pm a, \quad u_n = u_s = \psi_n = \psi_s = w = 0,$$

$$x = \pm a, \quad u_n = u_s = M_n = \psi_s = w = 0$$

where the origin of the coordinate frame is laid at the center of the square plate, which also applies to the next example.

Owing to the symmetry of the problem only one quadrant of the plate is modelled by 8 constant boundary elements and three meshes of the internal cell (3×3, 4×4 and 5×5). Table 1 shows the central deflection (w_m/t) of the plate and compares the result with the finite strip solution.[36]

Table 1. The central deflection (w_m/t) for example 1

Load Q		0.91575	4.5788	6.8681	9.1575
Finite strip		0.019908	0.098873	0.14694	0.19361
	3×3 cells	0.019903	0.098611	0.14571	0.19095
BEM	4×4	0.019906	0.098623	0.14592	0.19127
	5×5	0.019907	0.098625	0.14598	0.19135

Example 2: *a compressive sandwich plate*. Consider a square ($2a$=59.7cm) simply supported sandwich plate with identical isotropic facings (E= 0.668×10^{6}kg/cm^2, μ=0.3, t=0.0533cm) and a 0.46 cm thick core (G_c= 0.134×10^{4}kg/cm^2), subjected to uniformly in-plane compress N_x (kg/cm) at the boundary $x = \pm a$. The displacement boundary conditions used are as follows

$$w = \psi_y = 0, \quad \text{on } x = \pm a$$

$$w = \psi_x = 0, \quad \text{on } y = \pm a$$

Assuming that a symmetrical buckling pattern occurred, only one quadrant of the plate needs to be considered. In the analysis, 16 constant boundary elements and three meshes of internal cell (3×3, 4×4 and 5×5) are used. The results for central deflection w_m vs load N_x are listed in Table 2 and comparison is made with the results reported by Schmit and Manforton[37].

Table 2. The central deflection w_m(cm) for example 2

Load N_x		54.7	57.2	61.6	66
Schmit-Manfor		0	0.457	0.813	1.02
	3×3 cells	0	0.432	0.770	0.968
BEM	4×4	0	0.445	0.789	0.991
	5×5	0	0.450	0.790	1.005

The above two examples indicate that the boundary element model in Section 2 can predict nonlinear behavior of thick plates without foundation. It can be seen from tables 1 and 2 that the numerical results obtained by the present method agree wellwith those existing ones. Further, the results showed that little sensitivity to the varying internal mesh. In the course of computation convergence was achieved with about 18 iterations for example 1 and 24 iterations for example 2 at each loading step.

Example 3: a circular plate on a Winkler-type foundation. Consider a uniformly loaded circular plate resting on a Winkler-type foundation, and with radius a and clamped movable edge (i.e. $w = \psi_n = \psi_s = N_n = N_{ns} = 0$). Some parameters of the problem are assumed as

$$a/h = 50, \quad \mu=0.3, \quad k_w = 100D/a^4, \quad Q = qa^4/Et^4 = 15$$

A quadrant of the plate is modelled by 25 internal cells and three meshes of boundary element (16, 20, 24), and the loading step is taken as ΔQ=1. Table 3 shows the deflection (w/t) along the radius of the plate and compares with that by Katsikadelis[31].

Table 3. Deflection (w/t) along radius r for example 3

r/a		0.098	0.304	0.562	0.800	0.960
Katsikadelis		1.108	0.961	0.592	0.179	0.009
	16 b.e.†	1.096	0.950	0.584	0.171	0.0085
Present	20	1.102	0.957	0.588	0.174	0.0088
	24	1.109	0.960	0.590	0.175	0.0088

†b.e.=boundary elements.

Example 4: a 60° *skew sandwich plate on a Winkler-type foundation.* The skew plate is clamped immovable on all edges (i.e. $u_n = u_s = \psi_n = \psi_s = w = 0$ on the whole boundary) shown in Fig. 2. Some initial data are

$$h = 25.4\text{mm}, \quad t = 0.635\text{mm}, \quad \mu = 0.32, \quad a = b = 508\text{mm}, \quad z = t + h$$

$$G_c = 6.89\text{MPa}, \quad Q = 12a^3(1-\mu^2)q/(Etz^2), \quad K = 12a^2k_w(1-\mu^2)/(Etz)$$

The plate under consideration is modelled by 8×8 internal cells, three boundary meshes of (48, 60, 80) boundary elements, respectively. The loading step is taken to be ΔQ=12.5. Table 4 compares the present results with those obtained by Ng and Das[38] in which the values were taken from Fig. 11 in their paper.

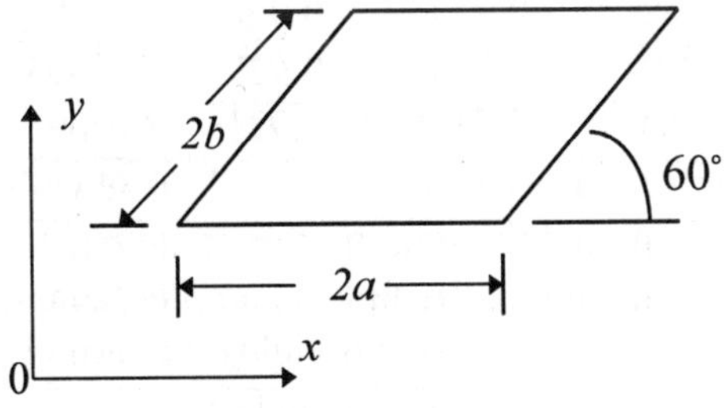

Fig. 2. A CI 60° skew plate on foundation.

Table 4. Central Deflection (w/h) for example 4 (K=20)

Load Q		25	50	75	100	125
Ng and Das		0.60	0.87	1.05	1.18	1.30
	48 b.e.	0.581	0.864	1.0395	1.1692	1.2723
Present	60	0.589	0.867	1.0412	1.1723	1.2802
	80	0.596	0.872	1.0521	1.1750	1.2837

Example 5: an annular plate on a Pasternak-type foundation. The annular plate is subjected to a uniform distributed load q ($Q=qa^4/Et^4$) and resting on a Pasternak-type foundation. The inner boundary of the plate was in a free edge condition while the out boundary condition was clamped immovable. Some initial data used in the example are given by

$$G_p a^2 / E = 1, \quad K = k_p a^4 / E = 5, \quad b/a = 1/3, \quad \mu = 1/3$$

where a and b are, respectively, outer and inner radii of the plate (see Fig. 3). In the example, a quarter of the plate is analyzed, and three internal meshes (16, 32 and 48 cells, see also Fig.3) as well as three meshes of boundary element (4, 6, 8 elements on inner boundary and 6, 8, 10 elements) have been used. The loading step is taken as ΔQ=5. Some results are listed in Tables 5 and 6.

Table 5. Maximum deflection (w/t) for example 5(8-10 b.e.)

Load Q		10	15	20	25	30
Smail		0.51	0.74	0.93	1.10	1.24
	16 cells	0.494	0.729	0.925	1.088	1.223
Present	32	0.497	0.733	0.929	1.093	1.228
	48	0.498	0.734	0.931	1.095	1.230

Table 6. Maximum deflection (w/t) for example 5 (48 cells)

Load Q		10	15	20	25	30
Smail		0.51	0.74	0.93	1.10	1.24
	10 b.e.	0.496	0.733	0.928	1.092	1.227
Present	14	0.497	0.733	0.929	1.093	1.228
	18	0.498	0.734	0.931	1.095	1.230

These three examples analyzed the problems of thick plate with elastic foundation. It is again found from Tables 3-6 that the present numerical results are in excellent agreement with the known ones. Among the three examples, example 3 provides the deflection over the entire plate, while example 5 yields the results with either varying internal meshes (16, 32 or 48 cells) or varying boundary meshes (4+6, 6+8, 8+10 boundary elements). As expected for all the three examples, it can be observed that the proposed formulation yields converging values along with refinement of the element meshes.

4.2 Postbuckling plates

As numerical illustration of the formulation given section 3, two simple examples have been considered. In order to allow for comparisons with those existing solutions, the obtained results are limited to a circular plate and a square plate subjected to uniform in-plane compressive load p_0 at the boundary $\partial\Omega$. The two examples are given as follows.

Example 6: a compressive circular plate. Consider a uniformly loaded circular with radius a and simply-supported movable edges (i.e. $M_n = w = N_n = N_{ns} = 0$). Some parameters for the problem are assumed as

$$\mu = 0.3, \quad E = 2\times10^6\,\mathrm{kg/cm^2}, \quad a/t = 48$$

In this case, its linear buckling load is p_{cr}=4.198D/a^2. In the analysis, a quadrant of the plate is modelled by 10 boundary elements and three internal meshes (see Fig. 4). The loading step is taken as $\Delta\beta=\Delta p_0/p_{cr}$=0.2. The numerical results describing the relationship between the maximum deflection (w_m/t) occurring at the center and the compressive load coefficient $\beta= p_0/p_{cr}(\beta>1)$ are shown in Table 7. Comparison is made with those given by Thompson and Hunt.[39]

Table 7. Maximum deflection (w_m/t) for example 6

β		1.2	1.4	1.6	1.8	2.0
Thompson *et al.*†		0.860	1.220	1.490	1.720	1.920
	20 cells	0.851	1.197	1.460	1.689	1.883
Present	36	0.857	1.206	1.471	1.698	1.894
	52	0.858	1.210	1.476	1.702	1.899

†Values obtained from Figure 68 on p.169.

It can be seen again from Table 7 that the present results agree well with those given by Thompson and Hunt[39]. In the course of computations, convergence was achieved with about 9 iterations at each loading increment. However, I found that the iterative method gave good results up to β=2 in the calculation, and the results became worse for higher values of β. The question is needed to further investigate mathematically.

Example 7: a compressive square plate. Consider a clamped movable square plate on an Winkler-type foundation subjected to a uniform in-plane compressive load p_0. The initial data used in the example are

$$\mu = 0.3, \quad E = 2\times10^6\,\mathrm{kg/cm^2}, \quad t/L = 1/50, \quad k_w = 5D\pi^4/L^4$$

where L is the length of the square plate. In the analysis, a 4×4 internal mesh and three boundary element meshes (8, 12 and 16 elements) have been used in a quarter of the square plate. The loading step is taken as 0.02p_{cr}(p_{cr}=7.229$D\pi^2/L^2$ in the example). Table 8 presents the results of maximum deflection (w_m/t) vs with the compressive load coefficient $\beta= p_0/p_{cr}(\beta>1)$.

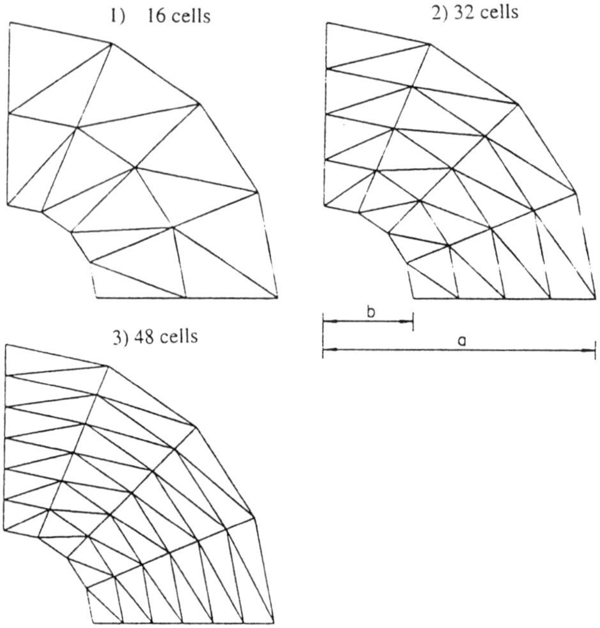

Fig. 3. Three internal meshes in example 5.

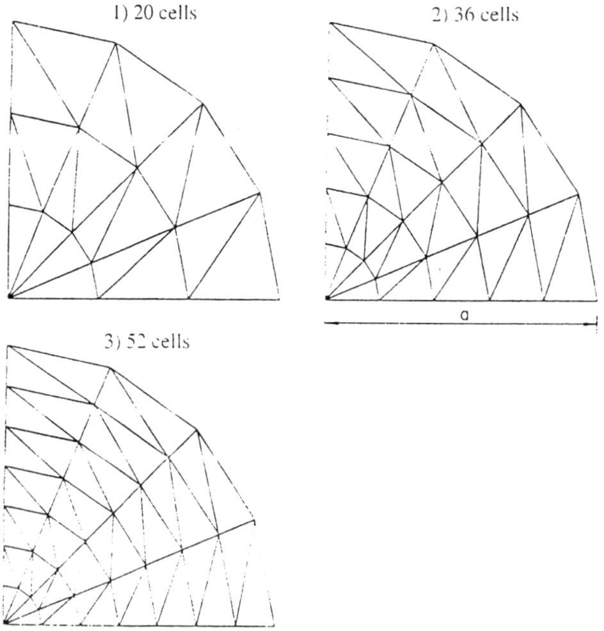

Fig. 4. Three internal meshes in example 6.

Table 8. Maximum deflection (w_m/t) for example 7

β		1.02	1.04	1.06	1.08	1.10
Naidu *et al.*†		0.3615	0.5128	0.6300	0.7298	0.8186
	8 b.e.	0.851	1.197	1.460	1.689	1.883
Present	12	0.857	1.206	1.471	1.698	1.894
	16	0.858	1.210	1.476	1.702	1.899

†The results were obtained by finite element method with a 2×2 mesh.

It can be also observed from Table 8 that convergence is very good along with refinement of the boundary element meshes. In our calculation, about 15 iterations were needed to achieve convergence at each loading step.

5 Concluding remarks

The boundary element formulation has been proposed for large deflection analysis of Reissner plates and postbuckling study of thin plates with (or without) foundation. The foundation may be of Winkler-type or Pasternak-type. The formulation was developed based on a variational approach. The approach showed that a nonlinear boundary integral equation can be exactly transformed from a modified variational functional. It also revealed the intrinsic relations between the variational principle and the boundary integral equation. In addition, the fundamental solutions corresponding the above two kinds of problem has been developed by way of the resolution of differential operator. Numerical computation has been carried out for large deflection and postbuckling analyses of plates with various shapes and also subject to various boundary conditions.

References

[1] Brebbia, C.A.(ed.). *New Developments in Boundary Element Methods*. Computational Mechanics Publications, Southampton, 1980

[2] Brebbia, C.A.(ed.). *Topics in Boundary Element Research*, Vol. **III**. Springer-Verlag, Berlin, 1987

[3] Beskos, D.E.(ed.). *Boundary Element Analysis of Plates and Shells*. Springer-Verlag, Berlin, 1991

[4] Alabadi, M.H. & Brebbia, C.A.(eds.). *Advanced Formulations in Boundary Element Methods*. Computational Mechanics Publications, Southampton, 1993

[5] Hansen, E.B. Numerical solution of integro-differential and singular integral equations for plate bending problems, *J. Elasticity*, **6**, 39-56, 1976

[6] Bezine, G. Boundary integral formulation for plate flexure with arbitrary boundary conditions, *Mech. Res. Comm*., **5**, 197-206, 1978

[7] Tottenham, H. The boundary element method for plates and shells, In: *Developments in Boundary Element Methods-1*(Edited by P.K. Banerjee & R. Butterfield), p.173-205, Applied Science Publishers, London, 1979

[8] Stern, M. & Lin, T.L. Thin elastic plates in bending, In: *Developments in Boundary Element Methods-4*(Edited by P.K. Banerjee & J.O. Watson), p.91-119, Applied Science Publishers, London, 1986
[9] Qin, Q.H. New algorithms for transient plate bending analysis by BEM, *Engineering Analysis with Boundary Elements*, **17**, 175-180, 1996
[10] Weeën F.Vander. Application of the boundary integral equation method to Reissner's plate model, *Int. J. Numer. Meth. Eng.*, **18**, 1-10, 1982
[11] Karam, V.J. & Telles, J.C.F. On the boundary elements for Reissner's plate theory, *Eng. Anal.*, **5**, 21-27, 1988
[12] Wang, J., Wang, X.X. & Huang, M.K. Fundamental solutions and boundary integral equations for Reissner's plates on two-parameter foundations, *Int. J. Solids Struct.*, **29**, 1233-1239, 1992
[13] Wang, J. & Schweizerhof, K. Boundary-domain element method for free vibration of moderately thick laminated orthotropic shallow shells, *Int. J. Solids Struct.*, **33**, 11-18, 1996
[14] Katsikadelis, J.T. & Armenakas, A.E. Plates on elastic foundation by BIE method, *J. Struct., Div., ASCE*, **110**, 1086-1105, 1984
[15] Costa, J.A. & Brebbia, C.A. The boundary element method applied to plates on elastic foundations, *Eng. Anal.*, **2**, 174-182, 1985
[16] Puttonen, J. & Varpasuo, P. Boundary element analysis of plates on elastic foundations, *Int. J. Numer. Meth. Eng.*, **23**, 287-303, 1986
[17] Kamiya, N. & Sawaki, Y. Integral equation formulationfor nonlinear bending of plates--formulation by weighted residual method, *Z. Angew. Math. Mech.*, **62**, 651-655, 1982
[18] Kamiya, N. & Sawaki, Y. An integral equation approach to finite deflection of elastic plates, *Int. J. Non-linear Mech.*, **17**, 187-194, 1982
[19] Kamiya, N., Sawaki, Y. & Nakamura, Y. Nonlinear bending analysis of heated sandwich plates and shells by the boundary element method, *Res. Mech.*, **8**, 29-38, 1983
[20] Kamiya, N. & Sawaki, Y. Finite deflection of plates, In *Topics in Boundary Element Research* (Edited by C.A. Brebbia), Vol. **1**, 204-224, 1984
[21] Kamiya, N. & Sawaki, Y. Boundary element analysis of nonlinear bending of sandwich plates and shallow shells, In: *Developments in Boundary Element Methods-4*(Edited by P.K. Banerjee & J.O. Watson), pp121-148, Applied Science Publishers, London, 1986
[22] Tanaka, M. Large deflection analysis of thin elastic plates, In: *Developments in Boundary Element Methods-3*(Edited by P.K. Banerjee & S. Mukherjee), p.115-136, Applied Science Publishers, London, 1984
[23] Ye, T.Q. & Liu, Y. Finite deflection analysis of elastic plate by the boundary element method, *Appl. Meth. Modeling*, **9**, 183-188, 1985
[24] Ye, J. Large deflection analysis of axisymmetric circular plates with variable thickness by the boundary element method, *Appl. Math. Modeling*, **15**, 325- 328, 1991

[25] Wang, X.X. Qian, J. & Huang, M.K. A boundary integral equation formulation for large amplitude nonlinear vibration of thin elastic plates, *Compu. Meth. Appl. Mech. Eng.*, **86**, 73-86, 1991
[26] Sawaki, Y., Takeuchi, K. & Kamiya, N. Finite deflection analysis of plates by the dual reciprocity boundary elements, In: *Boundary Elements*-XI (Edited by C.A. Brebbia & J.J.Connor), p.239-250, 1989
[27] Lei, X.Y., Huang, M.K. & Wang, X.X. Geometrically nonlinear analysis of a Reissner type plate by the boundary element method, *Compu. & Struct.*, **37**, 911-916, 1990
[28] Ye, J. Nonlinear bending analysis of plates and shells by using mixed spline boundary element and finite element, *Int. J. Numer. Meth. Eng.,* **31**, 1283- 1294, 1991
[29] Qin, Q.H. & Huang, Y.Y. BEM of postbuckling analysis of thin plates, *Appl. Math. Modeling*, **14**, 544-548, 1990
[30] Huang, Y., Zhong, W. & Qin, Q.H. Postbuckling analysis of plates on an elastic foundation by the boundary element method, *Compu. Meth. Appl. Mech. Eng.*, **100**, 315-323, 1992
[31] Katsikadelis, J.T. Large deflection analysis of plates on elastic foundation by the boundary element method, *Int. J. Solids Struct.*, **27**, 1867-1878, 1991
[32] He, X.Q. & Qin, Q.H. Nonlinear analysis of Reissner's plate by the variational approaches and boundary element methods, *Appl. Math. Modeling*, **17**, 149-155, 1993
[33] Qin, Q.H. Nonlinear analysis of Reissner plates on an elastic foundation by the BEM, *Int. J. Solids Struct.,* **30**, 3101-3111, 1993
[34] Kerr, A.D. Elastic and viscoelastic foundation models, *J. Appl. Mech.*, **31**, 491-498, 1964
[35] Simpson, H.C. & Spector, S.J. On the positive of the second variation of finite elasticity, *Arch. Rational Mech. Anal.*, **98**, 1-30, 1987
[36] Azizian, Z.G. & Dawe, D.J. Geometrical nonlinear analysis of rectangular Mindlin plates using finite strip method, *Compu. & Struct.*, **21**, 423-436, 1985
[37] Schmit, Jr. L.A. & Manforton, G.R. Finite deflection discrete element analysis of sandwich plates and cylindrical shells with laminated faces, *AIAA J.*, **8**, 1454-1461, 1970
[38] Ng, S.F. & Das, B. Finite deflection of skew sandwich plates on elastic foundations by the Galerkin method, *J. Struct. Mech.*, **14**, 355-377, 1986
[39] Thompson, J.M.T & Hunt, G.W. *A general theory of elastic stability*, Wiley, London, 1973
[40] Naidu, N.R., Raju, K.K. & Rao, G.V. Postbuckling of a square plate resting on an elastic foundation under biaxial compression, *Compu. & Struct.*, **37**, 343- 345, 1990

Appendix I: Notation

C	$5Et/12(1+\mu)$ for a homogenerous plate; $G_c(h+t)$ for a sandwich plate
C_w	a part of boundary $\partial\Omega$ of the solution domain Ω, on which deflection w is prescribed; C_R, etc. are defined similarly
D	$Et^3/12(1-\mu^2)$ for a homogenerous plate; $E(h+t)^2t/2(1-\mu^2)$ for a sandwich plate
E	modulus of elasticity
G	$E/2(1+\mu)$
G_c	core shear modulus
G_p	shear modulus of Pasternak-type foundation
h	core thickness
k_p	reaction coefficient of Pasternak-type foundation
k_w	reaction coefficient of Winkler-type foundation
M_{ii}	bending moment
M_{ij}	twisting moment($i\neq j$)
N_{ij}	membrane force tensor
n_i	components of the outward normal to the boundary $\partial\Omega$
q	lateral distributed load
Q_i	transverse shear force
r	$(x^2+y^2)^{1/2}$
s_i	components of the tangent to the boundary $\partial\Omega$
t	plate thickness, or face-sheet thickness in sandwich plate
$\mathbf{U}$	diaplacement vector, $\{u_1\ u_2\ \psi_1\ \psi_2\ w\}$
u_i	in-plane displacements
w	lateral deflection
δ	variational symbol
δ_{ij}	Kronecker delta
θ	arctg(y/x)
μ	Poisson's ratio
∇^2	$\partial^2/\partial x^2+\partial^2/\partial y^2$
ψ_i	average rotations normal to the plate mid-plane
$(\bar{\ })$	over a symbol denotes prescribed value

Appendix II: Fundamental solutions for out-of-deformation

$$\psi^*_{n1} = [S(Z_3)+E(Z_1,Z_2)]\cos\beta\cos(\beta-\phi)+[T(Z_3)-F(Z_1,Z_2)]\cos(2\beta-\phi)$$
$$\psi^*_{s1} = [S(Z_3)+E(Z_1,Z_2)]\cos\beta\sin(\beta-\phi)-[T(Z_3)+F(Z_1,Z_2)]\sin(2\beta-\phi),$$
$$w^*_1 = [A\sqrt{C_2}K_1(Z_2)(1-DC_2/C)+B\sqrt{C_1}Y_1(Z_1)(1+DC_1/C)]\cos\beta,$$
$$\psi^*_{n2} = [S(Z_3)+E(Z_1,Z_2)]\sin\beta\cos(\beta-\phi)+[T(Z_3)-F(Z_1,Z_2)]\sin(2\beta-\phi)$$
$$\psi^*_{s2} = [S(Z_3)+E(Z_1,Z_2)]\sin\beta\sin(\beta-\phi)+[T(Z_3)+F(Z_1,Z_2)]\cos(2\beta-\phi)$$
$$w^*_2 = [A\sqrt{C_2}K_1(Z_2)(1-DC_2/C)+B\sqrt{C_1}Y_1(Z_1)(1+DC_1/C)]\sin\beta,$$

$$\psi^*_{n3} = -[A\sqrt{C_2}K_1(Z_2)+B\sqrt{C_1}Y_1(Z_1)]\cos(\beta-\phi)$$
$$\psi^*_{s3} = -[A\sqrt{C_2}K_1(Z_2)+B\sqrt{C_1}Y_1(Z_1)]\sin(\beta-\phi)$$
$$w^*_3 = AC_2K_0(Z_2)(1-DC_2/C)+BC_1Y_0(Z_1)(1+DC_1/C)$$

where β and ϕ are shown in Fig. A1, and

$$A = \frac{1}{2\pi D(C_1+C_2)}, \qquad B = -\frac{1}{4D(C_1+C_2)},$$
$$S(Z_3) = -\frac{1}{\pi D(1-\mu)}\left[K_0(Z_3)+\frac{2}{Z_3}(K_1(Z_3)-1/Z_3)\right],$$
$$T(Z_3) = \frac{1}{\pi D(1-\mu)}\left[K_0(Z_3)+\frac{1}{Z_3}(K_1(Z_3)-1/Z_3)\right],$$
$$E(Z_1,Z_2) = BC_1Y_0(Z_1)-AC_2K_0(Z_2),$$
$$F(Z_1,Z_2) = BC_1Y_1(Z_1)/Z_1+AC_2K_1(Z_2)/Z_2,$$
$$Z_1 = r\sqrt{C_1}, \quad Z_2 = r\sqrt{C_2}, \quad Z_3 = \lambda r$$

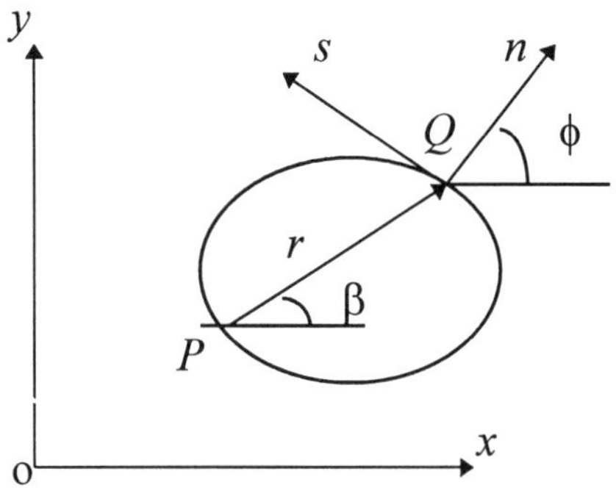

Fig. A1 The definitions of β and ϕ.

Chapter 9

Analysis of plates with variable thickness. An analog equation solution

M.S.Nerantzaki & J.T. Katsikadelis
Department of Civil Engineering, National Technical University of Athens, Zografou Campus, GR-157 73 Athens, Greece
Email: jkats@central.ntua.gr

Abstract

A BEM-based method, the analog equation method (AEM), is presented as it is employed to solve static and dynamic problems for plates of variable thickness subjected to inplane forces. The linear buckling problem is also treated using both the static and dynamic criteria. According to AEM the fourth order hyperbolic partial differential equation with variable coefficients is converted to a quasi-static linear problem for plates with constant stiffness subjected to an "appropriate" fictitious time dependent load under the same boundary conditions. The fictitious load is established using a technique based on BEM. The static problem results as a special case of the dynamic one, when the inertia and damping forces are neglected. The method is illustrated by applying it to several example problems including static plate bending problems, free and forced vibrations as well as buckling problems. Linear and exponential plate thickness variation laws are considered. The accuracy of the results is validated by comparing them with existing ones from analytic or other numerical methods.

1 Introduction

The study of plates with variable thickness is pursued in various engineering disciplines, such as civil engineering, aerospace engineering and the design of machines. Although there is an extensive literature on a static and dynamic analysis of plates with constant thickness, a rather limited amount of technical literature is available on the solution of problems dealing with plates of non-uniform thickness.

The reason for this is that in the case of plates with variable thickness, the governing differential equation is found to have variable coefficients, and this fact increases the difficulty of the solution. Prior to the advent of computers, plates with variable thickness could be analyzed only for certain simple geometries, boundary conditions, and thickness variation laws. The existing analytic solutions are limited to circular andannular plates with linear varying thickness along the radius subjected to axisymmetric loading, as well as to rectangular plates with unidirectional thickness variation subjected to constant or hydrostatic loading and boundary conditions amenable to use Levy-type solutions. Approximate methods, such as the Galerkin method and the Rayleigh–Ritz method, have also been used to treat this problem. Timoshenko & Woinowsky-Krieger,[1] Bares,[2] Bulson[3] and some recent papers by Katsikadelis & Nerantzaki[4–7] include literature surveys on plates with variable thickness analyzed by analytical and/or approximate methods; therefore no attempt to review the literature will be performed here.

An arbitrary thickness profile can be treated only by numerical methods. The finite difference method (FDM), the finite element method (FEM) and the boundary element method (BEM) are candidates to treat the problem at hand.

Although the FDM can solve static and dynamic problems for plates with variable thickness, its efficiency is drastically restricted when the geometry of the plate and the boundary conditions are not simple. The FEM can adequately solve static and dynamic problems for plates with variable thickness. Certain commercial computer codes for structural analysis include plate elements with variable thickness. However, to the authors' knowledge, no publications have appeared using FEM when inplane forces are taken into account. With regards to BEM, although there is considerable application of this method to the analysis of plates with constant thickness [e.g. Jaswon & Maitai,[8] Bezine,[9] Stern,[10] Hartmann & Zotemantel,[11] Katsikadelis & Armenakas,[12] Paris & deLeon[13]] little work has been published on plates with variable thickness using BEM. This is obviously because a fundamental solution for the governing equation cannot be established, at least in a form that could be useful to develop a pure boundary element method. A first attempt to use BEM for plates with variable thickness has been made by Katsikadelis & Sapountzakis,[14,15] who employed the fundamental solution of the plate with constant thickness and treated the term involving derivatives up to the third order as an unknown field quantity. This unknown term was established by augmenting the set of the boundary integral equations with a domain integral equation. Apparently, the developed procedure was not a pure BEM, since it required domain discretization. The use of a Gauss integration on the whole domain of the plate proposed by these authors alleviated the method from discretizing the domain into cells. Thus, this method retained most of the advantages of a BEM solution over the pure domain discretization method. Recently, a more effective BEM-based methods for the analysis of plates with variable thickness has been developed by Nerantzaki & Katsikadelis.[4–7] The method uses the analog equation concept as developed by Katsikadelis.[16] This method is more versatile and can treat the static and dynamic problem including also inplane forces. According to this method, to which we refer as the analog equation method (AEM), the fourth order hyperbolic type partial differential equation with variable coefficients is replaced by a quasi-static plate bending equation with constant thickness subjected to a fictitious time dependent load under the same boundary conditions. Subsequently, the deflections and all the derivatives involved in the governing differential equation are expressed in terms of the fictitious load. After substitution of these quantities into the differential equation, a semi-discretized equation of motion is derived from which the fictitious load is established by numerical evaluation. The static problem results as a special case of the dynamic one by neglecting the inertia and damping forces.

The method is not a pure BEM as it requires domain discretization to evaluate the domain integrals. However, this method can be developed to a boundary-only method, if the domain integrals containing the fictitious load are converted to boundary ones using the dual reprocity method.

The application of the method is illustrated by solving several example problems of plates with variable thickness including static, dynamic and buckling problems. The obtained numerical results are compared with those obtained from analytic or other numerical methods to validate the accuracy of the method.

2 Governing equations

Consider a thin elastic plate of variable thickness, $h = h(x,y)$, occupying the two-dimensional multiply connected region Ω of the x,y plane, bounded by the $K+1$ curves $\Gamma_0, \Gamma_1, \Gamma_2, \ldots, \Gamma_K$. The curves Γ_i $(i = 0,1,2,\ldots,K)$ may be piece-wise smooth (Figure 1).

Assuming that there is no abrupt variation in thickness, the expressions for bending and twisting moments derived for plates of constant thickness apply with sufficient accuracy to this case also and the equilibrium of a plate element subjected to a distributed transverse load $g(\mathbf{x},t)$ $\mathbf{x}: \{x,y\} \in \Omega$, $t \geq 0$ and inplane forces $N_x = N_x(\mathbf{x})$, $N_y = N_y(\mathbf{x})$ and $N_{xy}(\mathbf{x})$ yields the following differential equation of motion in terms of the deflection $w(\mathbf{x},t)$.

$$\begin{aligned} &D\nabla^4 w + 2\frac{\partial D}{\partial x}\frac{\partial}{\partial x}\nabla^2 w + 2\frac{\partial D}{\partial y}\frac{\partial}{\partial y}\nabla^2 w + \nabla^2 D \nabla^2 w \\ &-(1-\nu)\left(\frac{\partial^2 D}{\partial x^2}\frac{\partial^2 w}{\partial y^2} - 2\frac{\partial^2 D}{\partial x \partial y}\frac{\partial^2 w}{\partial x \partial y} + \frac{\partial^2 D}{\partial y^2}\frac{\partial^2 w}{\partial x^2}\right) \\ &-\left(N_x\frac{\partial^2 w}{\partial x^2} + 2N_{xy}\frac{\partial^2 w}{\partial x \partial y} + N_y\frac{\partial^2 w}{\partial y^2}\right) + c\frac{\partial w}{\partial t} + \rho h\frac{\partial^2 w}{\partial t^2} = g(\mathbf{x},t) \end{aligned} \tag{1}$$

where $D = Eh^3 / 12(1-\nu^2)$ is the variable flexural rigidity of the plate, $c = c(\mathbf{x})$ the distribution of the damping coefficient and ρ the mass density. Moreover, the deflection w must satisfy the following boundary conditions on the boundary $\Gamma = \bigcup_{i=0}^{i=K}$ and initial conditions inside Ω

$$\alpha_1 w + \alpha_2 V(w) = \alpha_3 \qquad \beta_1 \frac{\partial w}{\partial n} + \beta_2 M(w) = \beta_3 \qquad \text{on } \Gamma \tag{2a,b}$$

$$w(\mathbf{x},0) = \overline{w}(\mathbf{x}), \qquad \frac{\partial w(\mathbf{x},0)}{\partial t} = \dot{\overline{w}}(\mathbf{x}) \quad \text{in } \Omega \tag{3a,b}$$

where $\alpha_i = \alpha_i(\mathbf{x},t)$, $\beta_i = \beta_i(\mathbf{x},t)$, $\mathbf{x} \in \Gamma$, are functions specified on Γ; $M(w)$ and $V(w)$ are the bending moment and the reactive force on the boundary; $\overline{w}(\mathbf{x})$ and $\dot{\overline{w}}(\mathbf{x})$ are the initial deflection and the initial velocity of the points of the middle surface of the plate, respectively. The boundary conditions (2a,b) are the most general linear boundary conditions for the plate bending problem. All types of conventional boundary conditions are derived from (2a,b) by specifying appropriately the functions α_i and β_i, including support excitation.

Taking into account that the flexural rigidity D is a function of the variables x and y and using intrinsic co-ordinates n and s, the operators M, V appearing in eqns (2a,b) may be written as

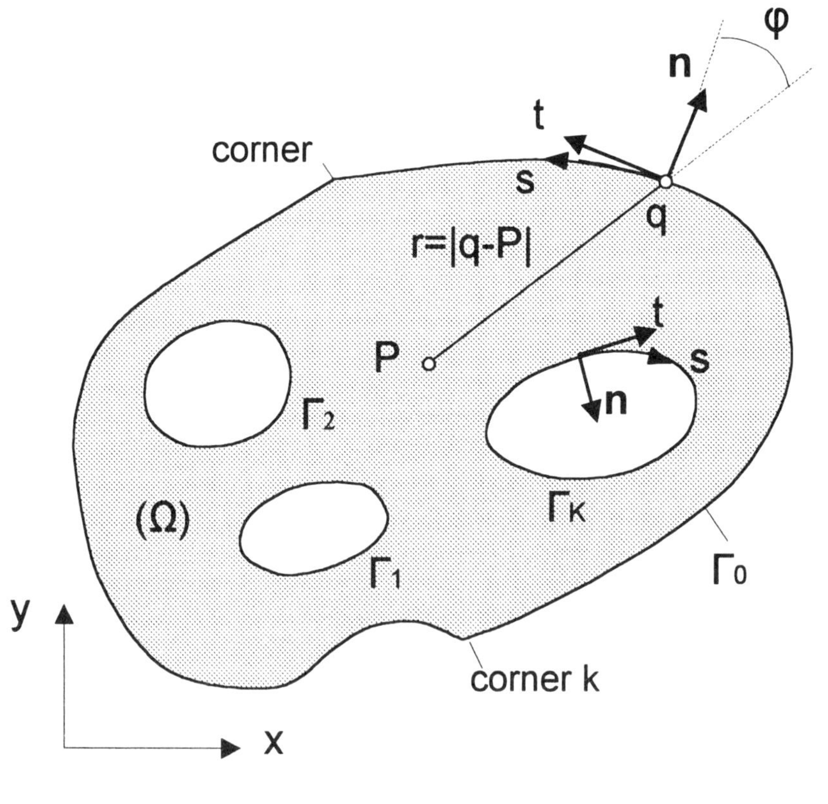

Figure 1. Geometry of the plate and notation.

$$M = -D\left[\nabla^2 + (\nu - 1)\left(\frac{\partial^2}{\partial s^2} + \kappa\frac{\partial}{\partial n}\right)\right] \tag{4a}$$

$$\begin{aligned} V = -D\left[\frac{\partial}{\partial n}\nabla^2 - (\nu - 1)\frac{\partial}{\partial s}\left(\frac{\partial^2}{\partial n \partial s} - \kappa\frac{\partial}{\partial s}\right)\right] + \frac{\partial D}{\partial s}(\nu - 1)\left(\frac{\partial^2}{\partial n \partial s} - \kappa\frac{\partial}{\partial s}\right) \\ - \frac{\partial D}{\partial n}\left[\nabla^2 + (\nu - 1)\left(\frac{\partial^2}{\partial s^2} + \kappa\frac{\partial}{\partial n}\right)\right] \end{aligned} \tag{4b}$$

in which $\kappa = \kappa(s)$ is the curvature of the boundary; $\partial / \partial s$ and $\partial / \partial n$ denote differentiation with respect to the arc length s of the boundary, and the outward normal n to it, respectively.

In the case of free or transversely elastically restrained edges, the boundary conditions (3a,b) must be supplemented with the corner condition

$$c_{1k} w + c_{2k} \|Tw\|_k = c_{3k}(t), \quad c_{2k} \neq 0 \tag{5}$$

where c_{ik} are specified constants at the corner $\mathbf{x}_k$ and T is the operator

$$T = D(1 - \nu)\left(\frac{\partial^2}{\partial s \partial n} - \kappa\frac{\partial}{\partial s}\right) \tag{6}$$

Thus, $T(w)$ is the twisting moment along the boundary and $\|Tw\|_k$ is its jump of discontinuity at the corner point $\mathbf{x}_k$. The stress resultants at a point inside Ω are given as

$$M_x = -D\left(\frac{\partial^2 w}{\partial x^2} + \nu\frac{\partial^2 w}{\partial y^2}\right) \tag{7a}$$

$$M_y = -D\left(\frac{\partial^2 w}{\partial y^2} + \nu\frac{\partial^2 w}{\partial x^2}\right) \tag{7b}$$

$$M_{xy} = D(1 - \nu)\frac{\partial^2 w}{\partial x \partial y} \tag{7c}$$

$$Q_x = -D\frac{\partial}{\partial x}\nabla^2 w - \frac{\partial}{\partial x}\left(\frac{\partial^2 w}{\partial x^2} + \nu\frac{\partial^2 w}{\partial y^2}\right) - \frac{\partial D}{\partial y}(1 - \nu)\frac{\partial^2 w}{\partial x \partial y} \tag{7d}$$

$$Q_y = -D\frac{\partial}{\partial y}\nabla^2 w - \frac{\partial D}{\partial y}\left(\frac{\partial^2 w}{\partial y^2} + \nu\frac{\partial^2 w}{\partial y^2}\right) - \frac{\partial D}{\partial x}(1 - \nu)\frac{\partial^2 w}{\partial x \partial y} \tag{7e}$$

Since the linear buckling problem is considered, the inplane forces N_x, N_y, N_{xy} are *a priori* known. They are given as

$$N_x = h\sigma_x = Ch\left(\frac{\partial u}{\partial x} + \nu\frac{\partial v}{\partial y}\right) \tag{8a}$$

$$N_y = h\sigma_y = Ch\left(\frac{\partial v}{\partial y} + \nu\frac{\partial u}{\partial x}\right) \tag{8b}$$

$$N_{xy} = hT_{xy} = C\frac{1-\nu}{2}h\left(\frac{\partial u}{\partial y} + \frac{\partial v}{\partial x}\right) \tag{8c}$$

in which $C = E/(1-\nu^2)$ and $u = u(x,y)$, $v = v(x,y)$ are the inplane displacement components which are established by solving independently the plane stress problem for plates with variable thickness, which, in absence of body forces, is described by the following Navier-type differential equations

$$\mu h\nabla^2 u + (\lambda+\mu)h\left(\frac{\partial^2 u}{\partial x^2} + \frac{\partial^2 v}{\partial x\partial y}\right) + \frac{\partial h}{\partial x}\left[\lambda\left(\frac{\partial u}{\partial x} + \frac{\partial v}{\partial y}\right) + 2\mu\frac{\partial u}{\partial x}\right] + \frac{\partial h}{\partial y}\mu\left(\frac{\partial u}{\partial y} + \frac{\partial v}{\partial x}\right) = 0 \tag{9a}$$

$$\mu h\nabla^2 v + (\lambda+\mu)h\left(\frac{\partial^2 u}{\partial x\partial y} + \frac{\partial^2 v}{\partial y^2}\right) + \frac{\partial h}{\partial x}\mu\left(\frac{\partial u}{\partial y} + \frac{\partial v}{\partial x}\right) + \frac{\partial h}{\partial y}\left[\lambda\left(\frac{\partial u}{\partial x} + \frac{\partial v}{\partial y}\right) + 2\mu\frac{\partial v}{\partial y}\right] = 0 \quad \text{in } \Omega \tag{9b}$$

under the boundary conditions

$$u = \bar{u}, \quad v = \bar{v}, \quad \text{on } \Gamma_1 \tag{10a}$$

$$u = \bar{u}, \quad f_y = \bar{f}_y, \quad \text{on } \Gamma_2 \tag{10b}$$

$$v = \bar{v}, \quad f_x = \bar{f}_x, \quad \text{on } \Gamma_3 \tag{10c}$$

$$f_x = \bar{f}_x, \quad f_y = \bar{f}_y, \quad \text{on } \Gamma_4 \tag{10d}$$

where $\bigcup_{i=1}^{i=4}\Gamma_i = \Gamma$ and λ and μ are the Lamé constants related to the elastic modulus E and Poisson's ratio ν as follows

$$\lambda = \frac{E\nu}{(1+\nu)(1-2\nu)}, \qquad \mu = \frac{E}{2(1+\nu)} \tag{11}$$

In the above relations quantities provided with an overbar designate prescribed quantities.

The boundary value problem (7a,b), (5a)–(5d) is solved using the FEM or the AEM[17].

The boundary forces f_x and f_y are given in terms of the displacements as

$$f_x = Ch\left[(u_x + \nu v_y)\cos a + \frac{1-\nu}{2}(u_y + v_x)\sin a\right] \tag{12a}$$

$$f_y = Ch\left[(\nu u_x + v_y)\sin a + \frac{1-\nu}{2}(u_y + v_x)\cos a\right] \tag{12b}$$

The plate problem described by eqn (1) includes as special cases all linear bending problems for plates with variable thickness, that is the static and dynamic problems with or without inplane forces including also the buckling problem. As it was mentioned in the introduction a D/BEM method has been developed by Katsikadelis & Sapountzakis for solving the static[14] and the dynamic[15] problem for plates with variable thickness. However, the analog equation method LAEM is presented in this chapter. This method is more versatile, and can be further developed as a boundary-only method.

3 The analog equation method

The initial-boundary value problem described by eqns (1), (2) and (3) is solved using the AEM. In the problem at hand this method is applied as follows.

Let w be the sought solution of eqn (1). This function is four times continuously differentiable with respect to the spatial co-ordinates x, y in Ω and three times on its boundary Γ. If the biharmonic operator is applied to this function we have

$$\nabla^4 w = q(\mathbf{x}, t) \tag{13}$$

Eqn (13) indicates that the solution of the original initial-boundary value problem can be obtained as the solution of a linear quasi-static plate bending problem having unit stiffness and subjected to the equivalent (fictitious) time-dependent load q under the given boundary and initial conditions.

According to the AEM, the unknown load distribution q can be established using the BEM. The direct BEM for plates (e.g. Bezine,[9] Stern,[10] Katsikadelis[18]) could be applied if the boundary terms in Rayleigh–Green identity were modified so that to include the boundary reaction defined by eqn (4b). However, this procedure is avoided because it would involve complicated singular and hypersingular kernels which would be difficult to manipulate and evaluate numerically. Therefore, an indirect BEM developed by Katsikadelis & Armenakas[12] has been employed in the development of AEM because of the simplicity of the kernels appearing in the boundary integrals.

According to this method, for any function w satisfying the non-homogeneous biharmonic eqn (13) the following integral representations are obtained.

$$\varepsilon w(\mathbf{x},t) = \int_\Omega \Lambda_4 q \mathrm{d}\Omega - \int_\Gamma (\Lambda_1 \Omega + \Lambda_2 X + \Lambda_3 \Phi + \Lambda_4 \Psi)\mathrm{d}s \tag{14}$$

$$\varepsilon \nabla^2 w(\mathbf{x},t) = \int_\Omega \Lambda_2 q \mathrm{d}\Omega - \int_\Gamma (\Lambda_1 \Phi + \Lambda_2 \Psi)\mathrm{d}s \tag{15}$$

where $\varepsilon = 2\pi,\ \pi,\ 0$ depending on whether the point $\mathbf{x}$ is inside the domain Ω, on the boundary Γ or outside Ω, respectively. Note that the boundary has been assumed to be smooth at the point $\mathbf{x}$. The kernels $\Lambda_i = \Lambda_i(r)$, with $|\xi - \mathbf{x}|$, and $\mathbf{x} \in \Omega, \xi \in \Gamma$ corresponding to the fundamental solution of eqn (13) are given as

$$\Lambda_1(r) = -\frac{\cos\varphi}{r}, \tag{16a}$$

$$\Lambda_2(r) = \ell n\, r + 1, \tag{16b}$$

$$\Lambda_3(r) = -\frac{1}{4}(2r\ell n\, r + r)\cos\varphi, \tag{16c}$$

$$\Lambda_4(r) = \frac{1}{4} r^2 \ell n\, r, \tag{16d}$$

φ = is the angle between r and the normal vector n

On the basis of eqns (4a,b) the boundary conditions (2a,b) are written as

$$\alpha_1 w - \alpha_2 \{ D \left[\Psi - (\nu - 1) \frac{\partial}{\partial s} \left(\frac{\partial X}{\partial s} - \kappa \frac{\partial \Omega}{\partial s} \right) \right] + \frac{\partial D}{\partial s} (\nu - 1) \left(\frac{\partial X}{\partial s} - \kappa \frac{\partial \Omega}{\partial s} \right)$$
$$- \frac{\partial D}{\partial n} \left[\Phi + (\nu - 1) \left(\frac{\partial^2 \Omega}{\partial s^2} + \kappa X \right) \right] \} = \alpha_3 \tag{17}$$

$$\beta_1 \frac{\partial w}{\partial n} - \beta_2 \{ D \left[\Phi + (\nu - 1) \left(\frac{\partial^2 \Omega}{\partial s^2} + \kappa X \right) \right] \} = \beta_3 \tag{18}$$

In eqns (14), (15), (17) and (18) the following notation has been used

$$\Omega = w(s), \; X = \frac{\partial w(s)}{\partial n}, \; \Phi = \nabla^2 w(s), \; \Psi = \frac{\partial \nabla^2 w(s)}{\partial n} \tag{19}$$

The integral representations (14) and (15) for $\mathbf{x} \in \Gamma$, together with the boundary conditions (17) and (18) constitute a set of four boundary equations with respect to the boundary quantities Ω, X, Φ, Ψ. Two of these equations are boundary integral equations and the remaining boundary differential equations. They are solved numerically. The boundary differential equations are solved using the finite difference method, whereas the integral equations are solved using the BEM. Constant boundary elements are employed. The domain integrals are evaluated using constant triangular or rectangular cells (Figure 2).

The above discretization yields the following set of linear equations.

$$\begin{bmatrix} [A_{11}] & [A_{12}] & [A_{13}] & [A_{14}] \\ [A_{21}] & [A_{22}] & [A_{23}] & [0] \\ [A_{31}] & [A_{32}] & [A_{33}] & [A_{34}] \\ [0] & [0] & [A_{43}] & [A_{44}] \end{bmatrix} \begin{Bmatrix} \{\Omega\} \\ \{X\} \\ \{\Phi\} \\ \{\Psi\} \end{Bmatrix} = \begin{Bmatrix} \{B_1\} \\ \{B_2\} \\ \{0\} \\ \{0\} \end{Bmatrix} + \begin{bmatrix} 0 \\ 0 \\ [C_3] \\ [C_4] \end{bmatrix} \{q\} \tag{20}$$

where $[A_{ij}]$ $(i, j = 1, 2, 3, 4)$ are $N \times N$ known coefficient matrices originating from the integration of the kernels on the boundary elements, $\{B_i\}$ $(i = 1,2)$ $N \times 1$ known constant column matrices and $[C_i]$ $(i = 3,4)$ $N \times M$ known coefficient matrices originating from the integration of the kernels over the domain cells. $\{\Omega\}$, $\{X\}$, $\{\Phi\}$, $\{\Psi\}$ are vectors including the N nodal values of the unknown boundary quantities, while $\{q\}$ is an $M \times 1$ vector including the values of the unknown fictitious loading at the nodal points inside Ω. N is the number of the boundary nodal points, whereas M is the number of the domain nodal points. From eqns (20) the

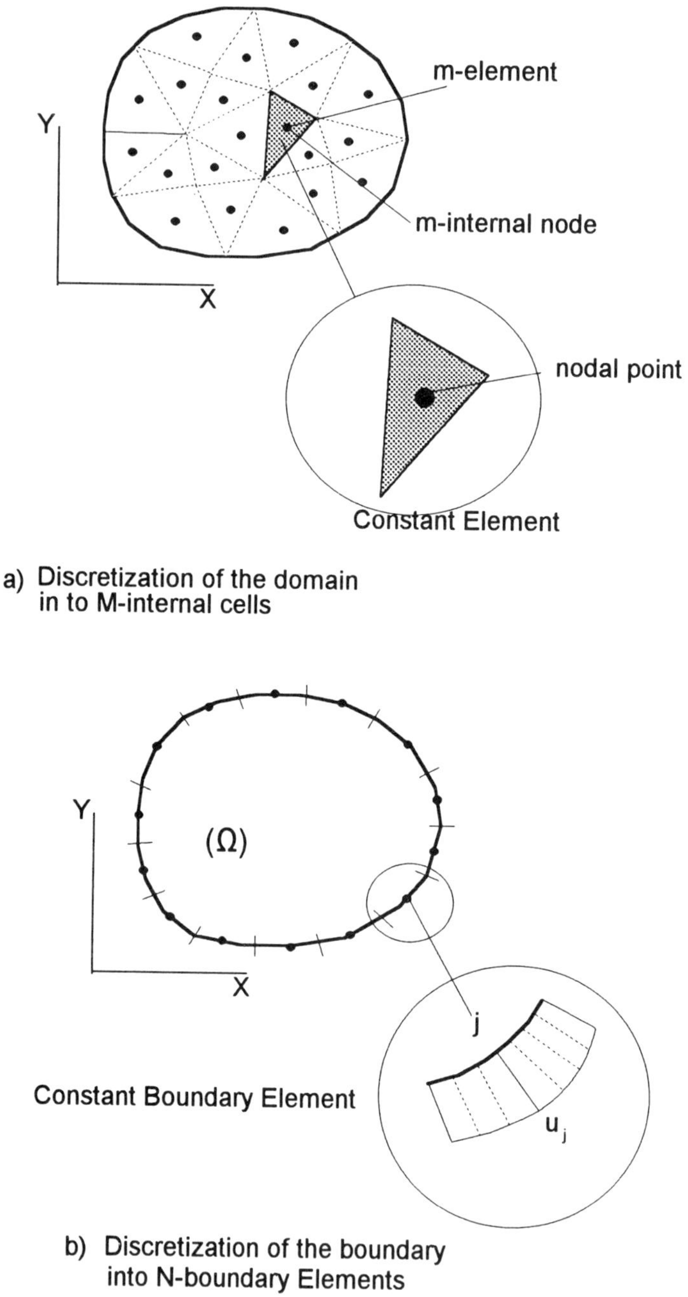

Figure 2. Boundary and domain discretization.

$$\{w\} = [G]\{q\} \tag{21}$$

boundary quantities Ω, X, Φ, Ψ are expressed in terms of the fictitious load vector $\{q\}$. Subsequently, their substitution into the discretized counterpart of eqn (14) yields where $\{w\}$ is a vector including the values of the deflection w at the M domain nodal points and $[G]$ is an $M \times M$ known coefficient matrix.

Subsequent differentiation of eqn (14), when $\varepsilon = 2\pi$, with respect to x and y gives

$$2\pi w_{xx}(\mathbf{x},t) = \int_R (\Lambda_4)_{xx} q \mathrm{d}\Omega - \int_\Gamma [(\Lambda_1)_{xx}\Omega + (\Lambda_2)_{xx} X + (\Lambda_3)_{xx} \Phi + (\Lambda_4)_{xx} \Psi)]\mathrm{d}s \tag{22a}$$

$$2\pi w_{yy}(\mathbf{x},t) = \int_R (\Lambda_4)_{yy} q \mathrm{d}\Omega - \int_\Gamma [(\Lambda_1)_{yy}\Omega + (\Lambda_2)_{yy} X + (\Lambda_3)_{yy} \Phi + (\Lambda_4)_{yy} \Psi)]\mathrm{d}s \tag{22b}$$

$$2\pi w_{xy}(\mathbf{x},t) = \int_R (\Lambda_4)_{xy} q \mathrm{d}\Omega - \int_\Gamma [(\Lambda_1)_{xy}\Omega + (\Lambda_2)_{xy} X + (\Lambda_3)_{xy} \Phi + (\Lambda_4)_{xy} \Psi)]\mathrm{d}s \tag{22c}$$

$$2\pi \nabla^2 w_x(\mathbf{x},t) = \int_R (\Lambda_2)_x q \,\mathrm{d}\Omega - \int_\Gamma [(\Lambda_1)_x \Phi + (\Lambda_2)_x \Psi]\mathrm{d}s \tag{22d}$$

$$2\pi \nabla^2 w_y(\mathbf{x},t) = \int_R (\Lambda_2)_y q \mathrm{d}\Omega - \int_\Gamma [(\Lambda_1)_y \Phi + (\Lambda_2)_y \Psi]\mathrm{d}s \tag{22e}$$

The derivatives of the kernels are given in the Appendix.

Eliminating the boundary quantities from the discretized counterparts of eqns (22a)–(22e) by means of eqns (20) and collocating at the M nodal points inside Ω yields

$$\{w_{xx}\} = [G_{xx}]\{q\} \tag{23}$$

$$\{w_{yy}\} = [G_{yy}]\{q\} \tag{24}$$

$$\{w_{xy}\} = [G_{xy}]\{q\} \tag{25}$$

$$\{\nabla^2 w_x\} = [G_{Lx}]\{q\} \tag{26}$$

$$\{\nabla^2 w_y\} = [G_{Ly}]\{q\} \tag{27}$$

where $[G_{xx}]$, $[G_{yy}]$, $[G_{xy}]$, $[G_{Lx}]$, $[G_{Ly}]$ are known $M \times M$ coefficient matrices. Note that eqns (21) and (23)– (27) are valid for homogeneous boundary conditions $(\alpha_3 = \beta_3 = 0)$. For non-homogeneous boundary conditions an additive, in general time dependent, vector will appear in the right hand side of these equations.

The final step of the AEM is to apply eqn (1) at the M nodal points inside the domain Ω. This yields

$$\begin{aligned}
&[D]\{\nabla^4 w\} + 2[D_x]\{\nabla^2 w_x\} + 2[D_y]\{\nabla^2 w_y\} + [\nabla^2 D]\{\nabla^2 w\} \\
&\quad - (1-\nu)([D_{xx}]\{w_{yy}\} - 2[D_{xy}]\{w_{xy}\} + [D_{yy}]\{w_{xx}\}) \\
&\quad - ([N_x]\{w_{xx}\} + 2[N_{xy}]\{w_{xy}\} + [N_y]\{w_{yy}\}) \\
&\quad + [c]\{\dot{w}\} + [\rho h]\{\ddot{w}\} = \{g\}
\end{aligned} \tag{28}$$

Subsequently, substituting eqns (13), (21) and (23)–(27) in eqn (28), we obtain

$$[M]\{\ddot{q}\} + [C]\{\dot{q}\} + [S]\{q\} = \{g\} \tag{29}$$

where $[M]$,$[C]$ and $[S]$ are known square matrices having dimensions $M \times M$ and are given by

$$[M] = [\rho h][G] \tag{30}$$

$$[C] = [c][G] \tag{31}$$

$$[S] = [K] - [B] \tag{32}$$

$$\begin{aligned}
[K] = [D] &+ 2[D_x][G_{Lx}] + 2[D_y][G_{Ly}] + [\nabla^2 D]([G_{xx}] + [G_{yy}]) \\
&- (1-\nu)([D_{xx}][G_{yy}] - 2[D_{xy}][G_{xy}] + [D_{yy}][G_{xx}])
\end{aligned} \tag{33a}$$

$$[B] = [N_x][G_{xx}] + 2[N_{xy}][G_{xy}] + [N_y][G_{yy}] \tag{33b}$$

Eqn (29) is the semidiscretized equation of motion of the plate with variable thickness with $[M]$,$[C]$ and $[S]$ representing the generalized mass, damping and stiffness matrices, respectively. Its solution yields $\{q\}$. The initial conditions for eqn (29) are obtained using eqn (21) as

$$\{q_o\} = [G]^{-1}\{\overline{w}\}, \quad \{\dot{q}_o\} = [G]^{-1}\{\dot{\overline{w}}\} \tag{34a,b}$$

In the above equations, the square matrices including the values of ρh, c as well as of D and its derivatives at the M nodal points inside Ω are diagonal matrices.

4 Evaluation of singular and hypersingular domain integrals

In evaluating the domain integrals in the discretized counterparts of eqns (14),(15) and (22a)–(22e) we come across to kernels, which behave as ℓnr, $1/r$ and $1/r^2$ for small values of the argument r. Thus, we have to evaluate singular and hypersingular domain integrals on the internal cells. This can be effectively done by converting the domain singular integrals into regular ones on the boundary of the cell using Green's reciprocal identity as follows.

The Green identity

$$\int_{\Omega^e}(u\nabla^2 U - U\nabla^2 u)\mathrm{d}\sigma = \int_{\Gamma^e}(uU_{,n} - Uu_{,n})\mathrm{d}s \tag{35}$$

For $u = 1$ and

$$\nabla^2 U = \upsilon^*, \quad \text{with} \quad \upsilon^* = \frac{1}{2\pi}(\ell n r + 1) \tag{36}$$

yields

$$\int_{\Omega^e}\upsilon^*\mathrm{d}\sigma = \int_{\Gamma^e} U_{,n}\mathrm{d}s \tag{37}$$

where Ω^e is the domain of the cell and Γ^e its boundary. The function U is obtained by integrating eqn (36). Thus, we have

$$U = \frac{1}{8\pi}r^2\ell n r \tag{38}$$

For the domain integrals involving derivatives, U must be replaced by its corresponding derivative in eqn (37). Thus

$$\int_{R^e}\upsilon^*_{,m}\mathrm{d}\sigma = \int_{\Gamma^e} U_{,mn}\mathrm{d}s, \qquad m = x, y, xx, xy, yy \tag{39}$$

The derivatives $U_{,mn}$ are given as

$$U_{xn} = -\frac{1}{8\pi}\{(2\ell n\, r + 1)\cos(\alpha + \varphi) + 2\cos\alpha\cos\varphi\} \tag{40}$$

$$U_{yn} = -\frac{1}{8\pi}\{(2\ell n\, r + 1)\sin(\alpha + \varphi) + 2\sin\alpha\cos\varphi\} \tag{41}$$

$$U_{xxn} = \frac{1}{4\pi r}(\cos\varphi - \sin 2\alpha \sin\varphi) \tag{42}$$

$$U_{xyn} = \frac{1}{4\pi r}\cos 2\alpha \sin\varphi \tag{43}$$

$$U_{yyn} = \frac{1}{4\pi r}(\cos\varphi + \sin 2\alpha \sin\varphi) \tag{44}$$

where

$$\alpha = \text{angle}(x, r) \quad \text{and} \quad \varphi = \text{angle}(n, r)$$

5 Applications

Following the procedure for AEM described in previous sections a computer program has been written in FORTRAN for solving static and dynamic problems of plates with variable thickness. For static problems the input data are the geometry of the plate described by the coordinates of the extreme points of the constant boundary elements, the material constants, the function of the thickness variation, the boundary conditions, the transverse loading and the inplane forces. For dynamic problems, in addition to the previous data, the mass density, the function of the damping distribution and the initial conditions (initial deflection and/or initial velocity distribution) are specified. To a given domain discretization the program computes the matrices $[G]$, $[G_{xx}]$, $[G_{xy}]$, $[G_{yy}]$, $[G_{Lx}]$ and $[G_{Ly}]$ defined by eqns (21), (23)–(27). The integrals on the kernels on the boundary element and on the domain cells are compluted numerically using one-and two-dimensional Gauss integration, respectively. Subsequently, it formulates the matrices $[M]$, $[C]$, $[K]$ and $[B]$ as required for the specific problem. Then the fictitious load is computed from eqn (29). Several example problems have been studied and the obtained results have been compared with those obtained from analytical or other numerical solutions to illustrate the efficiency and the accuracy of the method.

The accuracy of the method is mainly due to the accuracy of the computed derivatives by means of eqns (23)–(27). This is demonstrated by computing the

Table 1. Derivatives of the deflection surface of a uniformly loaded ($g = 1$) simply supported square ($a = b = 5$) plate. Upper row: AEM. Lower row: analytic

x=y	w_{xx}	w_{yy}	w_{xy}	$\nabla^2 w_x$	$\nabla^2 w_y$
0.25	–0.05509	–0.05509	1.08683	–0.35268	–0.35730
	–0.05461	–0.05461	1.09983	–0.35665	–0.35665
.75	–0.29535	–0.29535	0.81758	–0.54386	–0.54813
	–0.29507	–0.29507	0.81722	–0.54764	–0.54764
1.25	–0.56633	–0.56633	0.47691	–0.50643	–0.50975
	–0.56609	–0.56609	0.47678	–0.50584	–0.50584
1.50	–0.78514	–0.78514	0.18694	–0.34819	–0.35034
	–0.78492	–0.78492	0.18687	–0.34912	–0.34912
1.75	–0.90557	–0.90557	0.02166	–0.12326	–0.12409
	–0.90528	–0.90528	0.02166	–0.12958	–0.12958

derivatives of the deflection surface of square ($a = b = 5$) simply supported plate subjected to a uniformly distributed load $g = 1$. The computed values are presented in Table 1 as compared with those computed from an analytical solution (Navier series solution). For the AEM solution the domain has been discretized into 100 square cells.

5.1 Static analysis of plates with variable thickness without inplane forces

In this case the deflection is independent of time. Hence, it is $\dfrac{\partial w}{\partial t} = 0$, $\dfrac{\partial^2 w}{\partial t^2} = 0$.

Moreover, it is $N_x = N_y = N_{xy} = 0$ and eqn (1) reduces to

$$D\nabla^4 w + 2\frac{\partial D}{\partial x}\frac{\partial}{\partial x}\nabla^2 w + 2\frac{\partial D}{\partial y}\frac{\partial}{\partial y}\nabla^2 w + \nabla^2 D \nabla^2 w$$
$$-(1-\nu)\left(\frac{\partial^2 D}{\partial x^2}\frac{\partial^2 w}{\partial y^2} - 2\frac{\partial^2 D}{\partial x \partial y}\frac{\partial^2 w}{\partial x \partial y} + \frac{\partial^2 D}{\partial y^2}\frac{\partial^2 w}{\partial x^2}\right) = g(\mathbf{x}) \quad \text{in } \Omega \tag{45}$$

Then eqn (29), from which the fictitious load is established, becomes

$$[K]\{q\} = \{g\} \tag{46}$$

where $[K]$ is given by eqn (33a).

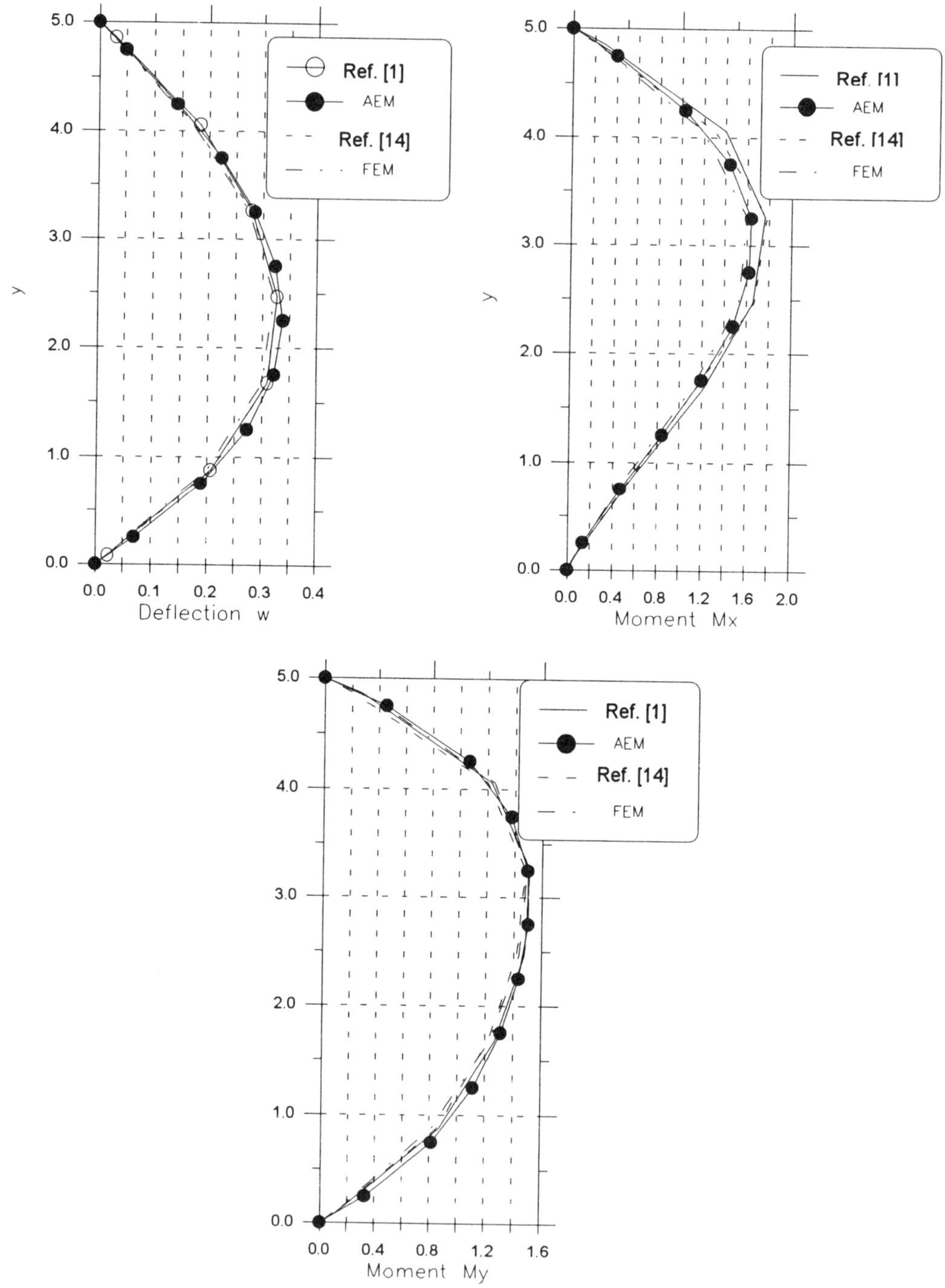

Figure 3. Deflections $w/(4g_o\alpha^4/\pi^5 D_o)$ and bending moments $M_x/(4g_o\alpha^2/\pi^3)$, $M_y/(4g_o\alpha^2/\pi^3)$ along the line $x=0$ in the plate of Example 1.

On the basis of eqn (46) numerical results for three example problems are presented. The results have been obtained using 60 constants boundary element and 100 square domain cells.

Example 1

A square simply supported plate with side length $\alpha = 5$ has been studied. The flexural rigidity is a linear function of y expressed in the form

$$D = D_o + D_1 y \quad 0 \leq y \leq 5 \tag{47}$$

The load density is chosen proportional to the flexural rigidity.

$$g = g_o(1 + \frac{D_1}{D_o} y) \tag{48}$$

The numerical results have been obtained for $D_1 = 7D_o / a$ and $v = 0.16$. They are shown in graphical form in Figure 3 as compared with those obtained from an analytical solution using two term series[1] as well from a FEM solution using 100 square elements and a BEM solution[14]. The results are in good agreement.

Table 2. Dimensionless deflections $\overline{w} = w / (ga^4 / D_o)$ and moments $\overline{Mx} / ga^2$, $\overline{Mxy} / ga^2$ along the line $x = y$ in a square simply supported plate with linear varying thickness . D_o is the stiffness at the center of the plate

	$\overline{w}(\times 10^3)$		$\overline{Mx} = \overline{My}(\times 10^2)$		$\overline{Mxy}(\times 10^2)$	
x/	AEM (Ref. 4)	FEM	AEM (Ref. 4)	FEM	AEM (Ref. 4)	FEM
−0.45	0.18	0.17	0.24	0.22	1.83	1.84
−0.35	1.31	1.29	1.32	1.30	1.53	1.51
−0.25	2.76	2.73	2.55	2.52	0.89	0.89
−0.15	3.89	3.81	3.55	3.51	0.32	0.32
−0.05	4.22	4.19	4.10	4.06	0.07	0.06
0.05	3.89	3.83	4.10	4.07	0.33	0.32
0.15	2.94	2.92	3.54	3.53	1.19	1.17
0.25	1.77	1.77	2.55	2.55	2.61	2.56
0.35	0.72	0.71	1.32	1.33	4.37	4.32
0.45	0.08	0.08	0.24	0.24	6.08	6.08

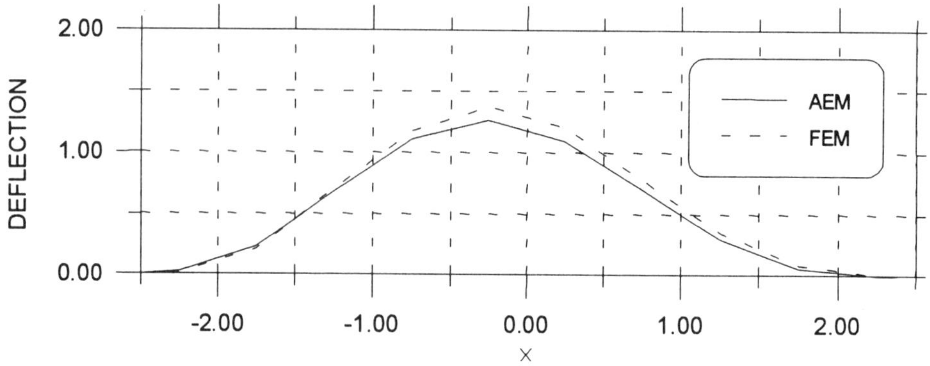

Figure 4. Deflections $\overline{w} = w / (g\alpha^4 / D_o)$ along the line $x = y$ in the plate of Example 3 (Ref. 4).

Example 2

A square simply supported plate with side length $\alpha = 5m$, elastic constants $\nu = 0.16, E = 2.1\times10^7 \text{kN/m}^2$, linearly varying thickness $h = 0.01(x+y)+0.15$ $(-2.5 \le x, y \le 1.5)$, subjected to uniform load $g = 10\text{kN/m}^2$ has been analyzed. Numerical results for the deflection and moments are shown in Table 2 as compared with those obtained from a FEM solution with 100 squares elements. The results are in good agreement.

Example 3

A square clamped plate with the same data as in Example 2 has been analyzed. Numerical results for the deflection along the line $x = y$ are shown in Figure 4 as compared with those obtained from a FEM solution with 100 square elements. The results are again in good agreement.

5.2 Dynamic analysis of plates with variable thickness without inplane forces

Free and forced vibrations of plates with variable thickness have been studied using the AEM.

5.2.1 Free vibrations

In absence of inplane forces $(N_x = N_y = N_{xy} = 0)$ and external transverse loading $(g = 0)$, eqn (1) becomes

$$D\nabla^4 w + 2\frac{\partial D}{\partial x}\frac{\partial}{\partial x}\nabla^2 w + 2\frac{\partial D}{\partial y}\frac{\partial}{\partial y}\nabla^2 w + \nabla^2 D\nabla^2 w$$
$$-(1-\nu)\left(\frac{\partial^2 D}{\partial x^2}\frac{\partial^2 w}{\partial y^2} - 2\frac{\partial^2 D}{\partial x\partial y}\frac{\partial^2 w}{\partial x\partial y} + \frac{\partial^2 D}{\partial y^2}\frac{\partial^2 w}{\partial x^2}\right) \tag{49}$$
$$+c\frac{\partial w}{\partial t} + \rho h\frac{\partial^2 w}{\partial t^2} = 0 \qquad \text{in } \Omega$$

the boundary conditions (2a,b) are also homogeneous, that is

$$\alpha_1 w + \alpha_2 V(w) = 0 \qquad \beta_1\frac{\partial w}{\partial n} + \beta_2 M(w) = 0 \quad on\ \Gamma \tag{50a,b}$$

The matrix $[B]$ vanishes and the equation of motion (29) becomes

$$[M]\{\ddot{q}\} + [C]\{\dot{q}\} + [K]\{q\} = \{0\} \tag{51}$$

in which $[M]$, $[C]$ and $[K]$ are defined by eqns (30), (31) and (33a).

The eigenfrequencies and mode shapes can be obtained by setting

$$q(\mathbf{x},t) = Q(\mathbf{x})e^{-i\omega t} \tag{52}$$

Substitution of eqn (52) into eqn (51) and taking into consideration undamped vibrations we obtain the following equation

$$([K] - \omega^2[M])\{Q\} = \{0\} \tag{53}$$

from which eigenfrequencies and eigenvectors are established numerically by solving the linear generalized eigenvalue problem for fully populated non-symmetric matrices. The mode shapes of the deflection surface are computed from eqn (21) as

$$\{w\} = [G]\{Q\} \tag{54}$$

On the basis of eqn (53) three example problems have been studied. The numerical results have obtained using 60 constant boundary elements and 100 rectangular constant domain cells.

Example 4

The free undamped vibrations of a simply supported square plate with side length a, $\nu = 0.30$ and linear thickness variation have been studied. The law of thickness variation is given as

Table 3. Eigenfrequencies $\Omega = \omega a^2 \sqrt{\rho h_o / D_o}$ for simply supported square plate with linearly varying thickness $h = h_o(1 + \alpha x / a + \beta y / b)$; $\beta = 0$; $\nu = 0.3$

	$\alpha = 0$			$\alpha = 0.2$			$\alpha = 0.4$		
n	AEM Ref. 5	Ref. 15	Ref. 19	AEM Ref. 5	Ref. 15	Ref. 19	AEM Ref. 5	Ref. 15	Ref. 19
1	19.8	19.7	19.7	21.8	21.7	21.7	23.7	23.6	23.6
2	49.8	49.3	49.4	54.6	54.0	–	59.2	58.6	–
3	49.8	49.3	49.4	54.6	54.0	–	59.3	58.6	–
4	80.0	78.6	79.0	87.8	86.3	–	95.4	93.8	–
5	100.2	97.5	99.0	109.7	106.7	–	118.4	114.9	–
6	100.2	97.9	99.0	110.0	107.4	–	119.3	116.4	–
7	130.6	125.7	128.3	143.3	138.1	–	155.6	150.0	–
8	130.6	125.7	128.3	143.5	138.1	–	156.1	150.0	–

$$h = h_o(1 + \alpha\frac{x}{a} + \beta\frac{y}{a}) \qquad 0 \le x,\ y \le a \tag{55}$$

The computed first eight eigenfrequencies for $\beta = 0$, $\alpha = 0,\ 0.2,\ 0.4$ are presented in Table 3 as compared with existing results from other approximate (Leissa[19]) and numerical (Katsikadelis & Sapountzakis[15]) solutions. The results are in good agreement. Moreover, the computed first eight eigenfrequencies for various values of the parameters α and β are given in Table 4.

Example 5

The free undamped vibrations of a clamped square plate with side length a, $\nu = 0.3$ and linearly varying thickness as in Example 4 have been studied. The computed first eight eigenfrequencies for $\beta = 0$, $\alpha = 0,\ 0.2,\ 0.4$ are presented in Table 5 as compared with those obtained from other approximate (Kuttler & Sigilito[20]) and numerical (Katsikadelis & Sapountzakis[15]) solutions. The results are in good agreement. Moreover, the computed first eight eigenfrequencies for various values of the parameters α and β are given in Table 6.

Example 6

A simply supported rectangular plate $(a \times b)$ with linearly varying thickness $h = h_o(1 + \alpha x / a)$ and $\nu = 0.3$ has been studied. The dependence of the computed fundamental eigenfrequencies on the plate aspect ratio a / b is presented in Table 7 as compared with an approximate solution (Leissa[19]).

5.2.2 Forced vibrations

In absence of inplane forces $(N_x = N_y = N_{xy} = 0)$ and $g(\mathbf{x},t) \neq 0$ the equation of motion (29) becomes

$$[M]\{\ddot{q}\} + [C]\{\dot{q}\} + [K]\{q\} = \{g\} \tag{56}$$

The initial conditions for $\{q\}$ are given by eqns (34a,b).

On the basis of eqn (56) two example problems have been studied. The numerical results have been obtained using 60 constant boundary elements and 100 rectangular constant domain cells.

Table 4. Eigenfrequencies $\Omega = \omega a^2 \sqrt{\rho h_o / D_o}$ for a simply supported square plate with linearly varying thickness $h = h_o(1 + \alpha x / a + \beta y / b)$; $\nu = 0.3$ (Ref. 5)

β	$\alpha = 0.2$	$\alpha = 0.4$	$\alpha = 0.6$
	23.7	25.6	27.5
	59.4	63.9	68.4
	59.6	64.4	69.0
$\beta = 0.2$	95.7	103.3	110.8
	119.5	128.4	136.6
	119.5	128.9	138.0
	155.9	168.7	179.9
	156.6	169.4	181.9
	25.6	27.5	29.4
	63.9	68.5	73.0
	64.4	69.3	74.0
$\beta = 0.4$	103.3	111.0	118.5
	128.4	137.9	146.2
	128.9	137.9	147.0
	168.7	180.1	191.0
	169.4	182.5	195.3
	27.5	29.4	31.3
	68.4	73.0	77.3
	69.0	74.0	78.8
$\beta = 0.6$	110.8	118.5	126.0
	136.6	146.2	155.6
	138.0	147.0	155.6
	179.9	191.0	203.5
	181.9	195.3	208.3

Table 5. Dimensionless eigenfrequencies $\Omega = \omega a^2 \sqrt{\rho h_o / D_o}$ for a clamped square plate with linearly varying thickness $h = h_o(1 + \alpha x / a + \beta y / b)$; $\beta = 0$; $\nu = 0.3$.

	$a = 0$			$a = 0.2$			$a = 0.4$		
n	AEM Ref. 5	Ref. 15	Ref. 20	AEM Ref. 5	Ref. 15	Ref. 20	AEM Ref. 5	Ref. 15	Ref. 20
1	36.0	35.9	36.0	39.8	39.4	39.5	43.7	42.7	42.9
2	72.4	73.1	73.4	79.5	78.6	80.3	86.1	85.2	87.2
3	72.4	73.1	73.4	79.6	78.8	80.4	86.7	85.9	87.4
4	109.7	107.4	108.2	120.5	117.9	118.6	131.2	128.0	–
5	133.4	129.0	131.6	146.3	141.0	–	158.3	151.3	–
6	133.7	129.9	132.2	146.7	142.3	–	159.3	153.9	–
7	164.4	157.3	165.0	180.8	172.7	–	196.2	187.8	–
8	164.4	157.3	165.0	181.0	172.8	–	197.2	188.0	–

Example 7

The undamped forced vibrations of a square simply supported elastic plate with with Poisson's ratio $\nu = 0.30$, side length a and linearly varying thickness in the x-direction as in Example 6 ($\alpha = 0.4$) have been studied. Two different loadings have been considered under zero initial conditions: (i) a suddenly applied constant load g_o; (ii) an impulsive load $g = g_o(1 - t / 0.5)\ H(0.5 - t)$; $H(.)$ is the Heaviside step function. The time history of the response ratios $R(t) = w(\mathbf{x},t) / w_{st}(\mathbf{x})$, where $w_{st}(\mathbf{x})$ is the static deflection, for the two points $x / a = y / a = 0.05$ and $x / a = y / a = 0.45$, is presented in Figures 5 and 6, respectively.

The graphical coincidence of the response ratio curves in both loading cases indicates that all points of the plate vibrate approximately in the fundamental mode implying that the contribution of higher modes is negligible.

In the first loading case, the dynamic magnification factor $D = \max R(t) = 2$ verifies the correctness of the results .

Example 8

The undamped forced vibrations of a square simply supported elastic plate with Poisson's ratio $\nu = 0.16$, side length $a = 5$ and linear variation of the thickness in x- and y-direction, under a sinusoidal impulsive load have been studied. The law of thickness variation is given by

Table 6. Eigenfrequencies $\Omega = \omega a^2 \sqrt{\rho h_o / D_o}$ for a clamped square plate with linearly varying thickness $h = h_o(1 + \alpha x / a + \beta y / b)$; $\nu = 0.3$ (Ref. 5)

β	$\alpha = 0.2$	$\alpha = 0.4$	$\alpha = 0.6$
	43.5	47.4	51.5
	86.4	93.0	99.4
	86.9	94.1	101.1
$\beta = 0.2$	131.4	142.1	152.8
	159.3	171.5	183.0
	159.6	172.3	184.6
	196.7	212.1	227.0
	197.7	214.1	230.1
	47.4	51.3	55.4
	93.0	99.6	105.9
	94.1	101.4	108.6
$\beta = 0.4$	142.1	152.9	163.6
	171.5	184.1	195.9
	172.3	184.7	197.0
	212.1	227.3	242.1
	214.1	230.7	247.1
	51.5	55.4	59.4
	99.4	105.9	112.2
	101.1	108.6	115.9
$\beta = 0.6$	152.8	163.6	174.3
	183.0	195.9	208.3
	184.6	197.0	208.9
	227.0	242.1	256.9
	230.1	247.1	263.8

$$h = 0.01(x + y) + 0.15 \quad 0 \le x, y \le 5 \tag{57}$$

while the impulsive load is given by

$$g = g_o \sin(20t) H(1.-t) \tag{58}$$

The time history of the response ratios $R(t) = w(\mathbf{x},t) / w_{st}(\mathbf{x})$ for two points, $x = y = 0.25$ and $x = y = 2.25$, is presented in Figure 7. The response ratio curves coincide graphically. This indicates that all points of the plate vibrate approximately in the fundamental mode as in example 7. In Figure 8 the time history of the deflection w

Table 7. Fundamental dimensionless eigenfrequency $\Omega = \omega a^2 \sqrt{\rho h_o / D_o}$ for linearly tapered rectangular $(a \times b)$ simply supported plates; $\nu = 0.3$, $h = h_o(1 + \alpha x / a)$

a/b	$\alpha = 0.2$		$\alpha = 0.4$		$\alpha = 0.6$	
	AEM Ref.6	Ref. 19	AEM Ref.6	Ref. 19	AEM Ref.6	Ref. 19
0.25	11.54	11.51	12.52	12.48	13.46	13.42
0.50	13.59	13.54	14.75	14.73	15.86	15.87
1.00	21.76	21.69	23.66	23.60	25.51	25.50
1.50	35.35	35.22	38.40	38.29	41.36	41.30
2.00	54.33	54.16	58.92	58.76	63.32	63.21

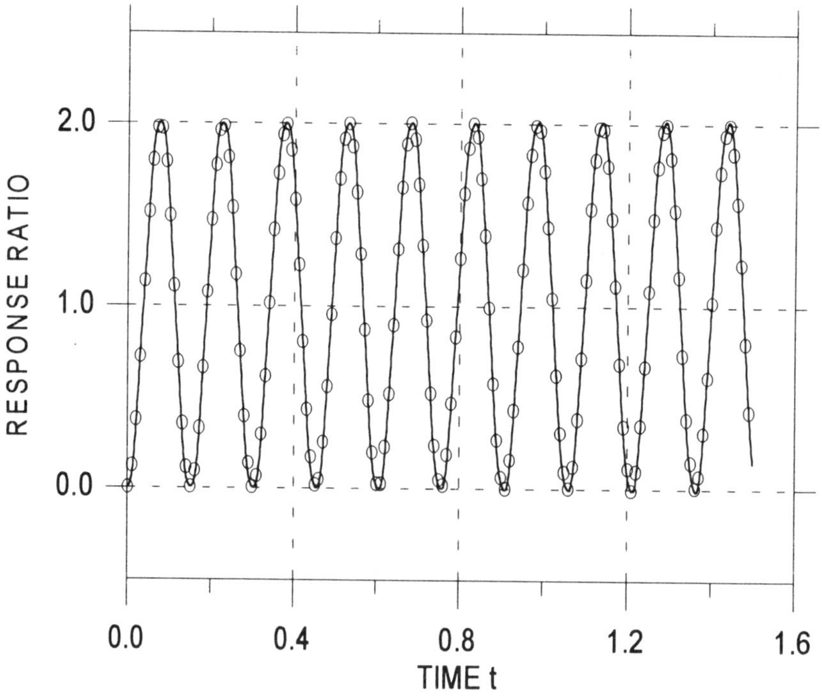

Figure 5. Time history of the response ratios of two points in a square simply supported plate with variable thickness $h = h_o(1 + 0.4x / a)$ $(\nu = 0.30)$ subjected to a suddenly constant load $g = g_o$ (Ref. 6).

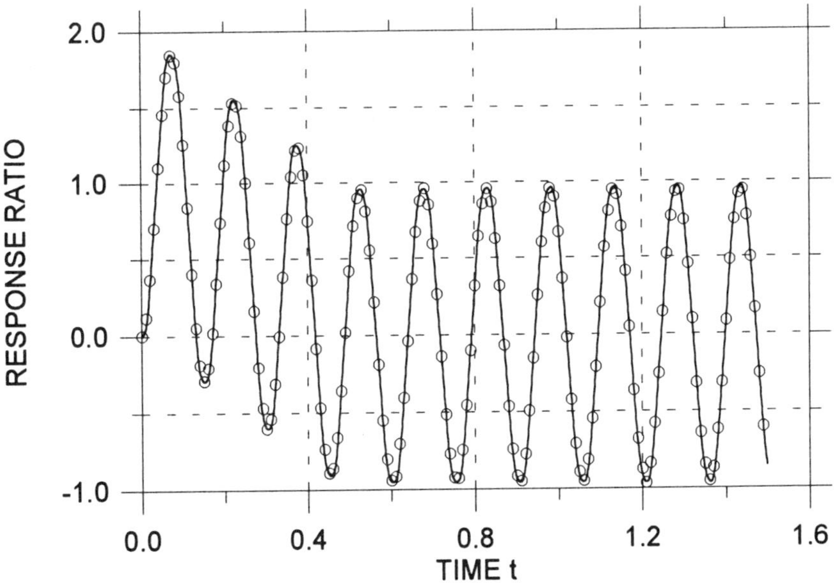

Figure 6. Time history of the response ratios of two points in a square simply supported plate with variable thickness, $h = h_o(1 = 0.4x / a)$ $(\nu = 0.30)$ subjected to a triangular impulse $g = g_o(1 - t / 0.5)\ H(0.5 - t)$ (Ref. 6).

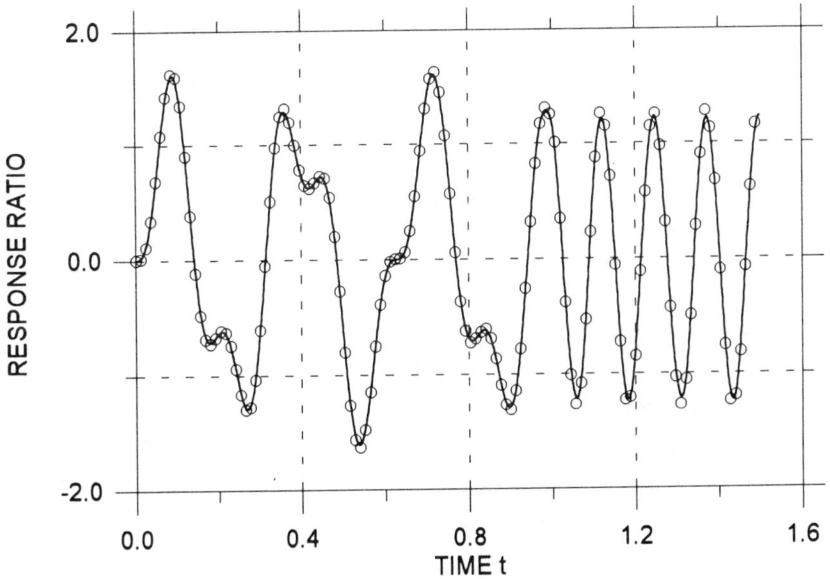

Figure 7. Time history of the response ratios of two points in a square simply supported plate with variable thickness, $h = 0.01(x + y) + 0.15$ subjected to sinusoidal impulse $g = g_o \sin(20t)H(1.-t)$ (Ref. 6).

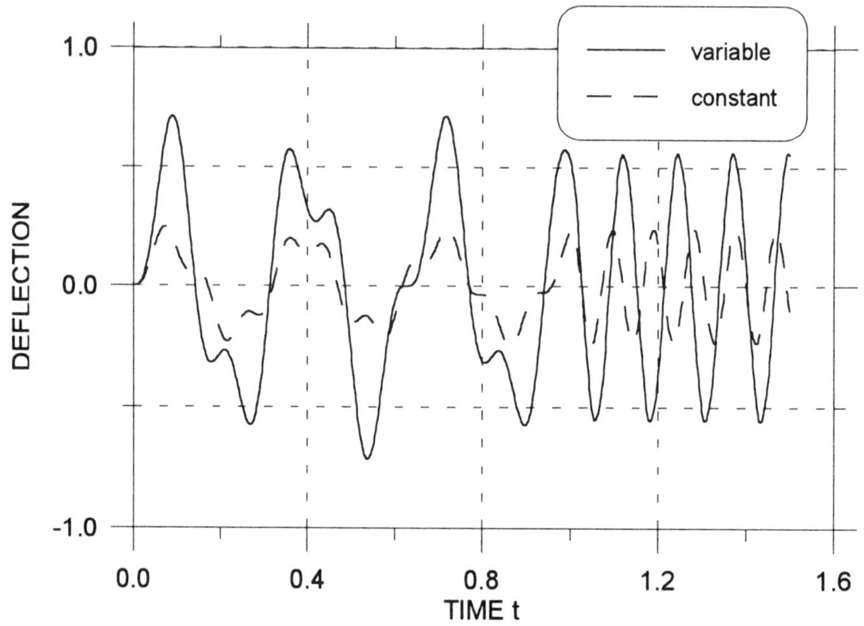

Figure 8. Time history of the deflection at point $x = y = 2.25$ in a square simply supported plate with variable thickness $h = 0.01(x + y) + 0.15$ and equivalent constant thickness $h_c = 0.20$ subjected to sinusoidal impulse $g = g_o \sin(20t)H(1.-t)$ (Ref. 6).

at point $x = y = 2.25$ of the plate with variable thickness is presented and compared with the deflection w_c of the same point if the material of the plate is uniformly distributed. This figure demonstrates the drastic difference of the response between constant and variable plate thickness.

5.3 Buckling of plates with variable thickness

Linear buckling of plates with variable thickness is considered. The differential equation for buckling is derived from eqn (1) by taking $g = 0$ and considering homogeneous boundary conditions. Thus, if the inplane forces are expressed in terms of one parameter, say k, the buckling problem is described by the following equations

$$\begin{aligned} &D\nabla^4 w + 2\frac{\partial D}{\partial x}\frac{\partial}{\partial x}\nabla^2 w + 2\frac{\partial D}{\partial y}\frac{\partial}{\partial y}\nabla^2 w + \nabla^2 D\nabla^2 w \\ &-(1-\nu)\left(\frac{\partial^2 D}{\partial x^2}\frac{\partial^2 w}{\partial y^2} - 2\frac{\partial^2 D}{\partial x\partial y}\frac{\partial^2 w}{\partial x\partial y} + \frac{\partial^2 D}{\partial y^2}\frac{\partial^2 w}{\partial x^2}\right) \\ -k&\left(N_x\frac{\partial^2 w}{\partial x^2} + 2N_{xy}\frac{\partial^2 w}{\partial x\partial y} + N_y\frac{\partial^2 w}{\partial y^2}\right) + c\frac{\partial w}{\partial t} + \rho h\frac{\partial^2 w}{\partial t^2} = 0 \quad \text{in } \Omega \end{aligned} \tag{59}$$

$$\alpha_1 w + \alpha_2 V(w) = 0 \qquad \beta_1\frac{\partial w}{\partial n} + \beta_2 M(w) = 0 \qquad \text{on } \Gamma \tag{60}$$

Eqn (29) becomes

$$[M]\{\ddot{q}\} + [C]\{\dot{q}\} + ([K] - k[B])\{q\} = \{0\} \tag{61}$$

By considering free undamped vibrations and using eqn (52) we obtain

$$(-\omega^2[M] + [K] - k[B])\{Q\} = \{0\} \tag{62}$$

If the dynamic criterion is employed the buckling parameters are established as the zeros of the curves $\omega^2 = f(k)$, which expresses the dependence of the eigenfrequencies on the inplane forces[21] and results from the vanishing of the determinant of eqn (62), that is

$$\det(-\omega^2[M] + [K] - k[B]) = 0 \tag{63}$$

If the static criterion is employed, the inertia term is neglected and eqn (62) becomes

$$([K] - k[B])\{Q\} = \{0\} \tag{64}$$

The solution of the eigenvalue problem (64) yields the buckling parameters k and the corresponding eigenvectors $\{Q\}$. The mode shapes of the deflection surface are established using eqn (21). Thus, we have

$$\{w\} = [G]\{Q\} \tag{65}$$

On the basis of eqn (64) the buckling parameters for certain rectangular plates with linear or exponential varying thickness have been computed. A few of these values have been also checked using eqn (63).

Example 9

A rectangular simply supported $(a \times b)$ with linearly varying thickness

$$h = h_o(1 + \alpha \frac{x}{a}), \quad \alpha = \frac{h_a}{h_o} - 1 \tag{66}$$

has been analyzed; h_o and h_a are the plate thickness at $x = 0$ and $x = a$.

The plate is uniformly compressed in the x direction, that is $N_x \neq 0$, $N_y = N_{xy} = 0$. The computed buckling parameters $k_M = N_x b^2 / \pi^2 D_M$, where $D_M = Eh_M^3 / 12(1 - \nu^2)$, $h_M = (h_o + h_a)/2$, are presented in Table 8 as compared with those given by Wittrik and Ellen[23] and Harik et al.[24]. They are in very good agreement, except for the case $a/b = 0.5$ in which the results given by Harik et al. deviate from those of AEM and Wittrik and Ellen.

Table 8. Values for buckling parameter k_M for a simply supported plate with linear variation in thickness and subjected to constant uniform force in the x-direction $(\nu = 1/3)$. Upper row: Ref. 24; middle row: Ref. 23; lower row: AEM (Ref. 7)

			h_a / h_o		
$\frac{a}{b}$	1.125	1.250	1.500	1.750	2.00
	6.199	6.246	6.416	6.359	6.405
0.5	6.225	6.162	5.965	5.722	5.463
	6.270	6.197	5.961	5.666	5.350
	3.970	3.878	3.720	3.560	3.317
1.00	3.966	3.882	3.638	3.364	3.100
	3.994	3.902	3.631	3.322	3.019
	4.144	3.862	3.373	2.967	2.642
1.50	4.146	3.861	3.339	2.908	2.557
	4.203	3.899	3.340	2.872	2.484
	3.857	3.593	3.062	2.624	2.273
2.00	3.859	3.591	3.048	2.597	2.236
	3.830	3.638	3.057	2.567	2.168

Table 9. Values for buckling stress parameter k_M for a simply supported plate with exponential variation in thickness and subjected to constant uniform force in the x-direction ($\nu = 1/3$). Upper row: Ref. 24; middle row: Ref. 23;. lower row: AEM (Ref. 7)

$\frac{a}{b}$	h_a / h_0: 1.125	1.250	1.500	1.750	2.00
	6.450	7.339	6.670	6.995	7.455
0.5	6.232	6.187	6.049	5.885	5.722
	6.280	6.224	6.050	5.838	5.615
	4.073	4.000	3.950	3.843	3.686
1.00	3.972	3.901	3.700	3.476	3.262
	4.000	3.922	3.698	3.451	3.212
	4.184	3.949	3.541	3.206	2.933
1.50	4.153	3.884	3.407	3.024	2.715
	4.210	3.924	3.417	3.008	2.678
	3.897	3.674	3.224	2.850	2.552
2.00	3.866	(3.616)	3.122	2.720	2.401
	3.925	3.644	3.141	2.712	2.370

Example 10

A rectangular simply supported plate with exponentially varying thickness

$$h = h_o e^{\frac{x}{a} \ell n \frac{h_\alpha}{h_o}} \tag{67}$$

has been analyzed. The plate is uniformly compressed in the x direction, as in Example 9. The buckling parameters k_M are presented in Table 9 as compared with those given by Witrik and Ellen[23] and Harik *et al.*[24] They are also in very good agreement, with the exception of the results of Harik *et al.* for $a/b = 0.5$.

6 Concluding remarks

This chapter has presented the Analog Equation method, a BEM-based method, as it is applied to analyze plates with variable thickness under static and dynamic loads including inplane forces. The main features of the method are:

(a) Plates with arbitrary shape and thickness variation subjected to any type of boundary conditions can be treated

(b) The general dynamic problem for plates with variable thickness under inplane forces is converted to a quasi-static bending problem for a plate with constant thickness without inplane forces which is solved by BEM using the static fundamental solution.
(c) The solution procedure is developed for the general dynamic problem including inplane forces. The static problems result as special cases by neglecting the inertia and damping forces.
(d) The solution procedure incorporates advantages of the known numerical methods: the FDM (collocation of the differential equation at the nodal points inside the domain and substitution of the derivatives by the values of a field variable, the fictitious load); the BEM (evaluation of the deflection and its derivatives by means of integral representations); the FEM (domain discretization in cells to approximate the domain integrals).
(e) Accurate results are obtained using a relatively small number of nodal points.
(f) Plates resting on elastic foundation can be easily treated if the appropriate subgrade reaction terms are included in eqn (1) [e.g. kw for Winkler-type foundation, $kw - G\nabla^2 w$ for Pasternak-type foundation[22]].
(g) The method can be developed as a pure BEM if the domain integrals containing the fictitious load are converted to boundary ones using the dual reciprocity method.

References

[1] Timoshenko, S.P. & Woinowsky-Krieger, S. *Theory of Plates and Shells*. McGraw-Hill, New York, 1959.

[2] Bares, R. *Tables for the Analysis of Plates, Slabs and Diaphragms*. Bauverlag GmbH, Berlin, 1979.

[3] Bulson, P.S. *The Stability of Flat Plates*. Elsevier, New York, 1959.

[4] Nerantzaki, M.S. Analysis of plates with variable thickness by the analog equation method, *Boundary Elements XVII*, eds C.A. Brebbia, S. Kim, T.A. Osswald & H. Power, Computational Mechanics Publications, Southampton, pp. 175–184, 1995.

[5] Nerantzaki, M.S. & Katsikadelis, J.T. Vibrations of plates with variable thickness: an analog equation solution, *Structural Dynamics, Eurodyn '96*, eds G. Augusti, C. Borri & P. Spinelli, Balkema Publishers, Rotterdam, **2,** 711–717, 1996.

[6] Nerantzaki, M.S. & Katsikadelis, J.T. An analog equation solution to dynamic analysis of plates with variable thickness, *Engineering Analysis with Boundary Elements*, **17**, 145–152, 1996.

[7] Nerantzaki, M.S. & Katsikadelis, J.T. Buckling of plates with variable thickness – an analog equation solution, *Engineering Analysis with Boundary Elements*, **18**, 149–154, 1996.

[8] Jaswon, M.A. & Maiti, M. An integral equation formulation of plate bending problems, *Journal of Engineering Mathematics*, **11**, 83–93, 1968.

[9] Bezine, G.P. Boundary integral formulation for plate flexure with arbitrary conditions, *Mechanics Research Communications*, **5**, 197–206, 1978.

[10] Stern, M. A general boundary integral formulation for the numerical solution of plate bending problems, *International Journal of Solids and Structures*, **15**, 769–782, 1979.

[11] Hartmann, F. & Zotemantel, R. The direct boundary element method in plate bending, *International Journal for Numerical Methods in Engineering*, **23**, 2049–2069, 1986.

[12] Katsikadelis, J.T. & Armenakas, A.E. A new boundary equation solution to the plate problem, *ASME, Journal of Applied Mechanics*, **56**, 364–374, 1989.

[13] Paris, F. & de Leon, S. Thin plates by the boundary element method by means of two poissons equations, *Engineering Analysis with Boundary Elements*, **17**, 111–122, 1996.

[14] Sapountzakis, E.J. and Katsikadelis, J.T. Boundary element solution for plates of variable thickness, *ASCE Journal of Engineering Mechanics*, **117**, 1241–1256, 1991.

[15] Katsikadelis, J.T. & Sapountzakis, E.J. A BEM solution to dynamic analysis of plates with variable thickness, *Computational Mechanics*, **7**, 369–379, 1991.

[16] Katsikadelis, J.T. The analog equation method – A powerful BEM-based solution technique for solving linear and nonlinear engineering problems, *Boundary Element Method* XVI, eds Brebbia, C.A., Computational Mechanics Publications, Southampton, pp.167–182, 1994.

[17] Katsikadelis, J.T. & Kandilas, C.B. Plane stress analysis of thin plates with variable thickness by the analog equation method, *Proc. of the 4th Nat. Congress on Mechanics*, eds P.S. Theocaris & E.E. Gdoutos, Demokritus University, Xanthi, pp.562–573, 1995.

[18] Katsikadelis, J.T. *The analysis of plates on elastic foundation by the boundary integral equation method*, Dissertation for the degree of Doctor of Philosophy, Polytechnic University of New York, 1982.

[19] Leissa, A.W. *Vibration of Plates* NASA SP–160, 1969.

[20] Kuttler, J.R. & Sigilito, V.G. Vibrational frequencies of clamped plates of variable thickness, *Journal of Sound and Vibration*, **862**, 181–189, 1983.

[21] Katsikadelis, J.T. & Sapountzakis, E.J. A BEM solution to dynamic analysis of plates subjected to inplane forces, *Proc. of the 1992 ESDA Conference, ASME PD–Vol.47–5, Structural Dynamics and Vibrations*, eds. Ertas A, Ovunc B., Konuk I., Istanbul, pp.41–48.

[22] Nerantzaki, M.S. Solving plate bending problems by the analog equation method, *Boundary Element Method XVI*, ed.C.A. Brebbia, Computational Mechanics Publications, Southampton, pp.283–291, 1994.

[23] Wittrik, W.H. & Ellen, C.H. Buckling of tapered rectangular plates in compression, *Aeronautical Quarterley*, **13**, 308–326, 1962.

[24] Harik, L.E., Liu, X. & Ekambaram, R. Elastic stability of plates with varying rigidities, *Computers and Structures*, **38**, 161–168, 1991.

Appendix

In this appendix the derivatives of the kernels functions $\Lambda_i(r)$ appearing in eqn (14) are given:

$$\nabla^2\Lambda_1 = \nabla^2\Lambda_2 = 0\ , \quad \nabla^2\Lambda_3 = \Lambda_1, \quad \nabla^2\Lambda_4 = \Lambda_2$$

$$\frac{\partial^2\Lambda_1}{\partial x^2} = -\frac{2}{r^3}\cos(2\omega - \varphi)\ , \qquad \frac{\partial^2\Lambda_1}{\partial y^2} = \frac{2}{r^3}\cos(2\omega - \varphi)$$

$$\frac{\partial^2\Lambda_1}{\partial x\partial y} = -\frac{2}{r^3}\sin(2\omega - \varphi)$$

$$\frac{\partial^2 \Lambda_2}{\partial x^2} = \frac{1}{r^2}(\sin^2 \omega - \cos^2 \omega) \ , \qquad \frac{\partial^2 \Lambda_2}{\partial y^2} = \frac{1}{r^2}(\cos^2 \omega - \sin^2 \omega)$$

$$\frac{\partial^2 \Lambda_2}{\partial x \partial y} = -\frac{\sin 2\omega}{r^2}$$

$$\frac{\partial^2 \Lambda_3}{\partial x^2} = \frac{\sin\varphi \cos\omega \sin\omega}{r} - \frac{\cos\varphi}{2r} \ , \qquad \frac{\partial^2 \Lambda_3}{\partial y^2} = -\frac{\sin\varphi \cos\omega \sin\omega}{r} - \frac{\cos\varphi}{2r}$$

$$\frac{\partial^2 \Lambda_3}{\partial x \partial y} = -\frac{\sin\varphi \cos 2\omega}{2r}$$

$$\frac{\partial^2 \Lambda_4}{\partial x^2} = \frac{1}{2}\ell nr + \frac{1}{4} + \frac{1}{2}\cos^2 \omega \ , \qquad \frac{\partial^2 \Lambda_4}{\partial y^2} = \frac{1}{2}\ell nr + \frac{1}{4} + \frac{1}{2}\sin^2 \omega$$

$$\frac{\partial^2 \Lambda_4}{\partial x \partial y} = \frac{1}{4}\sin 2\omega$$

$$\frac{\partial \Lambda_1}{\partial x} = -\frac{\cos(\omega - \varphi)}{r^2} \ , \qquad \frac{\partial \Lambda_1}{\partial y} = -\frac{\sin(\omega - \varphi)}{r^2}$$

$$\frac{\partial \Lambda_2}{\partial x} = -\frac{\cos\omega}{r} \ , \qquad \frac{\partial \Lambda_2}{\partial y} = -\frac{\sin\omega}{r}$$

where ω is the angle between the x axis and the vector r and φ is the angle between the outward normal n and the vector r.

Chapter 10

Stability

S. Syngellakis
Department of Mechanical Engineering, University of Southampton, Southampton, SO17 1BJ, UK
Email: ss@soton.ac.uk

Abstract

The boundary element method (BEM) has been applied to a wide range of linear and non-linear problems in the context of the general plate stability theory. A formulation for the bi-harmonic equation can yield the pre-buckling membrane state of stress for non-uniform distributions of edge loads. General procedures have been developed leading to a standard eigenvalue problem for the critical load. Gains in both efficiency and accuracy have been achieved by introducing and assessing the performance of both continuous and discontinuous models of various orders of approximation for boundary as well as domain unknowns. Dual reciprocity, whereby irreducible domain integrals are transformed into boundary integrals, has been applied by modelling the buckling mode either as a set of discrete nodal values to which curvatures are related through a Fourier analysis or as summation of continuous functions satisfying the plate support conditions. BEM predictions of critical loads for a large number of variously shaped, loaded and supported plates have been compared with published, exact or approximate, analytical and experimental results. This has led to optimum choice of various modelling parameters and established the reliability of the analyses. Finally, the non-linear post-critical plate behaviour has also been modelled by BEM and the resulting formulation validated through the simulation of well-documented experiments. This type of analysis couples an incremental solution for the membrane forces with one for the non-linear domain deflection initiated by imperfections.

1 Introduction

Thin plates resist very efficiently forces acting in their middle plane but, due to their slenderness, loss of stability at loads below their ultimate strength is a possibility that must be envisaged in engineering practice. Since the first experimental observation of the plate buckling phenomenon almost 150 years ago, the problem has been extensively investigated both analytically and experimentally.[1]

Closed form solutions for the critical load have been derived in the simplest cases of elastic plates with regular shapes, certain combinations of support conditions and uniform loading.[2–4] Rigorous analytical methods combined with computational techniques yielded very accurate numerical answers for a wide variety of plate geometry, as well as loading, material and boundary conditions. More recently, approximate methods implemented through computer codes have provided solutions of even wider applicability.

The advantages of the boundary element algorithms in terms of computer memory requirements, speed and input data structure have been demonstrated in a wide range of applications. BEM relies on the existence of a boundary integral equation that reduces the dimensions of a problem by one thus leading to its more efficient formulation and solution. Such a genuine boundary element formulation for the plate buckling problem has only been attempted in the very special cases of plates subjected to uniform uni-axial or bi-axial compression.[5,6] The obstacle to a general BEM solution is the unavailability of a fundamental solution valid for any membrane stress distribution. The alternative approach adopted by several investigators was the use of the fundamental solutions of the plate bending problem. The resulting integral equations contain an irreducible domain integral arising from in-plane loading and depending on the generated non-uniform membrane stress distribution as well as the unknown curvatures. According to one approach, the curvatures are modelled as additional nodal unknowns in a discretized domain.[7–9] This procedure has been extended to the stability analysis of orthotropic plates.[10] In all such reported applications, constant boundary elements were used and results were obtained for uniformly loaded, simply-supported or clamped rectangular plates.

An alternative formulation is based on a transformation of the integral equation of the problem so that the deflection replaces the curvatures in the domain integral.[11,12] Thus no additional integral equations are required for the generation of a consistent equivalent system of linear equations. The deflection is the only domain unknown which needs to be modelled over two-dimensional cells. This approach was extended and refined into a fully developed and computer implemented boundary element solution of the general problem.[13] Its accuracy was improved and its scope widened by introducing general boundary and domain discretization schemes, using higher-order continuous and discontinuous interpolation models over boundary elements and domain cells as well as dealing with any combination of edge conditions focusing, in particular, to the treatment of singular integrals along free edges. Non-uniform pre-buckling membrane states were admitted as input loading to the computer program by a parallel development of a similar boundary element formulation for the bi-harmonic equation governing the stress function from which the plane stress distribution due to any edge loading can be determined.

The basic advantage of BEM is to a certain extent compromised by the presence of irreducible domain integrals in the integral equation governing the plate buckling problem. For this reason, considerable research effort has been devoted to transforming such integrals into boundary ones. Dual reciprocity as a means of achieving this transformation was first introduced to a boundary element formulation by Brebbia and Nardini.[14] A version of the technique based on the Fourier series approximation of the known forcing function[15] can be applied to a wide range of problems and it has been extended to plate buckling.[16] The purpose of converting the integration from a domain to a boundary one is to avoid the domain discretization and thus achieve a more efficient numerical procedure for the evaluation of the critical loads.

All developed formulations can be reduced to a classical eigenvalue problem for the determination of the buckling load factor by easily accessible solution routines. The method has undergone extensive validation through its application to a wide variety of plate geometry as well as loading and support conditions and comparison of its predictions with exact or other approximate analytical or experimental results.

The extension of BEM to non-linear problems has not been achieved without penalties. As pointed out above, plate buckling formulations require domain discretization and modelling which reduce, to a certain extent, the efficiency of the method. A post-buckling analysis imposes similar requirements in addition to numerical complexities arising from non-linearity and membrane-flexural deformation coupling. Incremental and iterative BEM procedures have been developed for the prediction of the non-linear plate response to in-plane edge loading. Such behaviour can either by initiated by imperfections[17,18] or predicted as bifurcation path from critical equilibrium.[19,20] The deflection can be the only domain unknown requiring modelling and a domain discretization scheme. Higher-order models for the deflection lead to the elimination of the domain curvatures from the discrete system of equations. This increases the efficiency of the method by reducing the overall number of unknowns without loss of accuracy.

2 Plate stability theory

Plates with initial imperfections loaded in their plane, undergo some deflection before the theoretical critical buckling load is reached. In fact, when these deflections are not negligible compared to the thickness of the plate, the strains in the middle plane of the plate cannot be ignored in the analysis. This coupling between bending and membrane action is modelled by the well known von Kármán equations:[2]

$$\nabla^4 F = -\frac{1}{2} E[K(\hat{W},\hat{W}) - K(W^i,W^i)] \tag{1}$$

$$D\nabla^4 W = N_{\alpha\beta}\hat{W}_{,\alpha\beta} = hK(F,\hat{W}) \tag{2}$$

having defined the bilinear form

$$K(f,g) = (\nabla^2 f)(\nabla^2 g) - f_{,\alpha\beta}\, g_{,\alpha\beta} \tag{3}$$

and

$$\hat{W} = W + W^i. \tag{4}$$

D is the rigidity of an isotropic plate given by

$$D = \frac{Eh^3}{12(1-\nu^2)} \tag{5}$$

while W is the plate deflection, $N_{\alpha\beta}$ are the membrane forces, W^i the initial plate imperfection, h the thickness, E Young's modulus and ν Poisson's ratio of the plate material. Greek subscripts indicate mid-plane co-ordinates and a comma followed by a subscript denotes partial differentiation with respect to the respective co-ordinate. The summation convention applies over repeated indices. The membrane forces are related to the stress function F by

$$N_{\alpha\beta} = h[(\nabla^2 F)\delta_{\alpha\beta} - F_{,\alpha\beta}] \tag{6}$$

where $\delta_{\alpha\beta}$ is the Kronecker delta. Then equilibrium is identically satisfied by the membrane stresses provided that body forces are neglected. A smooth variation in thickness can be modelled as a variable rigidity $D(x_\alpha)$ leading to the addition of the terms

$$2D_{,\alpha}(\nabla^2 W)_{,\alpha} + \nu\nabla^2 D\nabla^2 W + (1-\nu)D_{,\alpha\beta}W_{,\alpha\beta} \tag{7}$$

to the left-hand side of eqn (2).

Using expressions (6) and referring to Fig. 1, showing the plate domain Ω bounded by the contour Γ, it is easily shown that, at any point $Q(\bar{x}_1, \bar{x}_2)$ along the boundary,

$$F = \int_O^Q [(x_1-\bar{x}_1)T_2 - (x_2-\bar{x}_2)T_1]\,d\Gamma \tag{8}$$

$$\frac{\partial F}{\partial n} = -s_1\int_O^Q T_1\,d\Gamma - s_2\int_O^Q T_2\,d\Gamma \tag{9}$$

where $\mathbf{T}(T_1,T_2)$ is the edge traction parallel to the middle plane and $\mathbf{s}(s_1,s_2)$ the unit tangent vector along Γ at Q. In deriving eqns (8) and (9), it has been assumed that F and both its first partial derivatives vanish at the origin O of the contour co-ordinate but this means that only constant and linear terms of F are omitted thus the stresses, given by eqn (6), are not affected. According to eqn (8), F can be physically interpreted as the resultant moment about Q of the traction acting over OP. Similarly, eqn (9) describes the normal derivative of F as the component of the resultant traction over OQ in the direction -s at Q.

In the case of perforated plates, additional compatibility and equilibrium conditions need to be imposed along the internal boundaries.[21] More specifically, the in-plane displacement **v** and rotation ω should satisfy the single-valuedness conditions

$$\oint_{\Gamma_k} d\mathbf{v} = \oint_{\Gamma_k} d\omega = 0 \tag{10}$$

where Γ_k is the boundary of the k-th hole. Also, for the stress function F to be single-valued, the traction acting on every Γ_k must constitute a self-equilibrium system, hence

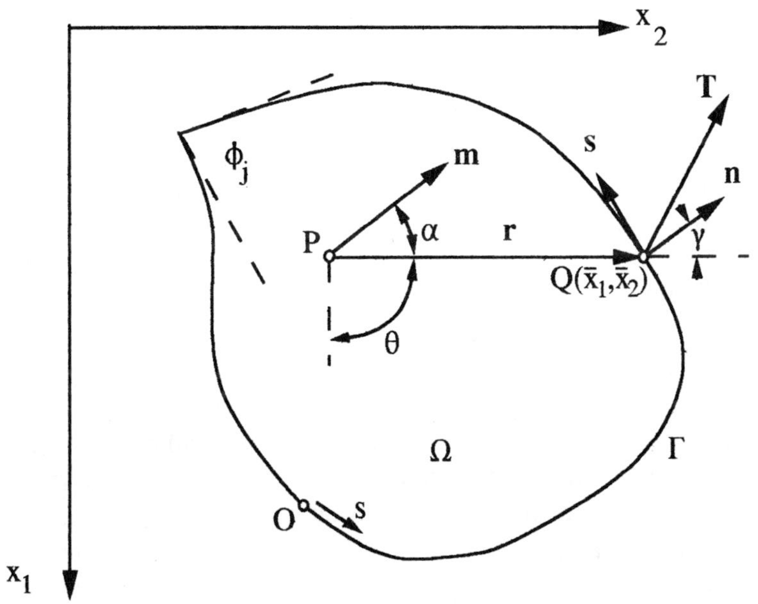

Figure 1. Plate orientation relative to a Cartesian frame of reference.

$$\oint_{\Gamma_k} \mathbf{T}\mathrm{d}\Gamma = \oint_{\Gamma_k} (x_1 T_2 - x_2 T_1)\mathrm{d}\Gamma = 0 \tag{11}$$

If condition (11) is not satisfied, then F is multi-valued. It was shown[21] that F can be separated into a single-valued part F^* and a multi-valued part $\hat{F}$ given by

$$\hat{F} = -\frac{1}{2\pi h}\sum_k (P_{1k}x_2 - P_{2k}x_1 + M_k)\tan^{-1}\frac{x_2 - x_{2k}}{x_1 - x_{1k}} \tag{12}$$

where $\mathbf{P}_k$ and M_k are, respectively, the out-of-balance resultant force and moment acting on Γ_k while $\mathbf{x}_k$ is any point in the region bounded by Γ_k. It is easily shown that $\nabla^4 \hat{F} = 0$; hence F^* replaces F as the unknown stress function in eqns (1) and (2) but needs to satisfy a much more complex set of boundary conditions than those described by eqns (8) and (9). The exact form of these conditions is discussed in detail by Cheng and Wang.[21]

2.1 Pre-buckling state

In the absence of imperfections, the unfactored edge traction $\mathbf{T}$ generates only a membrane state of stress. The problem is reduced to finding the stress function F satisfying the compatibility condition (1) simplified, in this case, to

$$\nabla^4 F = 0 \tag{13}$$

and boundary conditions (8)–(9). Alternatively, the plane stress problem of two-dimensional elasticity may be solved. The latter approach is certainly more convenient when displacement is specified on the boundary.

2.2 Critical equilibrium

It is assumed that Γ is smooth apart from a finite number N_k of corner points. The membrane state of equilibrium becomes unstable at a certain intensity of the factored edge in-plane load $\lambda\mathbf{T}$. Then a second stable equilibrium state exists, associated with a lateral deflection pattern $w(x_\alpha)$, which can be asymptotically identified through the solution of a series of linear variational principles.[22] In the context of Kirchhoff's thin plate theory, the critical load factor λ_c and the associated buckling mode $w(x_\alpha)$ can be determined from the first such variational equation

$$\Pi(w,\delta w)=D\int_{\Omega}[\nu(\nabla^2 w)(\nabla^2\delta w)+(1-\nu)w_{,\alpha\beta}\,\delta w_{,\alpha\beta}]d\Omega + \lambda_c\int_{\Omega}N_{\alpha\beta}w_{,\alpha}\,\delta w_{,\beta}\,d\Omega = 0 \quad (14)$$

where δ is the variational symbol. Integration of (14) by parts and application of Green's theorem leads to the integral equation:

$$\int_{\Omega}(D\nabla^4 w - \lambda_c N_{\alpha\beta}w_{,\alpha\beta})\delta w d\Omega + \int_{\Gamma}\left[(V+\lambda_c T_{\alpha}w_{,\alpha})\delta w - M_n\frac{\partial(\delta w)}{\partial n}\right]d\Gamma + \sum_{j=1}^{N_k}C_j\delta w_j = 0 \quad (15)$$

where the shear force V and bending moment M_n along the boundary are given by

$$V(w) = -D\,\frac{\partial}{\partial n}\left[\frac{\partial^2 w}{\partial n^2}+(2-\nu)\frac{\partial^2 w}{\partial s^2}\right] \quad (16)$$

$$M_n(w) = -D\left(\frac{\partial^2 w}{\partial n^2}+\nu\frac{\partial^2 w}{\partial s^2}\right) \quad (17)$$

and force C_j is equal to discontinuity jumps of the twisting moment

$$M_{ns}(w) = -D(1-\nu)\frac{\partial^2 w}{\partial n \partial s} \quad (18)$$

at corner j.

The condition that eqn (15) be satisfied for an arbitrary δw yields the field equation

$$D\nabla^4 w - \lambda_c N_{\alpha\beta}w_{,\alpha\beta} = 0 \quad (19)$$

over the domain Ω and the boundary conditions,

$$\text{either} \quad M_n=0 \quad \text{or} \quad \frac{\partial w}{\partial n} = \bar{\theta}(s) \quad (20)$$

$$\text{either} \quad V+\lambda_c T_{\alpha}w_{,\alpha}=0 \quad \text{or} \quad w = \bar{w}(s) \quad (21)$$

on Γ and

$$\text{either} \quad C(s_j)=0 \quad \text{or} \quad w_j = \bar{w}(s_j);\ j=1,\ldots,N_k \quad (22)$$

at the corners, where $\overline{\theta}(s)$, $\overline{w}(s)$ are, respectively, prescribed values of the slope and deflection along the whole or part of the boundary. The boundary element method is applied to the numerical solution of the boundary value problem defined by eqn (19) and conditions (20–(22).

3 Integral equations

3.1 Pre-buckling stresses

Introducing the stress function together with a fundamental solution u_i of eqn (13) into the familiar Rayleigh–Green identity for the biharmonic operator leads to

$$kf_i(P) + I_b^f(u_i,F) = 0; \; i=1,2 \tag{23}$$

with the definitions:

$$f_1 = F, \qquad f_2 = \frac{\partial F}{\partial n} \tag{24}$$

and

$$I_b^f(u_i,F) = \int_\Gamma \left[\frac{\partial \nabla^2 F}{\partial n} u_i - \nabla^2 F \frac{\partial u_i}{\partial n} + \frac{\partial F}{\partial n} \nabla^2 u_i - F \frac{\partial \nabla^2 u_i}{\partial n} \right] d\Gamma \tag{25}$$

The coefficient k is equal to unity if P is a domain point but is given by

$$k = \frac{\varphi}{2\pi} \tag{26}$$

if P is on Γ, where φ is the angle between the tangents to Γ at P. It is obvious that $k = 0.5$ along a smooth boundary. Since the Laplacian of F and its normal derivative are both unknown along the boundary, two fundamental solutions of eqn (13) are required in this boundary element formulation. These solutions are:

$$u_1(P,Q) = \frac{r^2}{8\pi} \ln r, \tag{27}$$

$$u_2(P,Q) = -\frac{r}{8\pi}(2\ln r + 1)\cos\alpha \tag{28}$$

and can be associated with stress fields due to body forces having point charge and dipole potentials, respectively, where r is the distance between the source point P and the field point Q and α the angle between the dipole orientation and line PQ. Explicit expressions for the kernels of integral (25) as well as of other, subsequently derived, boundary or domain integrals are given in Appendix A.

A boundary element solution of integral equations (23) can thus be generated based on modelling $\nabla^2 F$ and its normal derivative along the plate boundary. Once the nodal boundary values of these two variables have been determined, the second-order partial derivatives of the stress function F at any internal point can be evaluated, through numerical integration, using

$$F_{,\alpha\beta} = I_b^f(F, u_{i,\alpha\beta}) \tag{29}$$

and then substituted into eqn (6) to yield the membrane forces at any point within the plate domain.

The membrane forces can, of course, be found from the established BEM formulation of the plane stress problem in terms of the in-plane displacements. The boundary integral equation for this problem is

$$kv_\alpha + \int_\Gamma n_\gamma N_{\gamma\beta}(u_{\alpha\beta}) v_\beta d\Gamma = \int_\Gamma u_{\alpha\beta} T_\beta d\Gamma \tag{30}$$

where v_α are the displacement components, $u_{\alpha\beta}$ Kelvin's two-dimensional fundamental solution and $N_{\gamma\beta}(u_{\alpha\beta})$ the plane-stress elasticity expressions of the corresponding membrane forces. Having determined the unknown boundary displacements and traction, a second integral equation, obtained by substitution of the domain displacement, given by eqn (30) with $k=1$, into the constitutive equations, yields the membrane forces at any domain point.

3.2 Buckling load

The formulation is based on the integral equation (15) in which δw is replaced by a weighting function u. A second such integral equation results from interchanging the roles of w and u in eqn (15). The difference of the left-hand sides of these two integral equations is

$$D\int_\Omega [(\nabla^4 w)u - (\nabla^4 u)w]\mathrm{d}\Omega \tag{31}$$

$$- \lambda_c\{ I_d^w[u,p(w)] - I_d^w[w,p(u)] + I_t(u,w)\} + I_b^w(u,w) + J(u,w) = 0$$

where

$$p(w) = N_{\alpha\beta} w_{,\alpha\beta} \tag{32}$$

$$I_d^w[u,p(w)] = \int_\Omega u p(w) \mathrm{d}\Omega \tag{33}$$

$$I_b^w(u,w) = \int_\Gamma \left[uV(w) - \frac{\partial u}{\partial n} M_n(w) + M_n(u) \frac{\partial w}{\partial n} - V(u)w \right] d\Gamma \tag{34}$$

$$I_t(u,w) = \int_\Gamma T_\alpha (uw,_\alpha - u,_\alpha w) d\Gamma \tag{35}$$

$$J(u,w) = \sum_{j=1}^{N_k} [u(s_j)C_j(w) - C_j(u)w(s_j)] \tag{36}$$

Integration by parts shows that the expression

$$I_d^w\,[u,p(w)] - I_d^w\,[w,p(u)] + I_t(u,w)$$

vanishes identically, thus eqn (31) is, essentially, the Rayleigh–Green identity for the flexural plate theory. This new form is suitable to the buckling problem since two of its domain integrals are eliminated by identifying $w(x_\alpha)$ with the buckling mode and taking into account eqn (19). The remaining domain integrals can be eliminated only if u were a fundamental solution of eqn (19). Such a solution can be found in the special case of uniform bi-axial compression, that is, with $N_{11} = N_{22} = N$, $N_{12} = 0$. Then eqn (19) becomes

$$D\nabla^4 w - \lambda_c N \nabla^2 w = 0 \tag{37}$$

and its fundamental solutions are[5]

$$u_1 = \frac{1}{2\pi a^2}[\ln r + K_0(ar)]\,, \quad u_2 = \frac{\partial u_1}{\partial n} \tag{38}$$

where $a = \sqrt{N/D}$ and K_0 is the modified Bessel function of order zero. The resulting boundary integral equations are:

$$-kDg_i(P) + I_b^w(u_i,w) + \lambda_c I_t(u_i,w) + J(u_i,w) = 0; \quad i = 1,2 \tag{39}$$

with

$$g_1 = w, \qquad g_2 = \frac{\partial w}{\partial n}$$

The same boundary integral equation can be derived for uniform uni-axial compression, that is, with $N_{11} = N$, $N_{22} = N_{12} = 0$ but the fundamental solution in this case exists only in integral form,[6] it needs therefore to be evaluated numerically. Such formulations, although of some historical interest, have a very limited scope and the disadvantage of dealing with partial derivatives and singularities of rather complex fundamental solutions.

Since closed form fundamental solutions of the buckling problem, valid for any membrane stress distribution, are not available, the fundamental solutions u_1 and u_2 of the plate bending problem have been very widely adopted. These are again given by eqns (27) and (28) but should now be interpreted as deflections at a field point Q of an infinite plate of unit rigidity,[23] the first due to a transverse unit point force at the source point P, the second due to a unit moment at P about the direction normal to $\mathbf{m}$ at an angle α to the position vector $\mathbf{r}$ of Q relative to P (Fig. 1). Integral equation (31) is thus transformed to

$$-kDg_i(P) + \lambda_c\{I_t(u_i,w) + I_d^w[w,p(u_i)]\} + I_b^w(u_i,w) + J(u_i,w) = 0; \quad i=1,2 \tag{40}$$

Due to the presence of the domain integral (33), depending on the unknown deflection, eqn (40) is not a proper boundary integral equation. Despite this mathematical complication, the boundary element methodology can still be applied to the present problem by introducing a simple model for the domain deflection, complementing the conventional boundary modelling.

An alternative form of the integral equation can be derived by applying integration by parts to (33):

$$-kDg_i(P) + \lambda_c I_d^w[u_i,p(w)] + I_b^w(u_i,w) + J(u_i,w) = 0; \quad i=1,2 \tag{41}$$

where now the domain integral $I_d^w[u_i,p(w)]$ depends on the three curvatures which replace the deflection as the domain unknowns. Thus the formulation needs to be complemented by three integral equations derived by simply differentiating (41) with $i=1$ and placing the source point only in the domain:

$$-Dw_{,\alpha\beta}(P) + \lambda_c I_d^w[u_{1,\alpha\beta},p(w)] + I_b^w(u_{1,\alpha\beta},w) + J(u_{1,\alpha\beta},w) = 0 \tag{42}$$

It is possible to combine (42) into a single integral equation for the equivalent lateral pressure p defined by eqn (32):

$$-Dp(P) + \lambda_c I_d^w[p(u_1),p(w)] + I_b^w[p(u_1),w] + J[p(u_1),w] = 0 \tag{43}$$

In the case of a plate on elastic foundation, buckling behaviour can be assessed by including the reaction of the foundation in eqn (41) through the domain integral

$$-\beta\int_\Omega wu\,d\Omega \tag{44}$$

where β is the coefficient of subgrade reaction. This means that deflection as well as curvatures have to be modelled as domain unknowns. The combination of integral equations (41) and (43) are also applicable to orthotropic plates[10]. The fundamental solution of the corresponding bending problem and its derivatives, appearing as kernels in these equations, have been presented by Shi and Bezine.[24]

A BEM solution for plate bending can be based on the Rayleigh–Green identity in the same way as for the membrane stress problem. In such a formulation, the second integral equation yields the Laplacian of the deflection at the source point with the same Laplacian and its normal derivative replacing the bending moment and shear force as boundary unknowns. The boundary conditions are, of course, expressed in terms of the new variables. This approach has been extended to the buckling of plates of variable thickness[22] where the equivalent lateral pressure p has been incremented by the terms (7). It has also been adopted in a BEM buckling analysis of perforated plates[21] where boundary integrals are taken over external as well all internal boundaries.

3.3 Dual reciprocity

The domain integrals $I_{\mathrm{d}}^{\mathrm{w}}[u_i,p(w)]$ in eqns (41) depend on the curvatures and therefore cannot be evaluated unless modelling over the domain is introduced. This leads to an additional set of N_{d} nodal domain unknowns W_j so that the deflection could be approximated as

$$w(x_1,x_2) = \sum_{j=1}^{N_{\mathrm{d}}} W_j G_j(x_1,x_2) \tag{45}$$

Substituting eqn (45) into eqn (33) gives

$$I_{\mathrm{d}}^{\mathrm{w}}[u_i,p(w)] = \sum_{j=1}^{N_{\mathrm{d}}} W_j \int_{\Omega} u_i p(G_j)\mathrm{d}\Omega \tag{46}$$

It is assumed that the functions G_j are such that the domain integrals in eqn (46) can be evaluated numerically over domain cells. This integration is transferred to the boundary of the plate by a technique based on a Fourier series approximation of $p(G_j)$. The general form of the two-dimensional series and its discrete transform is presented in Appendix B.

For plates with arbitrary shape, the deflection itself can be approximated by a Fourier series with coefficients depending on its nodal values. An alternative model can be built from a set of trigonometric functions exactly satisfying the boundary conditions but is only applicable to rectangular plates. Following Tang,[15] both models ultimately lead to a Fourier series representation of the equivalent lateral pressure.

Having defined suitable deflection functions, their second partial derivatives can be combined with the known membrane force distribution $N_{\alpha\beta}$ to yield discrete values of $p(G_k)$ at any number of points within the domain of the plate. Hence, it is possible to transform each of these variables into its equivalent $N{\times}M$ Fourier series using expressions (B2) to compute the Fourier coefficients of each such expansion. If K_{jk} is defined as the $N_\mathrm{d}{\times}N_\mathrm{f}$ array containing these coefficients, then

$$p(G_j) = \sum_{k=1}^{N_\mathrm{f}} K_{jk}\Phi_k \tag{47}$$

where $\Phi_k(x_\alpha)$ is the k-th trigonometric function in the series. Equation (47) allows the transformation of the plate buckling domain integral into a boundary one using the Rayleigh–Green identity for the bi-harmonic operator. Let a function $F_k(x_\alpha)$ be related to $\Phi_k(x_\alpha)$ by

$$\nabla^4 F_k(x_\alpha) = \Phi_k \tag{48}$$

Since u_i $(i=1,2)$ are fundamental solutions of eqn (48), the Rayleigh–Green identity gives

$$\int_\Omega u_i \nabla^4 F_k \mathrm{d}\Omega = kf_i + I_\mathrm{b}^\mathrm{f}(u_i,F_k) \tag{49}$$

where $f_1=F_k(P)$ and $f_2=\partial F_k(P)/\partial n$. Thus the integrals $I_\mathrm{d}^\mathrm{w}[u_i,p(w)]$ can be transformed into boundary integrals provided a set of functions f_{nm}^j can be found such that

$$\nabla^4 f_{nm}^j(x_\alpha) = \phi_{nm}^j \tag{50}$$

where f_{nm}^j and ϕ_{nm}^j are, respectively, alternative representations of F_k and Φ_k, as indicated in Appendix B. Particular solutions of eqn (50) are

$$f_{00}^1 = \frac{1}{48}[(x_1-\bar{x}_1)^4 + (x_2-\bar{x}_2)^4] \tag{51}$$

$$f_{nm}^j = A_{nm}^2\,\phi_{nm}^j \tag{52}$$

where

$$A_{nm} = \frac{1}{\pi^2\left[\left(\frac{n}{a}\right)^2 + \left(\frac{m}{b}\right)^2\right]} \tag{53}$$

and $\bar{x}_1$, $\bar{x}_2$ are arbitrary constants. Explicit expressions of the derivatives of f^j_{nm} appearing in the boundary integral of eqn (49) are given in Appendix A. Tang[15] used this method to transform integrals containing the Poisson operator ∇^2. His approach is extended here to integrals containing the biharmonic operator ∇^4. It is thus possible to evaluate the integrals $I^{\mathrm{w}}_{\mathrm{d}}[u_i,p(w)]$ using boundary discretization only.

The accuracy of transformation (49) was tested numerically by evaluating both sides of this equation and comparing the results. Boundary and domain integrals were calculated for various functions Φ_k using Gaussian quadrature in both cases. It is noted that, in these calculations, the only approximation is due to the discretization. Perfect agreement was observed between domain and boundary integral values. A high percentage error was noted only when the integral value was itself too small. However, the actual difference between the two values was still small. A similar evaluation of overall accuracy of the transformation technique was applied to the integral $I^{\mathrm{w}}_{\mathrm{d}}[p(w),u_i]$ for several plate buckling examples. The error variation with the number of points of the discrete Fourier analysis showed a clear convergence of the transformed integrals to their true values. This convergence was, in most cases, faster than that noted in the respective Fourier approximations of $p(w)$ given by eqn (47).

3.4 Post-buckling

A step-wise non-linear solution can be obtained whereby the edge traction is incremented by t_α resulting in increments w, f and $n_{\alpha\beta}$ of W, F and $N_{\alpha\beta}$, respectively. These incremental variables are governed by

$$\nabla^4 f = -E[K(\hat{W},w) + \frac{1}{2}K(w,w)] \tag{54}$$

$$D\nabla^4 w = N_{\alpha\beta}w_{,\alpha\beta} + n_{\alpha\beta}(\hat{W}_{,\alpha\beta} + w_{,\alpha\beta}) \tag{55}$$

which are directly deduced from eqns (1) and (2). It should be noted that the quadratic terms in the incremental quantities have been retained in eqns (54) and (55).

The non-linear BEM formulation is based on the combination of Rayleigh–Green identities for the biharmonic and the flexural plate operator. The integral equations are derived, as previously, using the two fundamental solutions (27) and (28) of the biharmonic operator. After multiplying eqns (54) and (55) by the fundamental solutions u_i, $i=1,2$, both sides of the resulting equations are integrated over the domain, then Green's theorem and membrane stress equilibrium are applied, and finally the domain integrals in eqn (55) are transformed so that plate curvatures are eliminated from this equation. Through this process the following integral equations are obtained:

$$kf_i(P) + I_b^f(u_i,f) = I_d^f(u_i,w) \tag{56}$$

$$-kg_i(P) + \frac{1}{D}\{I_b^w(u_i,w) + I_t(u_i,w) + I_d^w[(w,p(u_i)] + J(u_i,w)\}$$
$$+ \frac{1}{D}\{\delta I_t(u_i,\hat{W}+w) + I_d^w[(\hat{W}+w,\delta p(u_i)]\} = 0; \; i=1,2 \tag{57}$$

where

$$I_d^f(u,w) = E\int_\Omega uK(W,w)\mathrm{d}\Omega \tag{58}$$

$$\delta p(u) = n_{\alpha\beta}u_{,\alpha\beta} \tag{59}$$

$$\delta I_t(u,w) = \int_\Gamma t_\alpha(uw_{,\alpha} - u_{,\alpha}w)\mathrm{d}\Gamma \tag{60}$$

while all other terms are as defined by eqns (32)–(36). The integral equations (56) and (57) are combined with boundary element modelling to generate a discrete system of equations. The boundary integrals $I_b^f(u_i,f)$ and $I_b^w(u_i,w)$ contain eight variables, four of which are known or can be evaluated from the specified support and loading conditions.

The in-plane displacements may replace the stress function as the variables coupled with the deflection.[19,20] Starting from the incremental form of the two-dimensional equilibrium equations, Kelvin's fundamental solution is used to generate an integral equation similar to eqn (30) with the addition of a domain integral depending on the first and second partial derivatives of the deflection increments.

4 Interpolation models

4.1 Boundary elements

It has been pointed out that, after accounting for the boundary conditions, only two from the four variables on which the boundary integrals depend, are unknown quantities which need to be approximated over boundary elements. Solution algorithms were developed for various orders and types of interpolations models. The more commonly used constant and linear continuous models are conceptually simple and easily programmable but they do not always lead to a sufficiently accurate and rapidly convergent solution. Their performance has been tested against that of the linear and quadratic discontinuous model which allows more flexibility in the variation of the approximated variables and can be designed for optimal performance.

Interpolation functions for discontinuous models are generated as polynomials in the natural co-ordinate ξ with origin at the midpoint of the element. These functions should satisfy

$$\phi_i(\xi_j) = \delta_{ij} \tag{61}$$

where δ_{ij} is the Kronecker delta and the range of indices depends on the order of approximation. In the case of linear interpolation, it is easily shown that the functions ϕ_i which satisfy conditions (61) are given by

$$\phi_1 = -\frac{\xi-\xi_2}{\xi_2-\xi_1}, \phi_2 = \frac{\xi-\xi_1}{\xi_2-\xi_1} \tag{62}$$

where ξ_1 and ξ_2 are the co-ordinates of the two nodes. Since there are two independent unknowns per node, the total number of boundary unknowns will be $N_b = 4N_e$ where N_e is the number of boundary elements.

A quadratic model can be generated by selecting the mid-point of the element as the third node. The interpolation functions are then given by

$$\phi_1 = \frac{\xi(\xi-\xi_2)}{\xi_1(\xi_1-\xi_2)}, \quad \phi_2 = \frac{-\xi(\xi-\xi_1)}{\xi_2(\xi_1-\xi_2)}, \quad \phi_3 = \frac{(\xi-\xi_1)(\xi-\xi_2)}{\xi_1\xi_2} \tag{63}$$

The more familiar continuous models result from setting $\xi_1=-1$ and $\xi_2=1$ in eqns (62) and (63).

If Z denotes a boundary unknown, its approximation over an element is written:

$$Z = \sum_k Z_k \phi_k \tag{64}$$

with the range of k depending on the order of the adopted model. Using a discontinuous boundary model bypasses the problem of normal slope discontinuity at the corners. An approximation is deduced for the tangential derivative appearing in the expressions of the jump term. With Z representing the normal derivative of the deflection over a simply supported or free boundary, eqn (64) can be substituted into expression (18) to give

$$M_{ns} = -D(1-\nu)\,\frac{2}{L}\sum_k Z_k \frac{\partial \phi_k}{\partial \xi} \tag{65}$$

over an element. The twisting moment M_{ns} at a corner can thus be expressed in terms of the boundary unknowns of the adjacent elements. In the case of linear element modelling, the approximation of M_{ns} near a corner has been further refined by assuming that its value at the centre of an element is given by expression (65) and varies quadratically over the three adjacent elements on either side of a corner.

4.2 Domain cells

The domain integral $I_d^w\,[(w,p(u_i)]$, appearing in eqns (40) and (57), is evaluated numerically by modelling only the deflection. This contrasts other BEM schemes, based on eqn (41), which admit additional unknowns representing the three domain curvatures at every nodal point. The domain of the plate may be discretized into N_c triangular or quadrilateral cells over which the deflection is given in terms of its nodal values W_i by

$$w = \sum_i \phi_i W_i \tag{66}$$

where ϕ_i are the adopted shape functions. The most common model for a domain unknown is the constant value over each cell. Linear continuous functions are also often used. Particular reference is made here to the less common discontinuous interpolation functions. Their linear form over triangular cells is

$$\phi_i = \sum_{j=1}^{3} \alpha_{ij}\zeta_j \tag{67}$$

where ζ_j, $j=1,2,3$, are the familiar area co-ordinates. The three internal nodes lie along the medians between vertices A, B, C and the centroid O as shown in Fig. 2(a). Their position is specified by the ratios

$$d_1 = \frac{(O1)}{(OA)}, \quad d_2 = \frac{(O2)}{(OB)}, \quad d_3 = \frac{(O3)}{(OC)} \tag{68}$$

If $d_1 = d_2 = d_3 = d$, the natural co-ordinates of nodal point 'k' must satisfy

$$\zeta_{jk} = \begin{cases} \dfrac{1+2d}{3} & \text{if } j = k \\ \dfrac{1-d}{3} & \text{if } j \neq k \end{cases} \tag{69}$$

The condition

$$\sum_{j=1}^{3} \alpha_{ij} \zeta_{jk} = \delta_{ik}$$

leads to the following expressions for the coefficients of the interpolation functions:

$$\alpha_{jk} = \begin{cases} \dfrac{2+d}{3d} & \text{if } j = k \\ -\dfrac{1-d}{3d} & \text{if } j \neq k \end{cases} \tag{70}$$

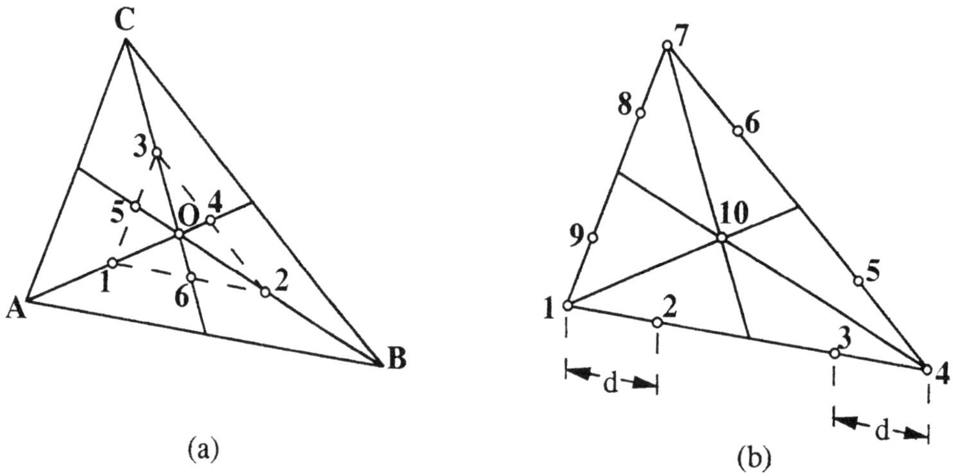

Figure 2. High-order domain cells for deflection modelling: (a) discontinuous quadratic; (b) continuous cubic.

The quadratic model is derived in the same way. With the position of the three additional nodes as shown in Fig. 2(a), their natural co-ordinates are given by

$$\zeta_{jk} = \begin{cases} \dfrac{1-d}{3} & \text{if } j=k-3 \\ \dfrac{2+d}{6} & \text{if } j\neq k-3 \end{cases} \tag{71}$$

with k=4,5,6. Imposing the necessary conditions on the corresponding interpolation functions leads to the expressions:

$$\phi_i = \frac{1}{9d^2}[18\zeta_i^2 - 3(4-d)\zeta_i + (2+d)(1-d)] \tag{72}$$

$$\phi_{i+3} = \frac{4}{9d^2}[9\zeta_j\zeta_k + 3(1-d)\zeta_i - (2+d)(1-d)] \tag{73}$$

with i=1,2,3 and the combination (i,j,k) representing all even permutations of 1,2,3. Again the more familiar continuous models are obtained by setting d=1 in eqns (71)–(73).

Along a free boundary, non-conformity between continuous boundary and domain modelling can be avoided if the same boundary key points are used to generate both domain and boundary discretization into cells and elements, respectively. In the case of discontinuous domain cells, the formulation can be further simplified by ignoring the boundary unknowns and extending the domain deflection model to the free boundary. Thus the deflection and its normal derivative at any point along a free edge can be easily expressed in terms of the nodal deflections of the cell containing that boundary point. This device, which has the additional advantage of reducing the overall number of unknowns, can only be used with the particular form of domain integral appearing in eqn (40). It should also be noted that a more accurate representation of the normal slope in terms of domain nodal deflections can be achieved through refinement of the domain mesh near the boundary.

It is noted that the domain integral $I_{\mathrm{d}}^{\mathrm{f}}(u_i,w)$ in the post-buckling eqn (56) depends on curvatures, but can still be treated numerically using the model for the deflection. This is achieved by adopting non-linear models in the form of high order polynomials or trigonometric functions. Then, the curvatures are obtained from eqn (66) as

$$w_{,\alpha\beta} = \sum \phi_{i,\alpha\beta} W_i \tag{74}$$

which leads to a direct relationship between nodal cell curvatures and deflections.

Continuous quadratic interpolation functions associated with six-node cells result in constant curvature over a cell. Since no great accuracy was expected from this rather crude model, the search for a more accurate solution led to the use of continuous cubic interpolation functions over the 10-node cell shown in Fig. 2(b) through which a linear variation of the curvature over the cell is achieved with the possibility of controlling its distribution by varying the position of the intermediate nodes on the sides of the cell relative to the vertices. Thus, the distance d of these nodes from the corner nodes becomes a key parameter affecting the accuracy of the formulation. Through finite difference considerations it can be shown that setting d equal to a quarter of the cell side results in the most smooth curvature modelling.

The estimates of the nodal curvatures predicted by such a model can be further improved by an averaging scheme. This is the consequence of having more than one curvature expression at every domain node belonging to more than one domain cell. Thus, averaging allows the adoption of a curvature model of the same order as that for the deflection.

Another deflection model that has been tried consists of trigonometric functions over the triangular cell. This has the generalized co-ordinate form

$$w = \sum_m \sum_n a_{mn} U_{1m}(x_1) U_{2n}(x_2) = \sum_k B_k \phi_k \tag{75}$$

and leads to the same order of non-linearity between curvatures and deflections. The model can be built up from products of either sinξ or cosξ with either sinη or cosη where

$$\xi = \frac{m\pi X_1}{a} \qquad \eta = \frac{n\pi X_2}{b}$$

with the origin of the local Cartesian frame of reference (X_1, X_2) on the boundary or in the domain of the cell and a, b arbitrarily chosen constant lengths which may serve as control parameters for testing the accuracy of the resulting formulation. The generalised co-ordinates B_k can be expressed in terms of the nodal deflection values by the standard procedure described in FEM textbooks.[26] In the case of plates of rectangular shape, it is particularly convenient to adopt approximation (75) with a and b equal to plate width and breadth, respectively, and U_{1n}, U_{2m} satisfying the boundary conditions. This simplifies matters considerably since there is no need for inter-element compatibility and the generalized co-ordinates a_{mn} are retained as the domain unknowns. Whatever the choice of the deflection model, the described method of modelling domain curvatures leads to a general matrix equation of the form

$$\{Y\} = [C]\,\{W\} \tag{76}$$

where $\{Y\}$ and $\{W\}$ are one-dimensional arrays containing the nodal domain curvatures and deflections, respectively.

4.3 Nodal deflection model

The non-zero values of the deflection of a plate at a discrete set of N_d points constitute the array of unknowns W_k. The deflection at any other point of the plate is related to these nodal values through the approximate two-dimensional Fourier analysis. Let a and b be the largest dimensions of the plate in the x_1 and x_2 directions, respectively. The domain Ω of $w(x_1,x_2)$ is extended to the rectangular domain Ω_f or $[-a, a]\times[-b,b]$. Then $w(x_1,x_2)$ can be assumed to be periodic with periods $2a$ and $2b$ in the x_1 and x_2 directions, respectively, and thus be approximated by an $N\times M$ Fourier series

$$w(x_1,x_2) = \sum_{j=1}^{N_f} B_j \Phi_j \tag{77}$$

The deflection w is thus considered as a Fourier expansion (B1) in the domain Ω_f with its values at all points outside the actual plate domain as well as on the laterally constrained boundaries set equal to zero. The coefficients B_j can be related linearly to the deflections at the discrete points through eqns (B2). Hence, a matrix $[A]$ of dimensions $N_f\times N_d$ can be constructed such that:

$$B_j = \sum_{k=1}^{N_d} A_{jk} W_k \tag{78}$$

Replacing B_j from eqn (78) in eqn (77) gives

$$w(x_1,x_2) = \sum_{k=1}^{N_d} \left(\sum_{j=1}^{N_f} A_{jk} \Phi_j \right) W_k \tag{79}$$

Since the elements of matrix A_{jk} are known, the deflection function has actually been defined in terms of its unknown values at some specified nodes in the domain of the plate. By setting

$$G_k(x_\alpha) = \sum_{j=1}^{N_f} A_{jk} \Phi_j \tag{80}$$

a formal equivalence between eqns (45) and (79) is established with the coefficients of the nodal deflections expressed as trigonometric series. It is apparent from eqn (80) that, when the in-plane stress distribution is uniform, $p(w)$ is the sum of simple trigonometric functions that are exactly reproduced by the Fourier analysis. In general, the membrane force distribution is non-uniform and, in most cases, can only be determined by some numerical method. Thus, functions $p(G_i)$ can only be specified as a set of discrete values over a two-dimensional domain. This leads to the numerical evaluation of coefficients K_{jk} in eqn (47) using the inverse formula (B2).

The nodal deflection model can be used for plates of any shape. Furthermore, even though the boundary conditions cannot be fully satisfied by the domain deflection function, zero deflection can be forced at the nodes on simply supported or clamped boundaries. It should be noted that boundary unknowns do not need to be specified along free boundaries since, in this case also, the boundary deflection and its normal derivative can be expressed in terms of the domain unknowns.

4.4 Trigonometric deflection model

Given a rectangular plate the sides of which are either simply supported, clamped or free, the deflection $w(x_\alpha)$ of the plate can be given by eqn (45) with the functions $G_j(x_\alpha)$ written in the form

$$G_j(x_\alpha) = U_{1n}(x_1)U_{2m}(x_2) \tag{81}$$

where $U_{1n}(x_1)$ and $U_{2m}(x_2)$ are trigonometric functions satisfying exactly the geometric and natural boundary conditions at the sides x_1=constant and x_2=constant, respectively. In this type of modelling, the unknown coefficients W_k are known as the generalised co-ordinates of the problem. Functions (81) as well as their first and second derivatives will also have to satisfy the condition of continuity over the domain of the plate.

Trigonometric functions $U_{1n}(x_1)$ or $U_{2m}(x_2)$, satisfying the simple support and clamped conditions can be easily found. However, the free edge condition cannot be satisfied by a function of a single space co-ordinate. In order to overcome this problem, the shear force terms with the Poisson's ratio have been disregarded. Thus, only the normal second and third derivatives of the function are required to vanish along free boundaries. In other words, the plate has been treated as a combination of two crossing beams as far as the boundary conditions of the deflection function are concerned. This assumption which has been adopted by other approximate analyses[27] is not expected to have a significant effect on the accuracy of the results.

The trigonometric deflection model allows the improvement of the accuracy by increasing N_d, that is, adding functions of higher order n and/or m. The model also has the advantage of allowing a selective choice of modes depending on the symmetry or anti-symmetry of geometry, loading and support conditions. In the case of simple boundary conditions and uniform loading, the trigonometric deflection model is expected to be computationally more efficient than the nodal one. However, it cannot be easily extended to non-rectangular plates because identifying deflection functions satisfying the boundary conditions would become difficult if possible at all. Moreover, the location of the source points for domain integral equations is not readily identifiable since there are no domain nodes but only generalized co-ordinates. These limitations are of course overcome by adopting the nodal deflection model.

5 The algebraic problem

5.1 Membrane forces

The application of the method to non-uniform fundamental equilibrium states requires the numerical solution of integral equations (23). Using any of the described models for the boundary unknowns, placing the source point at all boundary nodes and performing the integration over all elements gives a consistent system of algebraic equations of the form

$$[H_{\mathrm{b}}^{\mathrm{f}}]\,\{Z_{\mathrm{f}}\} = \{G_{\mathrm{f}}\} \tag{82}$$

where $\{Z_{\mathrm{f}}\}$ is the array of the unknown boundary nodal values of $\nabla^2 F$ and its normal derivative, $[H_{\mathrm{b}}^{\mathrm{f}}]$ a square coefficient matrix depending on the adopted modelling, and $\{G_{\mathrm{f}}\}$ a column matrix containing the weighted integrals of the given boundary values of F and its normal derivatives. Solving eqn (82) for $\{Z_{\mathrm{f}}\}$ allows the determination of $N_{\alpha\beta}$ at any internal point through integral equation (29). The membrane forces are actually computed at the nodes generated by the domain deflection model which is thus used to approximate their variation within cells.

5.2 Buckling mode

5.2.1 Deflection formulation

Boundary and domain models are substituted into eqn (40), the source point placed at all boundary nodes and integration is carried out over all elements leading to the following system of N_{b} linear algebraic equations

$$([H_{\mathrm{b}}^{\mathrm{w}}] + \lambda_{\mathrm{c}}\,[H_{\mathrm{t}}])\,\{Z_{\mathrm{w}}\} = ([H_{\mathrm{f}}] + \lambda_{\mathrm{c}}[H_{\mathrm{d}}])\{W\} \tag{83}$$

where $\{Z_{\mathrm{w}}\}$ and $\{W\}$ are the arrays containing the boundary and domain unknowns, respectively, the elements of matrix $[H_{\mathrm{b}}^{\mathrm{w}}]$ result from integrals $I_{\mathrm{b}}^{\mathrm{w}}(u_i,w)$ and jump terms $J(u_i,w)$, while matrices $[H_t]$ and $[H_d]$ are the consequence of integrals $I_t(u_i,w)$ and $I_d^w[w,p(u_i)]$, respectively. Matrix $[H_f]$ arises only in the presence of free boundaries along which edge deflection and normal slope are expressed in terms of domain unknowns. An additional system of N_d equations,

$$([D_b^w] + \lambda_c[D_t])\{Z_w\} = ([D_f] + \lambda_c[D_d])\{W\} \tag{84}$$

is obtained by applying eqn (40) with $k=i=1$ and placing the source point at all domain nodes. Then, integrals $I_b^w(u_1,w)$, $I_t(u_1,w)$ and $I_d^w[w,p(u_1)]$ would generate the elements of matrices $[D_b^w]$, $[D_t]$ and $[D_d]$, respectively, with the jump terms contributing to $[D_b^w]$ and the free edges to $[D_f]$, the latter reducing to the unit matrix for any other combination of edge support.

The critical load factor is the smallest eigenvalue of the system of eqns (83) and (84). Two numerical procedures are possible for its evaluation. The first involves elimination of the boundary unknowns from the above system and its transformation to

$$[A(\lambda_c)]\{W\} = \{0\} \tag{85}$$

where

$$[A(\lambda_c)] = ([D_b^w] + \lambda_c[D_t])([H_b^w] + \lambda_c[H_t])^{-1}([H_f] + \lambda_c[H_d]) - ([D_f] + \lambda_c[D_d])$$

Thus λ_c is found as the smallest root of the determinant of $[A]$. A numerical search of a change of sign of this determinant for a range of values of λ can give the approximate location of a root which can then be evaluated to the desired degree of accuracy by the secant method. The efficiency of this approach is severely undermined by the need to invert a $N_b \times N_b$ matrix for every new value of the load factor. Alternatively, the system of eqns (84) and (85) can be reduced to the standard eigenvalue problem

$$([A] - \lambda_c[B])\{X\} = \{0\} \tag{86}$$

where

$$[A] = \begin{bmatrix} [H_b^w] & -[H_f] \\ [D_b^w] & -[D_f] \end{bmatrix}, [B] = \begin{bmatrix} -[H_t] & [H_d] \\ -[D_t] & [D_d] \end{bmatrix}, \{X\} = \begin{Bmatrix} \{Z_w\} \\ \{W\} \end{Bmatrix} \tag{87}$$

from which λ_c^{-1} can be evaluated as the largest eigenvalue of the $[A]^{-1}[B]$. Since the inversion of the $(N_b+N_d)\times(N_b+N_d)$ matrix $[A]$ is performed only once while efficient routines yielding the largest eigenvalue of a matrix as well as the associated eigenvector are readily available, this method of solution proved to be more direct and efficient than the first. It is worth noting that, in the absence of any free boundaries, $[H_f] = [0]$, $[D_f] = [I]$ and it can be shown that, as a consequence,

$$[A]^{-1}[B] = \begin{bmatrix} -[H_b^w]^{-1}[H_t] & [H_b^w]^{-1}[H_d] \\ -[D_b][H_b^w]^{-1}[H_t]+[D_t] & [D_b][H_b^w]^{-1}[H_d]-[D_d] \end{bmatrix} \tag{88}$$

For fully clamped plates, both matrices $[H_t]$ and $[D_t]$ also vanish and the eigenvalue problem (86) reduces further to

$$[\,([D_b^w][H_b^w]^{-1}[H_d] - [D_d]) - \lambda_c^{-1}\;[I]\,]\,\{W\} = \{0\} \tag{89}$$

Thus, numerical efficiency is considerably improved by implementing the special formulations (88) and (89) derived above.

5.2.2 Curvatures formulation

A similar modelling and integration procedure, applied to eqns (41) and (42), leads to the system

$$[H_b^w]\{Z_w\} = \lambda_c \sum_{k=1}^{3} [H_{dk}]\{Y_k\} \tag{90}$$

$$[D_{bj}^w]\{Z_w\} = \{Y_j\} + \lambda_c \sum_{k=1}^{3} [D_{djk}]\{Y_k\}\,;\, j=1,2,3 \tag{91}$$

where $\{Y_j\}$ are one-dimensional arrays containing the nodal values of curvatures with the subscript values $j=1,2,3$ corresponding to $w_{,11}$, $w_{,22}$, $w_{,12}$. The above system can be reduced to an eigenvalue problem similar in form to eqn (89) where the components of the eigenvectors are the nodal curvatures grouped into a single array. An alternative, smaller system of equations is obtained if eqn (42) is replaced by eqn (43). Then, the buckling eigenvector consists of the discrete values of the equivalent lateral pressure $p(w)$.

5.2.3 Elastic foundation

The addition of the domain integral (44) in eqns (41) and (43) modifies the corresponding algebraic systems to:

$$[H_b^w]\{Z_w\} = \lambda_c[H_d]\{P\} - \beta[H_e]\{W\} \tag{92}$$

$$[D_{b2}^w]\{Z_w\} = ([I] + \lambda_c[D_{d2}])\{P\} - \beta[D_{e2}]\{W\} \tag{93}$$

where $\{P\}$ is the array containing the nodal values of p and subscript 2 indicates kernels depending on second partial derivatives of the fundamental solution. For the formulation to become consistent, the system of equations

$$[D_b^w]\{Z_w\} = \lambda_c[D_d]\{P\} + ([I] - \beta[D_e])\{W\} \tag{94}$$

is obtained from (41) with $i=1$ and the source point placed at all domain nodes $(k=1)$. The coefficients of matrices $[H_e]$, $[D_{e2}]$ and $[D_e]$ are essentially the same as the corresponding coefficients of $[H_d]$, $[D_{d2}]$ and $[D_d]$, respectively, apart from those associated with positive deflection values which must be set equal to zero. The problem is therefore non-linear and can only be solved by an iterative scheme whereby the buckling mode is initially obtained with β set equal to zero allowing the matrices $[H_e]$, $[D_{e2}]$ and $[D_e]$ to be initialized and then modified at each solution step until full plate-foundation interface compatibility is achieved.

5.2.4 Assembled plate structures

Interconnected plates of various orientations forming space structures are subjected to the combined action of in-plane and lateral forces thus coupling of membrane and flexural deformation should be taken into account in the assessment of overall buckling.[9] Integral equations (30), (41) and (42) are written for each plate in terms of the variables referred to a local frame of reference and the algebraic systems of equations for the nodal unknowns are obtained using boundary element and domain cell modelling as already described.

The local variables are next transformed into a new set referred to a global frame of reference and consisting of the three components of displacement and force as well as edge slope and moment, a total of eight variables. Along the interface between two plate components, the conditions of deformation compatibility and force equilibrium are imposed on the global variables. This leads to a consistent system of equations which is combined with homogeneous conditions along the structure boundary to generate an eigenvalue problem similar in form to eqn (89).

5.2.5 Dual reciprocity

The derivation of the transformation (49) means that the domain integral (46) is replaced by a summation of boundary integrals. Thus eqn (41) becomes a genuine boundary integral equation in which the source point P needs to be placed at all boundary as well as domain nodes in order to generate a consistent system of equations. In the case of trigonometric deflection model, the source point can be placed at the maxima of the shape functions adopted.

The linear system of equations obtained from eqn (41) by integration can be written in the form:

$$[H_b^w]\{Z_w\} = ([H_f] + \lambda_c[H_d])\,\{W\} \tag{95}$$

$$[D_b^w]\{Z_w\} = ([D_f] + \lambda_c[D_d])\,\{W\} \tag{96}$$

where the coefficient matrices $[H_b^w]$, $[H_f]$, $[H_d]$ correspond to boundary positions of the source point, the first arising from the evaluation of $I_b^w(u_i,w)$ as well as the jump terms $J(u_i,w)$ over simply supported and clamped sides, the second arising from the evaluation of $I_b^w(u_i,w)$ as well as the jump terms $J(u_i,w)$ over free sides, the third arising from the evaluation of $I_d^w[u_i,p(w)]$. Matrices $[D_b^w]$, $[D_f]$, $[D_d]$ are similarly obtained but correspond to domain positions of the source point.

Solving eqn (95) for $\{Z_w\}$, substituting in eqn (96) and rearranging gives

$$([A] - \lambda_c[B])\{W\} = \{0\} \tag{97}$$

where

$$[A] = [D_b^w][H_b^w]^{-1}[H_f] - [D_f]$$

$$[B] = -[D_b^w][H_b^w]^{-1}[H_d] + [D_d]$$

Thus an eigenvalue problem has been obtained which can now be solved by any standard technique. Vector $\{W\}$ will contain either the generalized co-ordinates or the nodal values of the deflection. In both cases the buckling mode of the plate can be found without actually using additional boundary integral equations as is usually required in BEM.

5.3 Post-buckling

Linear discontinuous boundary elements with two internal nodes per element can be used to model any boundary unknown quantity Z according to eqn (64). This leads to $4\times N_b$ algebraic equations, obtained by placing the source point on all N_b boundary nodes and evaluating the integrals of eqns (56) and (57). The general form of these equations is

$$[H_b^f]\{Z_f\} - \delta\lambda\{E_b\} = [H_d^f]\{Y\} + \{E_d^f\} \tag{98}$$

$$([H_b^w] + (\lambda+\delta\lambda)[H_t])\{Z_w\} = [H_d^w]\{W\} + [H_n]\{N\} + \{E_t\} + \{E_d^w\} \tag{99}$$

where λ is the load factor generating the current load from a reference state, and $\{Z_f\}$, $\{Z_w\}$, $\{N\}$ are one-dimensional arrays containing the boundary unknowns in $I_b^f(u_i,f)$, the boundary unknowns in $I_b^w(u_i,w)$, and the incremental membrane forces, respectively. The coefficient arrays on the left-hand side of eqns (98) and (99) are all fixed and need to be calculated only once throughout the incremental solution process.

An additional system of N_d algebraic equations is obtained by placing the source point on the N_d domain nodes and applying the first of eqns (57) with $k=1$. This operation results in the system

$$[\,[D_b^w] + (\lambda+\delta\lambda)[D_t]\,]\{Z_w\} = [D_d^w]\{W\} + [D_n]\{N\} + \{G_t\} + \{G_d^w\} \quad (100)$$

Integral expressions for the membrane forces are obtained by introducing the first of eqns (57) with $k=1$ into the incremental version of eqns (6). The number of nodes at which the membrane stresses are determined depends on the desired accuracy. The final result is a matrix equation of the form

$$\{N\} = [D_b^f]\{Z_f\} + [D_d^f]\{Y\} + \delta\lambda\{G_b\} + \{G_d^f\} \quad (101)$$

Quadratic continuous interpolation functions over the domain cells each with six nodes are used to model the in-plane stress distribution for the determination of the elements of $[H_d^w]$ and $[D_d^w]$.

At the beginning of each load step the solution is carried out without the one-dimensional arrays $\{E_d^f\}$, $\{E_d^w\}$, $\{G_d^w\}$ and $\{G_d^f\}$ containing the integrals of the quadratic terms in the unknowns. The arrays $\{Z_f\}$, $\{N\}$, and $\{Y\}$ are eliminated from the formulation using matrix eqns (76), (98), and (101). This operation is carried out without severe computational penalty because it does not involve matrix inversions. The final system of equations contains only the elements of $\{Z_w\}$ and $\{W\}$ as unknowns. After all incremental quantities are determined, the result is further corrected by including the integrals of the quadratic terms and solving the system iteratively within the current load step.

An alternative solution algorithm does not require detailed modelling of domain deflection but relies heavily on iteration.[19] A deflection shape has to be assumed at the start of the first load step and entered into all domain integrals which then can be evaluated. The boundary integral equations can thus be solved by BEM as ordinary static problems first for the membrane displacements and then for the plate deflection. The latter is compared to the initial estimate to assess convergence and either this process is repeated with modified deflection values or the solution moves to the next load step. It is worth noting that the coefficients of boundary unknowns depend only on the plate properties and are therefore evaluated once in this iterative approach.

5.4 Singular integrals

A typical boundary integral over an individual boundary element Γ_e has the form

$$I^e = \sum_k Z_k \int_{\Gamma_e} \Lambda \phi_k d\Gamma \tag{102}$$

where Λ would be the kernel paired with variable Z in an integral equation. The integration over elements which do not contain the source point is performed numerically using Gaussian quadrature. Gaussian integration with 4 points over a boundary element and 12 points over a domain cell has been found to provide satisfactory accuracy. As the distance r between the source point and the field point becomes smaller, kernels containing the factors $\ln r$, r^{-1} and r^{-2} approach indeterminate values. Consequently, the standard Gaussian quadrature for numerical integration becomes less accurate and a larger than the usual number of integration points may be needed.

Analytical integration is possible and indeed essential over elements containing the source point not only for achieving better accuracy but also for avoiding elaborate numerical schemes to cope with the singularities of certain kernels. With the source point identified with internal node 'm' which is taken as the origin of a local co-ordinate s, the integrals over element 'e' in the case of the linear interpolation model reduce to the form

$$I^e = \frac{2I_1^e}{L(\xi_2 - \xi_1)} (Z_2 - Z_1) + I_0^e Z_m \tag{103}$$

where

$$I_0^e = \int_{\Gamma_e} \Lambda ds , \quad I_1^e = \int_{\Gamma_e} \Lambda s ds \tag{104}$$

Analytical expressions of non-vanishing integrals (104) for various kernels Λ have been published.[13] It should be pointed out that, in the presence of a free boundary along which w is a boundary unknown, singular integrals due to $\Lambda = V(u_2)$ arise and therefore the given expressions represent their principal value.

It is not however possible to evaluate analytically integrals arising from the transformation of $I_d^w [u_i, p(w)]$ because the boundary integrals resulting from dual reciprocity contain the trigonometric functions of the Fourier series and their derivatives. A technique proposed by Telles[28] employs a special purpose variable transformation that would automatically consider a higher number of integration points near singularities, hence increasing the accuracy of the Gaussian quadrature. For the r^{-2} singularity, which is too strong to be treated by the Telles's method, the principal values of the singular integrals are evaluated by a general numerical technique developed by Kutt[29,30] for functions of the form $f(x)/x^n$.

Special numerical integration schemes to deal with singularities of type r^{-2}, occurring when the source point is placed on a free boundary, are avoided in formulations where the unknown deflection and normal slope along a free boundary can be expressed in terms of the modelled domain deflection. In the case of continuous domain interpolation models, cells bordering the free edge can be modified so that nodes, which would otherwise be on the boundary, are moved within the cell. The free boundary is thus discretized into elements only for the purpose of carrying out the Gaussian integration with the expressions for w and $\partial w/\partial n$ at the Gaussian points obtained from the assumed deflection over the adjacent domain cell.

6 Results and discussion

6.1 Programming features

All developed formulations were programmed on computers with some of the resulting codes featuring automatic mesh generation schemes for both the contour and the area of the plate. For plates of polygonal shape, each edge is divided into a specified number of segments not necessarily of the same length. Using the constant boundary element model, it was shown[6] that reducing the element length near the corners improves the representation of the corner forces in terms of the adjacent nodal values of the normal slope. This scheme proved not essential with higher-order element models. The generation of triangular domain cells as well as boundary elements, with their internal nodes, may be initiated from a set of key points along the plate boundary. Domain cells containing the boundary elements along free boundaries, if any, need to be identified so that their modelling be used to approximate the free boundary unknowns.

Modelling can be restricted to half or a quarter of a plate by accounting for any uni-axial or bi-axial symmetry or anti-symmetry of geometry and loading. This programming feature decreases further the number of unknowns, thus improving the speed of execution and the accuracy of the results, it should however be used with caution since the buckling mode of a plate with symmetric shape and load may not necessarily be symmetric. It is therefore suggested that all possible combinations of symmetry and anti-symmetry conditions are tried along the symmetry axes as a means of identifying the minimum buckling load. Any in-plane loading can be considered with the resulting in-plane stresses specified at certain points of the domain or boundary of the plate.

The data structure is rather simple as expected in a BEM program in which no domain discretization is needed. The input data consist of the geometric and material properties of the plate, the number of boundary elements on each side of the plate and either the number and type of trigonometric deflection functions or just the number of domain nodes.

6.2 Buckling loads

The performance of the various boundary and domain models was assessed by analysing a large number of plate shapes, with various loading and support conditions. A selection of analysed plate examples is shown in Fig. 3 where clamped, simply supported and free boundaries are indicated by double solid, single solid and broken lines, respectively. In all these examples, the plates have an aspect ratio *a*/*b* of 1. In the case of uniform, either normal or shearing, traction *T*, the critical value of load factor λ, defined by

$$\lambda = \frac{b^2}{\pi^2 D} T, \tag{105}$$

was computed. In the case of non-uniform loading, the load factor is given by

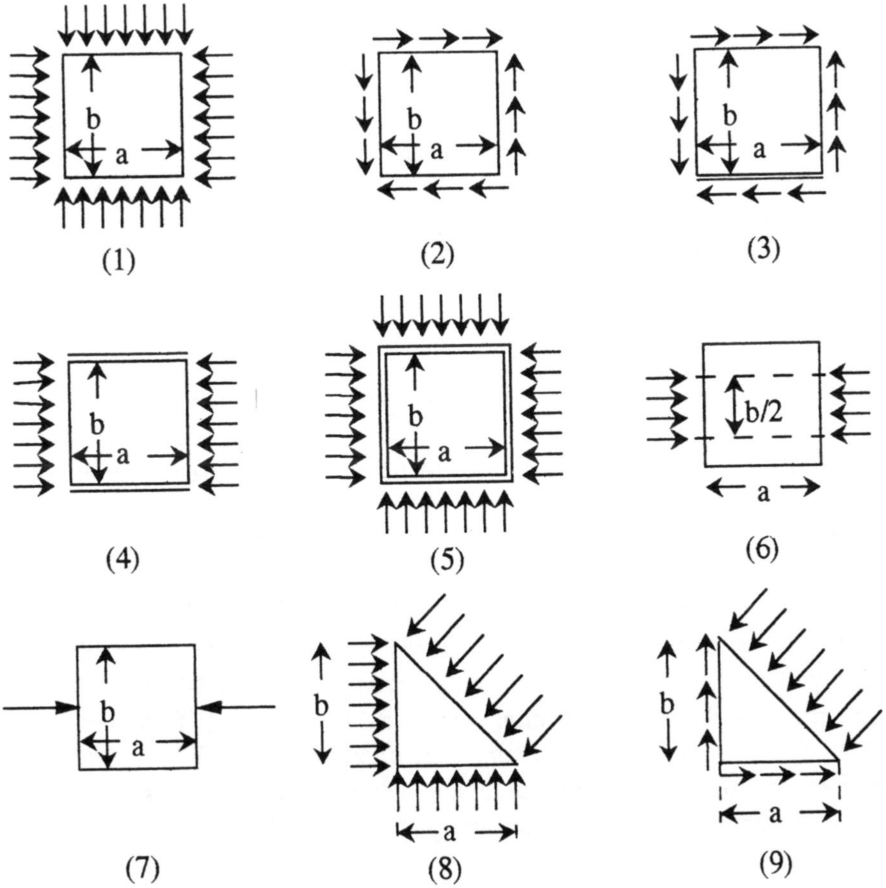

Figure 3. Analysed plate buckling examples.

$$\lambda = \frac{b}{\pi^2 D} P \tag{106}$$

where P is the total compressive force in the direction of loading.

The optimum internal nodal position in discontinuous cells and elements was identified through a systematic assessment of the numerical error in the calculation of the critical load of plates with simple geometry, load and support conditions. With the position of the boundary nodes kept constant, the critical load of several plates was determined for the range of parameter d, defining the nodal position within a domain cell according to eqns (68). The percentage error was found to be minimum for d=0.5 in all analysed cases for both the linear and the quadratic domain model. A similar parametric study using a symmetrical arrangement of internal boundary element nodes indicated that the best choice for their natural co-ordinates is $|\xi_1| = |\xi_2| = 0.5$.

A number of plates were analysed using various combinations of boundary and domain models. For a meaningful comparison of the performance of the various orders of approximation considered, the generated number of unknowns was kept, as much as possible, within a reasonably narrow range. It was noted that the consistent combination of linear discontinuous boundary elements and domain cells, although not always the most accurate, gives the best overall performance with regard to efficiency and reliability. The convergence of the formulation with linear discontinuous modelling was verified by letting the number of boundary elements or domain cells increase independently and assessing the percentage error of the solution.

A selection of results for plates with various combinations of boundary conditions and non-uniform membrane stress distribution is presented in Table 1. These can be compared to the corresponding published critical values obtained by other exact or approximate methods. The accuracy achieved by the BEM solution can be considered satisfactory. It should be noted that a degree of approximation was involved in some of the earlier predictions, especially those for non-uniformly loaded plates. This must be taken into account in the assessment of the BEM results.

Table 1 also contains critical loads obtained by the algorithm based on dual reciprocity using both the nodal and trigonometric deflection models. It was numerically demonstrated that the results given by the nodal deflection model converge towards the actual buckling factor as the number of domain nodes is increased. The most interesting results were those obtained for triangular plates. The degree of convergence and accuracy achieved in these cases confirm the validity of the approach of expanding the plate domain for the purpose of Fourier analysis.

Table 1. BEM predictions of critical loads

Plate example	BEM			Published	
	Domain integration	Dual reciprocity		λ_c	Reference
		Nodal model	Trigonometric model		
1	1.999	2.002	1.998	2.0	2
2	9.346	9.343	9.292	9.34	2
3	10.81	10.985	10.6	10.98	31
4	7.76	7.707	7.669	7.69	2
5	5.287	5.348	5.34	5.3	2
6	3.095	3.107	3.108	3.05	32
7	0.66	0.68	0.68	0.67	33
8	5.05	5.05	-	5.11	34
9	11.56	11.8	-	11.55	34

Existing geometric and loading symmetries have been taken into consideration when choosing the trigonometric deflection shapes of the plate. It should be noted that careful selection of the set of shape functions speeds considerably the convergence of the solution towards its exact value. Among an arbitrarily chosen set of functions, those with a major contribution to the buckling load can be identified from the solution for the deflection amplitudes and then selected to represent the deflection in a re-run of the code implementing the analysis. This procedure enhances the accuracy of the solution algorithm.

The presented results demonstrate the validity of dual reciprocity for both the nodal and the trigonometric deflection models. However, the accuracy of the results depends on the order of the Fourier approximation or the choice of deflection functions. Comparison shows that, for the adopted values of these parameters, the nodal performed consistently better than the trigonometric model.

With or without dual reciprocity, the boundary element formulations based on modelling the deflection of the plate rather than the curvatures have the advantage of yielding directly the plate buckled shape, normalized with respect to the maximum deflection. Thus, there is no need for further use of integral equations. The deflected shape given by either approach using various boundary and domain models was always found in excellent agreement with published predictions[2] for a number of examples.

The performance of a BEM computer code was compared to that of a finite element program using conforming quadrilateral linear elements with four degrees of freedom per node (w, $w_{,1}$, $w_{,2}$, $w_{,12}$).[35] Four examples were analysed using both programs, run on the same computer, with respective meshes designed to yield a comparable degree of accuracy. The results confirmed that the finite element formulation handles more efficiently problems of a certain size due to the symmetry and sparseness of its stiffness matrix, while economy is achieved by the boundary element method through a reduction of the size of the problem despite the presence of irreducible domain integrals. It is natural to expect that removal of the latter would improve greatly the performance of the BEM solution of this problem. The scheme based on the concept of dual reciprocity,[16] whereby these domain integrals are transformed into boundary ones without, however, avoiding the need for domain unknowns, obviates domain discretization and integration, but there is still need for multiple boundary integration. The efficiency of this scheme is not therefore much greater than that of the original formulation using domain discretization which has the additional advantage of greater versatility in modelling plates of arbitrary shape. There is certainly potential for further refinements of the described numerical algorithms that would improve the degree of their accuracy over a wider range of applications.

6.3 Post-buckling

The effectiveness of the numerical algorithm described by eqns (98)–(101) was assessed by comparing its predictions with a selection of experimental data taken from the literature. Tests have made a significant contribution to the assessment of the effect of various design parameters on practical buckling behaviour. Review articles[1] give an appreciation of the various techniques and rig designs employed by a large number of researchers in order to simulate as realistically as possible all manifestations of the phenomenon in the laboratory. Experiments showed quite early that the load carrying capacity of plates is considerably higher than the predictions of the linearized theory. The existence of secondary buckling along the fundamental equilibrium load–deformation curve was also originally observed[36] and then thoroughly investigated[37] experimentally.

Control of the in-plane boundary conditions was achieved through an innovative experimental arrangement[38] using flexible strips to apply uniform strain and, therefore, stress distribution. Accurate simulation of the out-of-plane boundary conditions[39] led to meaningful comparisons of experimental results with theoretical predictions. Load eccentricity was examined through tests whose other objective was the validation of a general analysis accounting for displacement-controlled loading and initial imperfections.[40]

Care was taken to test the algorithm on a selection of examples with various geometrical, support or loading characteristics so that the versatility of the method could be demonstrated.[17,18] Results are shown here for the experimental set-up which simulates more closely traction-specified in-plane boundary conditions.[38] The plate data which were introduced into the numerical solutions were: $a=b=508$ mm, $h=6.35$ mm, $E=18$ GPa, $\nu=0.25$, maximum $W^i=0.349$ mm. The simply supported square plate was subjected to a total compressive force P, uniformly distributed over two opposite sides. The initial imperfection was estimated by fitting the initial theoretical load–deformation curve to the experimental data. The predictions from all tested deflection models are plotted in Fig. 4 together with the experimental results. This diagram gives a clear indication of the degree of accuracy that can be achieved by the various models.

As expected, the quadratic polynomial model proved inadequate. This was based on a boundary and domain discretization into 16 elements and 32 cells, respectively. The averaging scheme for the curvature distribution improved its accuracy but not to an acceptable degree. Satisfactory results were finally achieved by the use of the cubic model based on a 18-cell discretization and combined with a search for the most suitable value for d, in this case 0.25.

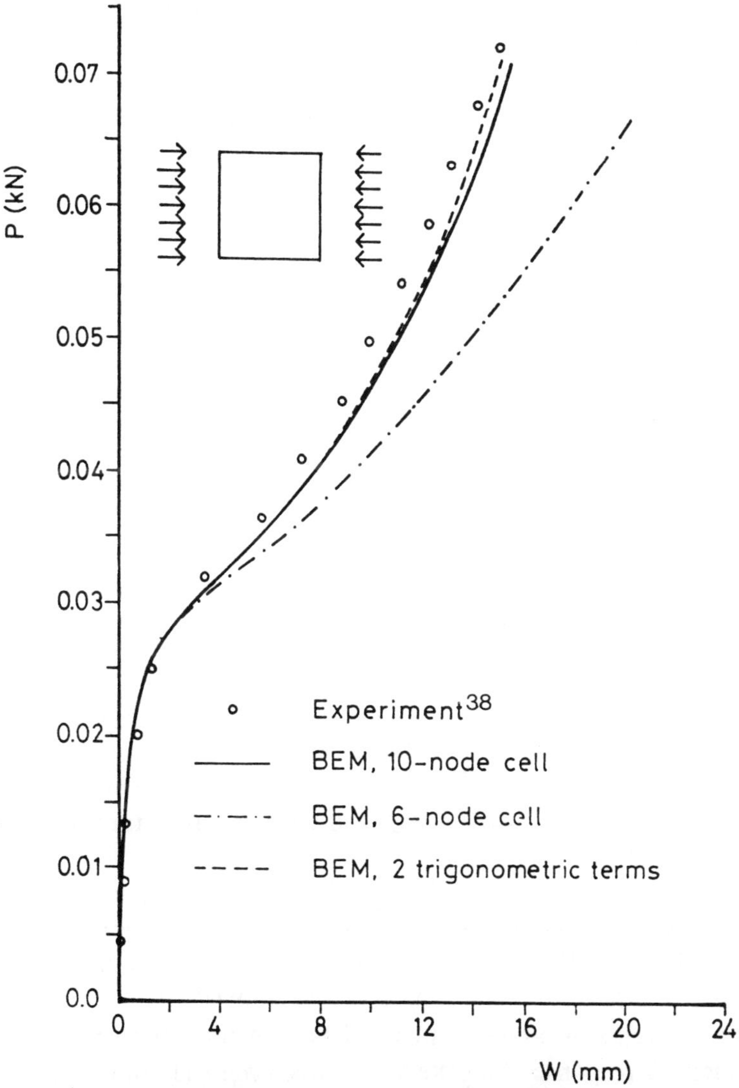

Figure 4. Measured non-linear plate deflection and BEM predictions for various domain models.

The square shape and simple support conditions of the plate suggested the initial representation of the deflection over the whole domain by a single trigonometric term. This resulted in a very good agreement between experiment and analysis which was enhanced in the region of advanced deformation when a second term was added to the expansion. As can be seen from Fig. 4, the result obtained using the trigonometric model also validated the formulation based on the cubic cell model.

In other experiments the results of which were simulated by the non-linear BEM analysis, kinematic in-plane restrictions were imposed along part or the whole of the boundary. The actual conditions had therefore to be replaced by the traction conditions which most closely approximate them so that the presented analysis could be applied. It has been pointed out that this inconsistency generates discrepancies between experiment and theory only for loads well above the critical. This conclusion was confirmed by the BEM results reported in the literature.[17,18]

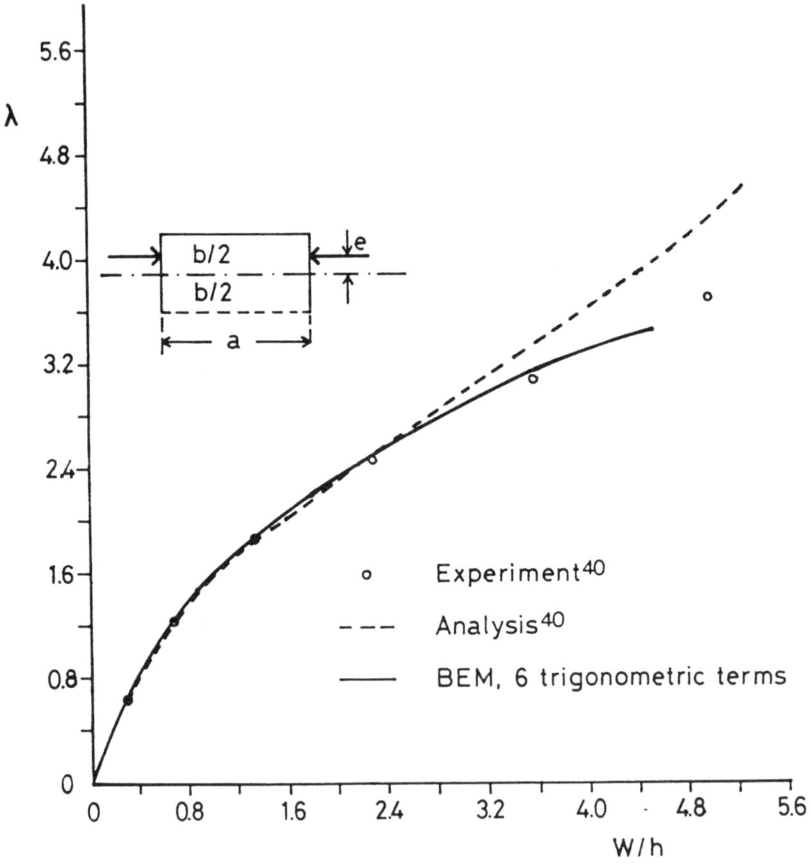

Figure 5. Experimental measurement and BEM prediction of post-buckling response of a plate with a free edge and non-uniform loading.

The interesting feature of the experiments conducted by Rhodes *et al.*[40] is the non-uniformity of the applied load and the internal membrane stresses from the start of the loading. Comparisons of BEM predictions with the results from one of the reported tests can be made by referring to Fig. 5. This particular example had one edge free and a load eccentricity of $e=0.3b$. The other problem data were: $a=508$ mm, $b=254$ mm, $h=1.24$ mm, $E=227.5$ GPa, $\nu=0.25$, maximum $W^i=0.088$ mm. The shown results were obtained using 16 boundary elements and 32 domain cells. Again the non-linear BEM analysis proved absolutely reliable for a significant part of the load-deflection curve. The number of load steps varied according to the length of the experimental record and the size of the load increment. This affects the running time of the developed computer code which also depends on the number of domain unknowns generated by the deflection model. It is therefore obvious that the more generally applicable cubic model is more computational demanding than the single-cell trigonometric model whose applicability is however restricted only to rectangular plates.

7 Conclusions

The comparison of BEM predictions with experimental and other analytical results has established the method as a reliable analytical tool for plate stability, while pointing to the direction for further development. The parameters affecting the efficiency of the numerical algorithms were investigated. The types of boundary elements, the number of nodes per element, the number of points for Gaussian quadrature, the evaluation of integrals with singularities, the boundary discretization schemes, the method of solution of the linear system of equations are all factors that affect the accuracy and speed of the BEM calculations of buckling loads. A systematic study of some of these factors[13] led to the acceptance of the linear discontinuous element with 2 nodes per element as the best choice for boundary meshing.

The transformation of the domain integrals into boundary ones results in a numerical algorithm for plate buckling having certain advantages over that based on domain meshing and integration over cells. Input files are considerably simplified. Apart from the physical properties of the plate and the number of boundary elements on each side of the plate, either the number of nodes or the type and number of deflection functions need to be specified. The time required for the solution of the system is reduced but this is outweighed by the time taken for the calculation of Fourier coefficients.

Dual reciprocity is formulated through the representation of the deflection as a discrete set of nodal values over a regular pattern of equidistant points within a rectangular area encompassing the plate domain. Any plate shape can therefore be analysed; a standard format for data input is required; there is a clear convergence of the solution as the number of domain nodes is increased. However, computer implementation requires considerable time and memory space. For rectangular plates with simple boundary conditions and loading, the trigonometric model yields very efficient solutions in terms of accuracy, computer memory and speed with only a small number of terms. It is applicable to plates with free boundaries and non-uniform plane stress distributions although the selection of suitable functions becomes more difficult and a larger number of them may be required. Moreover, choosing the position of the source point to generate domain equations has a strong effect on performance.

The extent to which initial imperfections influence the critical equilibrium and the post-buckling behaviour of the plates has been the subject of extensive experimental research. It was shown through numerical examples how successfully can a measured non-linear plate behaviour be simulated by a BEM algorithm. Applied to plates of arbitrary shape and loading, the non-linear BEM analysis can thus provide reliable information about post-buckling stability, snap-through and strength in general which is very important to design.

Alternative formulations of the integral equations may lead to a wider applicability of the elastic BEM analysis to plates with any combination of out-of-plane and in-plane support conditions. The iterative procedure obviating deflection modelling has been tested only on circular plates[19,20] and its potential deserves to be explored further. A long term objective can be the extension of the non-linear analysis to elasto-plastic plate buckling. Another stability problem of practical significance concerns stiffened plates which can be analysed as a collection of interconnected panels or as orthotropic elements.

References

[1] Walker, A.C. A brief review of plate buckling research, *Behaviour of Thin-Walled Structures*, eds. J. Rhodes & J. Spence. Elsevier Applied Science, London, pp. 375–398, 1984.

[2] Timoshenko, S. & Gere, J.M. *Theory of Elastic Stability*. 2nd edn. McGraw–Hill, New York, 1961.

[3] Bulson, P.S. *The Stability of Flat Plates*. Chatto & Windus, London, 1970.

[4] Allen, H.G. & Bulson, P.S., *Background to Buckling*. McGraw–Hill, London, 1980.

[5] Manolis, G.D., Beskos, D.E. & Pineros, M.F. Beam and plate stability by boundary elements. *Computers & Structures*, **22**, 917–923, 1986.

[6] Kawabe, H. Plate buckling analysis by the boundary element method, *Theory and Applications of Boundary Element Methods*, eds. M.Tanaka & Q. Du. Pergamon, Oxford, pp. 367–374, 1987.

[7] Bezine, G., Cimetiere, A. & Gelbert, J.P. Unilateral buckling of thin elastic plates by the boundary integral equation method, *International Journal for Numerical Methods in Engineering*, **21**, 2189–2199, 1985.

[8] Costa, J.A. & Brebbia, C.A. Elastic buckling of plates using the boundary element method, *Boundary Elements VII*, eds. C.A. Brebbia & G. Maier. Springer–Verlag, Berlin, pp. 4.29–4.42, 1985.

[9] Tanaka, M. & Miyazaki, K. Elastic buckling analysis of assembled plate structures by boundary element method, *Boundary Elements VIII*, eds. M. Tanaka & C.A. Brebbia. Springer–Verlag, Berlin, pp. 547–559, 1986.

[10] Shi, G. Flexural vibration and buckling analysis of orthotropic plates by the boundary element method, *International Journal of Solids & Structures*, **26**, 1351–1370, 1990.

[11] Syngellakis, S. & Kang, M. A boundary element solution of the plate buckling problem, *Engineering Analysis*, **4**, 75–81, 1987.

[12] Liu, Y. Elastic stability analysis of thin plate by the boundary element method – a new formulation, *Engineering Analysis*, **4**, 160–164, 1987.

[13] Syngellakis, S. & Elzein, A. Plate buckling loads by the boundary element method, *International Journal for Numerical Methods in Engineering*, **37**, 1763–1778, 1994.

[14] Brebbia, C.A. & Nardini, D. Dynamic analysis in solid mechanics by an alternative boundary element procedure, *Soil Dynamics and Earthquake Engineering*, **2**, 228–233, 1982.

[15] Tang, W. *A Generalized Approach for Transforming Domain Integrals into Boundary Integrals in Boundary Element Methods*, Ph.D. Thesis, Computational Mechanics Institute, Southampton, 1987.

[16] Elzein, A. & Syngellakis, S. Dual reciprocity in boundary element formulations of the plate buckling problem, *Engineering Analysis with Boundary Elements*. **9**, 175–184, 1992.

[17] Syngellakis, S., Elzein, A. & Walker, A.C. Comparison between experimental and boundary element predictions of plate buckling behaviour, *Engineering Analysis with Boundary Elements*, **9**, 103–108, 1991.

[18] Elzein, A. & Syngellakis, S. High-order elements for the BEM stability analysis of imperfect plates, *Advances in Boundary Elements*, Vol. 3: Stress Analysis, eds. C.A. Brebbia & J.J. Connor. Springer–Verlag, Berlin, pp. 269–289, 1989.

[19] Qin, Q. & Huang. Y. BEM of postbuckling analysis of thin plates, *Applied Mathematical Modelling*, **14**, 544–548, 1990.

[20] Tanaka, M., Matsumoto, T. & Zheng, Z. An incremental approach to the pre/post buckling problem of thin elastic plates via boundary-domain element method, *Boundary Element Methods*, eds. M. Tanaka & Z. Yao. Elsevier Science, Oxford, pp. 3–14, 1996.

[21] Cheng, C.J. & Wang, R. Boundary integral equations and the boundary element method for buckling analysis of perforated plates, *Engineering Analysis with Boundary Elements*, **17**, 57-68, 1996.

[22] Budiansky, B. Theory of buckling and post-buckling behavior of elastic structures, *Advances in Applied Mechanics*, Vol. 14, ed. C.S. Yih. Academic Press, New York, pp. 1-65, 1974.

[23] Timoshenko, S. & Woinowsky-Krieger, S. *Theory of Plates and Shells*, 2nd edn. McGraw-Hill, New York, 1959.

[24] Shi, G. & Bézine, G. A general boundary element formulation for the anisotropic plate bending problems, *Journal of Composite Materials*, **11**, 345-364, 1988.

[25] Nerantzaki, M.S. & Katsikadelis, J.T. Buckling of plates with variable thickness - an analog equation solution, *Engineering Analysis with Boundary Elements*, **18**, 149-154, 1996.

[26] Zienkiewicz, O. C. *The Finite Element Method*, 3rd edn. McGraw-Hill, London, 1977.

[27] Hambly, E.C. *Bridge Deck Behaviour*. Chapman & Hall, London, 1976.

[28] Telles, J.C.F. A self adaptive coordinate transformation for efficient numerical evaluation of general boundary element integrals, *International Journal for Numerical Methods in Engineering*, **24**, 1-15, 1987.

[29] Kutt, H.R. *Quadrature formulae for finite-part integrals*. National Research Institute for Mathematical Sciences, CSIR Special Report, WISK 178, Pretoria, 1975.

[30] Kutt, H.R. *On the numerical evaluation of finite-part integrals involving an algebraic singularity*. National Research Institute for Mathematical Sciences, CSIR Special Report, WISK 179, Pretoria, 1975.

[31] Cook, I.T. & Rockey, K.C. Shear buckling of rectangular plates with mixed boundary conditions, *Aeronautical Quarterly*, **14**, 349-356, 1963.

[32] Khan, M.Z. & Walker, A.C. Buckling of plates subjected to localised edge loading, *The Structural Engineer*, **50**, 225-232, 1972.

[33] Alfutov, N.A. & Balabukh L.I. On the possibility of solving plate stability problems without a preliminary determination of the initial state of stress, *Journal of Applied Mathematics & Mechanics*, **31**, 730-736, 1967.

[34] Wittrick, W.H. Symmetrical buckling of right-angled isosceles triangular plates, *Aeronautical Quarterly*, **5**, 131-143, 1954.

[35] Desai, C.S. & Abel, J.F. *Introduction to the Finite Element Method*. Van Nostrand Reinhold, New York, 1972.

[36] Stein, M. *Loads and deformations of buckled rectangular plates*. Technical Report R-40, NASA, Langley Research Center, Va, 1959.

[37] Uemura, M. & Byon, O-I. Secondary buckling of a flat plate under uniaxial compression part 2. Analysis of clamped plate by F.E.M. and comparison with experiments, *International Journal of Non-Linear Mechanics*, **13**, 1-14, 1978.

[38] Hoff, N.J., Boley, B.A. & Coan, J.M. The development of a technique for testing stiff panels in edgewise compression, *Proceedings of the Society for Experimental Stress Analysis*, **5**, 68–74, 1948.

[39] Yamaki, N. Experiments on the postbuckling behavior of square plates loaded in edge compression, *Transactions ASME, Journal of Applied Mechanics*, **28**, 238–244, 1961.

[40] Rhodes, J., Harvey, J.M. & Fok, W.C. The load-carrying capacity of initially imperfect eccentrically loaded plates, *International Journal of Mechanical Sciences*, **17**, 161–175, 1975.

[41] Tolstov, G.P. *Fourier Series*. Prentice-Hall, London, 1962.

Appendix A

The most common kernels in various boundary and domain integrals are:

$$\frac{\partial u_1}{\partial n} = \frac{r}{8\pi}(2\ln r + 1)\cos\gamma$$

$$\frac{\partial u_1}{\partial s} = -\frac{1}{8\pi}(2\ln r + 1)\sin\gamma$$

$$u_{1,\alpha\beta} = \frac{1}{8\pi}(2\ln r + 1)\delta_{\alpha\beta} + \frac{r_{,\alpha} r_{,\beta}}{4\pi}$$

$$\nabla^2 u_1 = \frac{1}{2\pi}(\ln r + 1)$$

$$\frac{\partial \nabla^2 u_1}{\partial n} = \frac{1}{2\pi r}\cos\gamma$$

$$\frac{\partial u_{1,\alpha\beta}}{\partial n} = \frac{1}{4\pi r}[n_\alpha r_{,\beta} + n_\beta r_{,\alpha} + (\delta_{\alpha\beta} - 2r_{,\alpha} r_{,\beta})\cos\gamma]$$

$$\nabla^2 u_{1,\alpha\beta} = \frac{1}{2\pi r^2}(\delta_{\alpha\beta} - 2r_{,\alpha} r_{,\beta})$$

$$\frac{\partial \nabla^2 u_{1,\alpha\beta}}{\partial n} = -\frac{1}{\pi r^3}[n_\alpha r_{,\beta} + n_\beta r_{,\alpha} + (\delta_{\alpha\beta} - 4r_{,\alpha} r_{,\beta})\cos\gamma]$$

$$M_n(u_1) = -\frac{D}{8\pi}\,[2(1+\nu)(\ln r+1) + (1-\nu)\cos 2\gamma]$$

$$M_{ns}(u_1) = \frac{D(1-\nu)}{8\pi}\,\sin 2\gamma$$

$$V(u_1) = -\frac{D}{4\pi r}\,[2+(1-\nu)\cos 2\gamma]\cos\gamma + \frac{D(1-\nu)}{4\pi R}\,\cos 2\gamma$$

$$\frac{\partial u_2}{\partial n} = -\frac{1}{8\pi}(2\ln r+1)\cos(\alpha-\gamma) - \frac{1}{4\pi}\cos\alpha\,\cos\gamma$$

$$\frac{\partial u_2}{\partial s} = -\frac{1}{8\pi}(2\ln r+1)\sin(\alpha-\gamma) + \frac{1}{4\pi}\cos\alpha\,\sin\gamma$$

$$u_{2,\alpha\beta} = -\frac{1}{4\pi r}\,[(\delta_{\alpha\beta} - 2r_{,\alpha}r_{,\beta})\cos\alpha + m_\alpha r_{,\beta} + r_{,\alpha}m_\beta]$$

$$\nabla^2 u_2 = -\frac{1}{2\pi r}\,\cos\alpha$$

$$\frac{\partial \nabla^2 u_2}{\partial n} = \frac{1}{2\pi r^2}\,\cos(\alpha+\gamma)$$

$$M_n(u_2) = \frac{D}{4\pi r}\,[(1+\nu)\cos\alpha + (1-\nu)\sin\alpha\,\sin 2\gamma]$$

$$M_{ns}(u_2) = \frac{D(1-\nu)}{4\pi r}\,\sin\alpha\,\cos 2\gamma$$

$$V(u_2) = \frac{D(1-\nu)}{8\pi r^2}\,[\cos(\alpha-3\gamma) - 2\cos(\alpha+3\gamma)] - \frac{D(5-\nu)}{8\pi r^2}\cos(\alpha+\gamma) - \frac{D(1-\nu)}{2\pi R r}\sin 2\gamma\,\sin\alpha$$

where R is the radius of curvature of the plate contour.

The derivatives of the functions f^j_{nm} appearing in the boundary integrals generated by dual reciprocity are

$$\frac{\partial f^1_{00}}{\partial n} = \frac{1}{12}\,[(x_1-\bar{x}_1)^3 \cos\psi + (x_2-\bar{x}_2)^3 \sin\psi]$$

$$\nabla^2 f^1_{00} = \frac{1}{4}\,[(x_1-\bar{x}_1)^2 + (x_2-\bar{x}_2)^2]$$

$$\frac{\partial \nabla^2 f^1_{00}}{\partial n} = \frac{1}{2}\,[(x_1-\bar{x}_1) \cos\psi + (x_2-\bar{x}_2) \sin\psi]$$

$$\frac{\partial f^j_{nm}}{\partial n} = \pi A^2_{nm} Z^j_{nm}\,(\mathrm{x}_\alpha)$$

$$\nabla^2 f^j_{nm} = A_{nm}\, f^j_{nm}$$

$$\frac{\partial \nabla^2 f^j_{nm}}{\partial n} = \pi A_{nm} Z^j_{nm}$$

with

$$Z^1_{nm} = -\frac{n}{a}\cos\psi \sin\xi \cos\eta - \frac{m}{b}\sin\psi \cos\xi \sin\eta$$

$$Z^2_{nm} = -\frac{n}{a}\cos\psi \sin\xi \sin\eta + \frac{m}{b}\sin\psi \cos\xi \cos\eta$$

$$Z^3_{nm} = \frac{n}{a}\cos\psi \cos\xi \cos\eta - \frac{m}{b}\sin\psi \sin\xi \sin\eta$$

$$Z^4_{nm} = \frac{n}{a}\cos\psi \cos\xi \sin\ + \frac{m}{b}\sin\psi \sin\xi \cos\eta$$

where ψ is the angle between the x_1 axis and the normal to the boundary and

$$\xi = \frac{n\pi x_1}{a}\,,\ \eta = \frac{m\pi x_2}{b}$$

Appendix B

Fourier series approximation

The basis of converting domain to boundary integrals is the double Fourier series approximation of a function in two variables defined over the rectangular domain

$[-a,a]\times[-b,b]$ and satisfying the necessary convergence criteria.[41] The truncated form of such an approximation can be written

$$f(x_1,x_2)=\sum_{n=0}^{N}\sum_{m=0}^{M}\sum_{j=1}^{4}\alpha_{nm}k_{nm}^{j}\phi_{nm}^{j} \tag{B1}$$

where

$$\begin{aligned}\phi_{nm}^{1} &= \cos\xi\,\cos\eta\\ \phi_{nm}^{2} &= \cos\xi\,\sin\eta\\ \phi_{nm}^{3} &= \sin\xi\,\cos\eta\\ \phi_{nm}^{4} &= \sin\xi\,\cos\eta\end{aligned}$$

and

$$\alpha_{nm}=\begin{cases}\frac{1}{4} \text{ for } n=m=0; n=0, m=M; n=N, m=0; n=N, m=M\\ \frac{1}{2} \text{ for } 0<m<M, n=0,N; 0<n<N, m=0,M\\ 1 \text{ for } 0<m<M, 0<n<N\end{cases}$$

The Fourier coefficients k_{mn}^{j} are obtained as double integrals over the rectangular domain Ω_f. The present application of the Fourier series however deals with a discrete number of points rather than a continuous field. If the value of the function $f(x_1,x_2)$ is given at a discrete set of $(2N)\times(2M)$ equidistant points over Ω_f, then the Fourier coefficients can be approximated in the following manner

$$k_{nm}^{j}=\frac{1}{NM}\sum_{p=-N+1}^{N}\sum_{q=-M+1}^{M}f(x_{1p},x_{2q})\phi_{nm}^{j}(x_{1p},x_{2q}) \tag{B2}$$

Since

$$\begin{aligned}\phi_{0m}^{j} &= \phi_{Nm}^{j} = 0 \text{ for } j=3,4\\ \phi_{n0}^{j} &= \phi_{nM}^{j} = 0 \text{ for } j=2,4\end{aligned}$$

the respective coefficients are also taken equal to zero.

The double Fourier series (B1) can be written in the more concise form

$$f(x_1,x_2)=\sum_{k=1}^{N_f}K_k\Phi_k \tag{B3}$$

with K_k representing the Fourier coefficients, Φ_k the system of trigonometric functions and $N_f=4NM$ the total number of terms in the truncated system. Numerical analysis becomes much more versatile with the use of the discrete series (B2) for calculating the Fourier coefficients since this approach obviates analytical integration which has to be performed for each single term of the model for the deflection.